ADVANCES IN FLAVONOIDS FOR HUMAN HEALTH AND PREVENTION OF DISEASES

AAP Advances in Nutraceuticals

ADVANCES IN FLAVONOIDS FOR HUMAN HEALTH AND PREVENTION OF DISEASES

Edited by

Nisha Sharma, PhD

Deepika Saini, PhD

Rajesh K. Kesharwani, PhD

Prakash Chandra Gupta, PhD

Raj K. Keservani, MPharm

AAP APPLE ACADEMIC PRESS

First edition published 2024

Apple Academic Press Inc.
1265 Goldenrod Circle, NE,
Palm Bay, FL 32905 USA

760 Laurentian Drive, Unit 19,
Burlington, ON L7N 0A4, CANADA

CRC Press
2385 NW Executive Center Drive,
Suite 320, Boca Raton FL 33431

4 Park Square, Milton Park,
Abingdon, Oxon, OX14 4RN UK

Library and Archives Canada Cataloguing in Publication

Title: Advances in flavonoids for human health and prevention of diseases / edited by Nisha Sharma, PhD, Deepika Saini, PhD, Rajesh K. Kesharwani, PhD, Prakash Chandra Gupta, PhD, Raj K. Keservani, MPharm.
Names: Sharma, Nisha (Director of the School of Pharmaceutical Sciences), editor. | Saini, Deepika, editor. | Kesharwani, Rajesh K., 1978- editor. | Gupta, Prakash Chandra (Lecturer in pharmaceutical sciences), editor. | Keservani, Raj K., 1981- editor.
Series: AAP advances in nutraceuticals.
Description: First edition. | Series statement: AAP advances in nutraceuticals | Includes bibliographical references and index.
Identifiers: Canadiana (print) 20230475310 | Canadiana (ebook) 20230475396 | ISBN 9781774913444 (hardcover) | ISBN 9781774913451 (softcover) | ISBN 9781003369813 (ebook)
Subjects: LCSH: Flavonoids. | LCSH: Flavonoids—Therapeutic use. | LCSH: Flavonoids—Health aspects.
Classification: LCC QK898.F5 A38 2024 | DDC 615.1—dc23

Library of Congress Cataloging-in-Publication Data

Names: Sharma, Nisha (Director of the School of Pharmaceutical Sciences), editor. | Saini, Deepika, editor. | Kesharwani, Rajesh K., 1978-editor. | Gupta, Prakash Chandra (Lecturer in pharmaceutical sciences), editor. | Keservani, Raj K., 1981- editor.

Title: Advances in flavonoids for human health and prevention of diseases / edited by Nisha Sharma, PhD, Deepika Saini, PhD, Rajesh K. Kesharwani, PhD, Prakash Chandra Gupta, PhD, Raj K. Keservani, MPharm.

Description: First edition. | Palm Bay, FL : Apple Academic Press, 2024. | Series: AAP advances in nutraceuticals | Includes bibliographical references and index.

Identifiers: LCCN 2023028261 (print) | LCCN 2023028262 (ebook) | ISBN 9781774913444 (hbk) | ISBN 9781774913451 (pbk) | ISBN 9781003369813 (ebk)

Subjects: LCSH: Flavonoids. | Flavonoids--Therapeutic use. | Flavonoids--Health aspects.

Classification: LCC QK898.F5 A38 2024 (print) | LCC QK898.F5 (ebook) | DDC 615.3/21--dc23/eng/20230724

LC record available at https://lccn.loc.gov/2023028261

LC ebook record available at https://lccn.loc.gov/2023028262

ISBN: 978-1-77491-344-4 (hbk)
ISBN: 978-1-77491-345-1 (pbk)
ISBN: 978-1-00336-981-3 (ebk)

ABOUT THE BOOK SERIES: AAP ADVANCES IN NUTRACEUTICALS

SERIES EDITORS:

Raj K. Keservani, MPharm
Associate Professor, Faculty of B. Pharmacy, CSM Group of Institutions, Prayagraj, Uttar Pradesh, India/Email: rajksops@gmail.com

Rajesh K. Kesharwani, PhD
Associate Professor, Department of Computer Application, Nehru Gram Bharati (Deemed to be University), Prayagraj, India

Anil K. Sharma, PhD, MPharm
Assistant Profesoor, Department of Pharmacy, School of Medical and Allied Sciences, GD Goenka University, Gurugram, India

In the modern era, mankind has witnessed a paradigm shift with respect to fundamental eating behavior. The lack of physical workouts and busy schedules at offices and in households have promoted the consumption of junk foods, which eventually results in numerous diseases and disorders since the nutritional content from fast food is inadequate. Plumpness and obesity have become a global health threat. The leading causes of death from most developing nations are noncommunicable diseases such as cardiovascular diseases, cancer, arthritis, osteoporosis, and liver toxicity. Patients suffering from such lifestyle ailments are a bit apprehensive towards prolonged use of costly modern therapeutics, which encourages instead the use of alternative approaches for management of such diseases and disorders.

The emerging sector of the nutraceutical industry encompasses products derived from nature, dietary supplements, and functional foods. Nutraceuticals are used for treatment and prevention of a broad range of diseases, such as the common cold, arthritis, sleep-related disorders, cancers, cardiovascular complications, metabolic disorders, and others.

The research on nutraceuticals is increasing day by day, considering the beneficial effects of food or food supplements in the management of diverse diseases. The issue of paramount concern is standardization and establishment of clinical efficacy of nutraceuticals, which is in fact a challenge for researchers around the globe.

This book series aims at realizing the significance of the variety of nutraceuticals for human well-being. The books in the series will also emphasize the role of dieticians and nutritionists for the prescription of judicious eating habits. The key food components such as carbohydrates, proteins, and lipids as well as micronutrients (vitamins, minerals) are demonstrated to maintain good health, obviating the need for medicines. Thus, nutraceuticals that are indeed derived from food ingredients are believed to be potential alternative therapeutics.

Asymptomatic diseases/disorders necessitate proper diagnosis. Stressed circumstances have been reported to cause weight loss. Workplace stress is a key variable in the genesis and propagation of metabolic disorders. The nuances of the underlying mechanisms are still to be deciphered. Poor lifestyle trends have been attributed along with the consumption of junk foods in several such instances. Sugar-rich carbonated beverages, the use of contraband drugs, and the consumption of liquor in excess are assigned as risk factors. This book series will cover advances and applications in addition to providing the basics of nutraceuticals.

The books in this series are valuable resources for industry professionals to design and develop quality products for end users. In addition, the book series will be fortified by ideas of innovation, concept building, manufacturing aspects, quality control, and regulatory status of nutraceuticals.

Ostensibly, business has seen a rise in the nutraceutical consumption in recent years. Pure, safe, efficacious products are the need of the hour globally. There are a number of books on nutraceuticals in the market, yet these are only a few books presenting certain aspects of nutraceuticals therein. This book series offers a holistic view of this promising strategy, which is picking up pace with time. This book series is a unique endeavor to bring manufacturing, research and development, and marketing strategies under a single umbrella.

CURRENT BOOKS IN THE SERIES

- **Micronutrients and Macronutrients as Nutraceuticals**
  Editors: Prakash Chandra Gupta, PhD, Sayan Bhattacharyya, MBBS, MD, Nisha Sharma, PhD, MPharm, Rajesh K. Kesharwani, PhD, and Raj K. Keservani, MPharm

- **Formulations, Regulations, and Challenges of Nutraceuticals**
Editors: Tingirikari Jagan Mohan Rao, PhD, Rajesh K. Kesharwani, PhD, Raj. K. Keservani, MPharm, and Anil K. Sharma, PhD
- **Food Supplements and Dietary Fibers in Health and Disease**
Editors: Bhushan R. Rane, PhD, MPharm, Raj K. Keservani, M Pharm, Durgesh Singh, DPhil, Nayan A. Gujarathi, PhD, and Ashish S. Jain, PhD
- **Nutraceuticals in Cancer Prevention, Management, and Treatment**
Editors: Raj K. Keservani, MPharm, Bui Thanh Tung, PhD, Sippy Singh, DPhil, and Rajesh K. Kesharwani, PhD
- **Advances in Flavonoids for Human Health and Prevention of Diseases**
Editors: Nisha Sharma, PhD, Deepika Saini, PhD, Rajesh K. Kesharwani, PhD, Prakash Chandra Gupta, and Raj. K. Keservani, MPharm
- **The Flavonoids: Extraction and Applications**
Editors: Deepika Saini, PhD, Rajesh K. Kesharwani, PhD, and Raj K. Keservani, MPharm
- **Immune-Boosting Nutraceuticals for Better Human Health: Novel Applications**
Editors: Urmila Jarouliya, PhD, Raj K. Keservani, MPharm, Rajesh K. Kesharwani, PhD, Virendra K. Patel, PhD, and Adi D. Bharti
- **Applications of Functional Foods in Disease Prevention**
Editors: Raj K. Keservani, MPharm, and Eknath D. Ahire, MSPharm
- **Plant Metabolites and Vegetables as Nutraceuticals**
Editors: Raj K. Keservani, MPharm, Bui Thanh Tung, PhD, Rajesh K. Kesharwani, PhD, MPharm, and Eknath D. Ahire
- **Preventive and Therapeutic Role of Vitamins as Nutraceuticals**
Editors: Khemchand R. Surana, Eknath D. Ahire, Raj K. Keservani, MPharm, and Rajesh K. Kesharwani, PhD
- **Nutraceuticals and Bone Health**
Editors: Deepak Sharma, PhD, Madan Mohan Gupta, PhD, Anil K. Sharma, PhD, MPharm, Raj K. Keservani, MPharm, and Rajesh K. Kesharwani, PhD

- **Flavonoids as Nutraceuticals**
  Editors: Rajesh K. Kesharwani, PhD, MTech, Deepika Saini, PhD, Raj K. Keservani, MPharm, and Anil Kumar Sharma, PhD, MPharm
- **Herbals as Nutraceuticals: Their Role in Healthcare**
  Editors: Raj K. Keservani, MPharm, Rajeshwar Kamal Kant Arya, PhD, MPharm, and Rajesh K. Kesharwani, PhD, MTech
- **Nutraceutical Fruits: Overview and Disease Prevention**
  Editor: Raj K. Keservani, MPharm
- **Nutrigenomics and Nutraceuticals**
  Editors: Raj K. Keservani, MPharm, Eknath D. Ahire, Shubham J. Khainar, PhD, Sanjay J. Kshirsagar, PhD, and Rajesh K. Kesharwani, PhD
- **The Nature of Nutraceuticals: History, Properties, Sources, and Nanotechnology**
  Editors: Rajesh K. Kesharwani, PhD, Prashant Kumar, PhD, and Raj K. Keservani, MPharm
- **Nutraceuticals for the Treatment and Prevention of Sexual Disorders**
  Editors: Raj K. Keservani, PhD, Sharangouda J. Patil, PhD, and Ivan Aranha, PhD
- **Antioxidants as Nutraceuticals**
  Editors: Nayan A. Gujaratih, PhD, Raj K. Keservani, PhD Rajesh K. Kesharwani, PhD, Bhushan R. Rane, PhD, and Yogeeta Sameer Goyal, PhD

ABOUT THE EDITORS

Nisha Sharma, PhD
Director, School of Pharmaceutical Sciences, C.S.J.M. University, Kanpur, Uttar Pradesh, India

Nisha Sharma, PhD, MPharm, is presently working as a Director of the School of Pharmaceutical Sciences, C.S.J.M. University, Kanpur, India. With a teaching experience of over 23 years, she has authored many research articles in journals and books. Dr. Sharma is a popular figure in the pharmaceutical field. She has chaired many scientific sessions at national and international conferences. She has served as a resource person for the All India Council for Technical Education (AICTE) faculty development program. Dr. Sharma received BPharm from J. S. S. College of Pharmacy, Mysore, India, in 1995, an MPharm from BITS Pilani, Rajasthan, India, in 1997, and a PhD from Bhagwant University, Ajmer, Rajasthan, India, in 2014. Her areas of research include phytochemistry, ethnopharmacology, and herbal formulations. She is associated with many professional bodies, like (IPA) Indian Pharmaceutical Association, (IPGA) Indian Pharmacy Graduates' Association, (IHPA) Indian Hospital Pharmacists' Association, (APTI) Association of Pharmaceutical Teachers of India and many others.

Deepika Saini, PhD
Assistant Professor, Department of Zoology, Chamanlal Mahavidyalaya, Haridwar, Uttarakhand, India

Deepika Saini, PhD, is currently working as an Assistant Professor in the Department of Zoology, Chamanlal Mahavidyalaya, Haridwar, Uttarakhand. She has a teaching experience of about five years. She has also worked as an adjunct faculty member in the Natural Science Department of the University of Maryland Eastern Shore, Salisbury (MD), USA. She has published in about 14 national and international publications. She has presented more than 14 papers in national and international seminars/conferences and has also attended many workshops all around India. She has also been awarded the Best Teacher Award 2020 by the Society of

Research in Biological Studies. Dr. Deepika is also providing her services as a reviewer and editorial board member to more than eight international and national journals including *The American Journal of Life Sciences, The International Journal of Zoological Studies, The International Journal of Biological Studies,* etc. She has successfully edited four books. During the pandemic year 2020, she organized five webinars with reputed institutes like the Zoological Survey of India and the Botanical Survey of India. She has organized two national conferences funded by the Uttarakhand Council of Science and Technology, Dehradun, and the Uttarakhand Council for Biotechnology, Dehradun.

Rajesh K. Kesharwani, PhD
Associate Professor, Department of Computer Application,
Nehru Gram Bharati (deemed to be university), Prayagraj,
Uttar Pradesh, India

Rajesh K. Kesharwani, PhD, MTech, is currently working as an Associate Professor in the Department of Computer Application at Nehru Gram Bharati (deemed to be university), Prayagraj, India. He has more than 11 years of research and nine years of teaching experience in various institutes in India, imparting bioinformatics and biotechnology education. He has received several awards, including the NASI-Swarna Jayanti Puruskar by The National Academy of Sciences of India. He has supervised one PhD, and more than 20 UG and PG students for their research work, has authored over 49 peer-reviewed articles, 20 book chapters, and 14 edited books with international publishers (e.g., Springer Nature, IGI Global, Taylor & Francis, etc.). He has been a member of many scientific communities as well as a reviewer for many international journals. He has presented many papers at various national and international conferences. Dr. Kesharwani received his PhD from the Indian Institute of Information Technology, Allahabad, and worked at NIT Warangal for two semesters. He has been a recipient of the Ministry of Human Resource Development (India) Fellowship and a Senior Research Fellowship from the Indian Council of Medical Research, India. His research fields of interest are medical informatics, protein structure and function prediction, computer-aided drug designing, structural biology, drug delivery, cancer biology, nano-biotechnology, and biomedical sciences.

Prakash Chandra Gupta, PhD

Assistant Professor, School of Pharmaceutical Sciences, C.S.J.M. University, Kanpur, Uttar Pradesh, India

Prakash Chandra Gupta, PhD, MPharm, is presently working as an Assistant Professor in the School of Pharmaceutical Sciences, C.S.J.M. University, Kanpur, India. He has more than 15 years of academic and research experience. His research interest includes the standardization of herbal medicine, pharmacological screening, and the development of herbal formulations. He has published many research and review articles in reputed national and international journals. He has received various research project grants from different funding agencies. Dr. Gupta received his B. Pharm. degree in 2004 from the Department of Pharmacy, Kumaun University, Nainital, Uttarakhand, India, and his M. Pharm. degree (Pharmacognosy) in 2006 from the Institute of Pharmacy, Barkatullah University Bhopal, Madhya Pradesh, India, and PhD degree in 2014 from the Department of Pharmaceutical Science, Bhagwant University, Ajmer, Rajasthan, India.

Raj K. Keservani, MPharm

Associate Professor, Faculty of B. Pharmacy, CSM Group of Institutions, Allahabad, Uttar Pradesh, India

Raj K. Keservani, MPharm, is working as an Associate Professor in the Faculty of B. Pharmacy at CSM Group of Institutions, Allahabad, India. He has more than 12 years of academic (teaching) experience in pharmaceutical education from various Indian institutes. He has supervised more than 25 UG students for their research work, has published 31 peer-reviewed papers in the field of pharmaceutical sciences in national and international journals, one patent, 43 book chapters, three co-authored books, and 19 edited books. He is also active as a reviewer for several international scientific journals. Mr. Keservani graduated with a pharmacy degree from the Department of Pharmacy, Kumaun University, Nainital (Uttarakhand), India. He received his Master of Pharmacy (MPharm) (specialization in pharmaceutics) from the School of Pharmaceutical Sciences, Rajiv Gandhi Proudyogiki Vishwavidyalaya, Bhopal, India. Mr. Keservani is a life member of the Society of Pharmaceutical Education and Research (SPER). His research interests include nutraceutical and functional foods, novel drug delivery systems (NDDS), transdermal drug delivery/drug delivery, health science, cancer biology, and neurobiology.

CONTENTS

CONTRIBUTORS

Eknath D. Ahire
Department of Pharmaceutics, MET's Institute of Pharmacy, Affiliated to Savitribai Phule Pune University, BKC, Adgaon, Nashik, Maharashtra, India

Pankaj Bhatt
Department of Pharmaceutics, KIET Group of Institutions (KIET School of Pharmacy), Muradnagar, Ghaziabad, Uttar Pradesh, India

Diwakar Chauhan
Department of Chemistry, School of Basic and Applied Sciences, Galgotias University, Greater Noida, Uttar Pradesh, India

Monika Chauhan
Department of Forensic Science, School of Basic and Applied Sciences, Galgotias University, Greater Noida, Uttar Pradesh, India

Manas Kumar Das
Director, Orlean College of Pharmacy, Greater Noida, Uttar Pradesh, India

Saumya Das
Associate Professor, Pharmacy Institute, Noida Institute of Engineering and Technology, Greater Noida, Uttar Pradesh, India

Deepa Devi
Scientist-B, Parasitology Division, ICMR–National Institute of Malaria Research, Delhi, India

Deepika Dhawan
Department of Dietetics and Applied Nutrition, Amity Medical School, Amity University Haryana, Gurugram, Haryana, India

Lovnish Gupta
Department of Forensic Science, School of Basic and Applied Sciences, Galgotias University, Greater Noida, Uttar Pradesh, India

Mohini R. Jagtap
JMCTE Institute of Pharmacy, Nashik, Maharashtra, India

Urmila Jarouliya
SOS in Biochemistry, Jiwaji University, Gwalior, Madhya Pradesh, India

Sagarika Kabra
School of Pharmaceutical Sciences, Maharishi University of Information Technology, Lucknow, Uttar Pradesh, India

Raj K. Keservani
Faculty of B.Pharmacy, CSM Group of Institutions, Prayagraj, Uttar Pradesh, India

Vipin Kesharwani
School of Pharmaceutical Sciences, Maharishi University of Information Technology, Lucknow, Uttar Pradesh, India

Shubham J. Khairnar
Department of Pharmacology, MET's Institute of Pharmacy, Affiliated to Savitribai Phule Pune University, BKC, Adgaon, Nashik, Maharashtra, India

Sanjay J. Kshirsagar
Department of Pharmaceutics, MET's Institute of Pharmacy, Affiliated to Savitribai Phule Pune University, BKC, Adgaon, Nashik, Maharashtra, India

Ajay Kumar
Department of Life Science, School of Basic and Applied Sciences, Galgotias University, Greater Noida, Uttar Pradesh, India

Vipin Kumar
Department of Pharmaceutical Sciences, Gurukul Kangri (Deemed to be University), Haridwar, Uttarakhand, India

Deepika Saini
Assistant Professor, Department of Zoology, Chaman Lal Mahavidhyalya, Landhaura, Haridwar, Uttarakhand, India

Pradip Kumar Saini
Department of Crop Physiology, Acharya Narendra Deva University of Agriculture and Technology, Kumarganj, Ayodhya, Uttar Pradesh, India

Pragati Saini
Department of Forensic Science, School of Basic and Applied Sciences, Galgotias University, Greater Noida, Uttar Pradesh, India

Bhupesh Chandra Semwal
Institute of Pharmaceutical Research, GLA University, Mathura, Uttar Pradesh, India

Kamal Kant Sharma
Research Scholar, Department of Botany and Microbiology, Gurukula Kangri (Deemed to be University), Haridwar, Uttarakhand, India

Akhand Pratap Singh
School of Pharmaceutical Sciences, Maharishi University of Information Technology, Lucknow, Uttar Pradesh, India

Neha P. Singh
Dev Bhoomi Institute of Pharmacy and Research, Manduwala, Dehradun, Uttarakhand, India

Richa Singh
Department of Dietetics and Applied Nutrition, Amity Medical School, Amity University Haryana, Gurugram, Haryana, India

Shivam Singh
Scientist B, Government District Hospital Cum Trauma Center, Waidhan, Singrauli, Madhya Pradesh, India

Suruchi Singh
School of Pharmacy, Glocal University, Mirzapur Pole, Saharanpur, Uttar Pradesh, India

Khemchand R. Surana
Department of Pharmaceutical Chemistry, Shreeshakti Shaikshanik Sanstha, Divine College of Pharmacy, Satana, Nashik, Maharashtra, India

Tarkeshwar
Department of Genetics and Plant Breeding, Acharya Narendra Deva University of Agriculture and Technology, Kumarganj, Ayodhya, Uttar Pradesh, India

ABBREVIATIONS

12-HETE	12-hydroxyeicosatetraenoic acid
6-OHDA	6-hydroxydopamine
ACC	acetyl CoA carboxylate
AD	Alzheimer's disease
AIBL	Australian imaging biomarkers and lifestyle study of ageing
AMI	acute myocardial infarction
AMPK	adenosine monophosphate-activated protein kinase
AP-1	activator protein-1
APP	amyloid precursor protein
AR	aldose reductase
ARA	arachidonic acid
ATP	adenosine triphosphate
BACE-1	beta-site APP cleaving enzyme-1
BBB	blood-brain barrier
BCRP	breast cancer resistance protein
BDNF	brain-derived neurotrophic factor
BP	blood pressure
C/EBP	CCAAT-enhancer-binding protein
CAP	Cb1 associated protein
CAT	catalase
CDKs	cyclin dependable kinases
CHD	coronary heart disease
CHF	congestive heart failure
COX	cyclooxygenase
CREB	cAMP response element-binding protein
CVD	cardiovascular disease
Cy-3-Glu	cyanidin-3-glucoside
DM	diabetes mellitus
ECG	epicatechin gallate
EGC	epigallocatechin
EGCG	epigallocatechin gallate

EGFR	epidermal growth factor receptor
ERK	extracellular signal-regulated kinase
FGFR2	fibroblast growth factor receptor 2
FOXO	forkhead box O
GDM	gestational diabetes mellitus
GHES	glucosyl hesperidin
GLUT1	glucose transporter 1
GP	glycoprotein
GPx	glutathione peroxidase
GR	glutathione reductase
GSPE	grape seed proanthocyanidin extract
GST	glutathione S-transferase
H_2O_2	hydrogen peroxide
HMG-CoA	3-hydroxy-3-methylglutaryl-coenzyme A
HOMA-IR	homeostatic model assessment of insulin resistance
HSD	3-hydroxysteroid dehydrogenase
HSV-1	herpes simplex virus type-1
HSV-2	herpes simplex virus type-2
IBD	irritable bowel disease
ICAM-1	intercellular adhesion molecule 1
IGF-1	insulin-like boom factor-1
IGFBP1	insulin-like growth factor binding protein 1
IHD	ischemic heart disease
IL-1β	inhibits interleukin beta
IL-6	interleukin 6
JNK	c-Jun N-terminal kinases
LDL	low-density lipoprotein
LDL-C	low-density lipoprotein cholesterol
LNAME	NG-nitro-L-arginine-methyl-ester
LOX	lipoxygenase
LPS	lipopolysaccharide
LU	lutein
MAPK	mitogen-activated protein kinase
MCI	moderate cognitive impairment
MCP-1	monocyte chemoattractant protein 1
MD	Mediterranean diet

MMP-1	matrix metalloproteinase 1
MnSOD	manganese superoxide dismutase
MS	multiple sclerosis
NAD	nicotinamide adenine dinucleotide
NADPH	nicotinamide adenine dinucleotide phosphate
NCDs	noncommunicable diseases
NF-κB	nuclear factor-kappa B
NO	nitric oxide
NSAIDs	nonsteroidal anti-inflammatory drugs
O_2	oxygen
OHDs	oral hypoglycemic drugs
PD	Parkinson's disease
PEPCK	phosphoenol pyruvate carboxy kinase
PGC-1α	peroxisome proliferator-activated receptor-c coactivator-1α
PGH2	prostaglandin-endoperoxide
PGs	prostaglandins
PI3K	phosphatidylinositide 3-kinases
PIH	pregnancy-induced hypertension
PIN	prostatic intraepithelial neoplasia
PKA	protein kinase A
PKD	polycystic kidney disease
PPAR	peroxisome proliferator-activated receptor gamma
PRFs	polyphenol-rich foods
PUD	peptic ulcer disease
RGC	retinal ganglia cells
RNS	reactive nitrogen species
ROS	reactive oxygen species
RPE	retinal pigment epithelial
SLC2	solute carrier 2
SOD	superoxide dismutase
SREBP1c	sterol regulatory element-binding protein 1c
sVCAM-1	soluble vascular cell adhesion molecule 1
T2DM	type-2 diabetes mellitus
TGF-β-I	transforming growth factor β-I
TIN	tubule interstitial nephritis
TNBC	triple-negative breast cancer

TNF-α	tumor necrosis factor α
UDP-GT	UDP-glucuronyl transferase
uPA	urokinase plasminogen activator
USDA	United States Department of Agriculture
VCAM-1	vascular cell adhesion molecule 1
VDAC	voltage-gated anion channel
VEGF	vascular endothelial growth factor

PREFACE

Recently, there has been an upsurge of interest in the therapeutic potential of medicinal plants, which might be due to their phenolic compounds, specifically flavonoids. Flavonoids are a group of phenolic compounds having pharmacological properties and are mostly found in the stem, fruits, vegetables, bark, grains, roots of plants, etc.

The versatile health benefits of flavonoids in various epidemiogenic studies need to be focused on. Nowadays, these natural compounds are seeking researchers' attention worldwide due to their potential as anti-depressants, immune-boosters, anti-oxidants, anti-mitogenic, and anti-inflammatory. Flavonoids are proving themselves a boon to agriculture as well.

This book provides an overview of flavonoids including their potentiality of treating human health issues and environment-friendly concepts related to agriculture. The upcoming generations, including academicians and researchers, would benefit from this endeavor of ours.

This book, *Advances in Flavonoids for Human Health and Prevention of Diseases,* comprises 11 chapters describing the recent advances in polyphenol (flavonoids) derivatives for human health and in managing and preventing diseases.

Polyphenols and polyphenol-rich diets assist in preventing the start and development of a number of chronic pathological illnesses, including cancer, diabetes, cardiovascular disease (CVD), and aging. This is discussed in Chapter 1 – "Nutritional Properties of Polyphenols," written by Eknath D. Ahire and his associates.

Suruchi Singh and her colleagues provide an overview of the benefits of natural food dyes anthocyanidins and anthocyanins as nutraceuticals and their functions as food pigments in Chapter 2 – "Phytonutrients Anthocyanidins and Anthocyanins: Dietary and Medicinal Pigments with Possible Health Benefits."

A growing amount of research suggests that flavonoid-rich diets are good for improving health, and the prevention of NCDs. This topic is discussed in Chapter 3 – "Management and Prevention of Diseases by Flavonoids" by Flavonoids, written by Eknath D. Ahire.

Various flavonoids such as tea flavonoids, vegetable flavonoids, berry flavonoids, cocoa flavonoids, citrus flavonoids, etc., help regulate various body functions to prevent and control cardiac diseases and are reported in Chapter 4 – "Flavonoids and Cardiovascular Diseases" written by Monika Chauhan and her associates.

Day-by-day, incidences of neuro-degenerative illnesses are increasing in the old age population. Various new technologies are helpful in preventing or treating dysfunctions in the brain as well as its related cognitive loss. This is discussed in Chapter 5 – "Neuroprotective Effects of Flavonoids" written by Vipin Kesharwani and his associates.

Dietary flavonoids are secondary plant metabolites found in a variety of fruits, vegetables, and seeds. In addition, its role in diabetes is discussed in Chapter 6 – "Overview and Potential Benefits of Flavonoids in Diabetes," written by Sagarika Kabra and her associates.

Richa Singh and Deepika Dhawan summarize that flavonoids have a scope of its use as an adjunct medicine for diabetic patients in the future in Chapter 7 – "Anti-Diabetics Benefits of Dietary Flavonoids."

Tarkeshwar and associates report the use of flavonoids in pregnancy-induced disorder in Chapter 8 – "Application of Flavonoids in Pregnancy Induced Disorders."

Chapter 9 – "Flavonoids and Their Anti-Cancer Properties," written by Kamal Kant Sharma and associates, emphasizes the anti-cancer activity, its biochemistry, and the most probable mechanisms of action of some therapeutically important flavonoids, which can be further used in the development of novel nutraceutical active anti-cancer remedies.

Chapter 10 – "Plant Metabolites as Immunomodulators," written by Saumya Das and associates, emphasize the need to evaluate the potential of Indian ayurvedic remedies as adjuvants to counteract the side effects of modern therapies.

CHAPTER 1

NUTRITIONAL PROPERTIES OF POLYPHENOLS

SHUBHAM J. KHAIRNAR[1], EKNATH D. AHIRE,[2] MOHINI R. JAGTAP[3], KHEMCHAND R. SURANA[4], SANJAY J. KSHIRSAGAR[2], and RAJ K. KESERVANI[5]

[1]*Department of Pharmacology, MET's Institute of Pharmacy, Affiliated to Savitribai Phule Pune University, BKC, Adgaon, Nashik, Maharashtra, India*

[2]*Department of Pharmaceutics, MET's Institute of Pharmacy, Affiliated to Savitribai Phule Pune University, BKC, Adgaon, Nashik, Maharashtra, India*

[3]*JMCTE Institute of Pharmacy, Wadala Road, Nashik, Maharashtra, India*

[4]*Department of Pharmaceutical Chemistry, Shreeshakti Shaikshanik Sanstha, Divine College of Pharmacy, Satana, Nashik, Maharashtra, India*

[5]*Faculty of B.Pharmacy, CSM Group of Institutions, Prayagraj, Uttar Pradesh, India*

ABSTRACT

Polyphenols are secondary plant metabolites that protect plants from UV radiation and disease attack. Over the last decade, there has been a lot of interest in the health implications of dietary plant polyphenolic compounds

Advances in Flavonoids for Human Health and Prevention of Diseases. Nisha Sharma, Deepika Saini, Rajesh K. Kesharwani, Prakash Chandra Gupta, & Raj K. Keservani (Eds.)

as antioxidants. According to epidemiological research and linked meta-analyzes, long-term plant polyphenol-rich diets protect against cancer, cardiovascular disease (CVD), diabetes, osteoporosis, and neurological illnesses. This chapter discusses the pharmacological functions of plant polyphenols in terms of their importance to human health. The research findings in this assessment offer updated information on polyphenol organic outcomes and their significance to human health. Polyphenols and polyphenol-rich diets assist in preventing the start and development of a number of chronic pathological illnesses, including cancer, diabetes, CVD, and aging.

1.1 INTRODUCTION

Polyphenolic compounds are naturally occurring compounds found in a variety of foods such as fruits, vegetables, cereals, and beverages. For example, grapes, apples, pears, cranberries, and berries can comprise 200–300 mg of polyphenols per 100 g fresh weight. Polyphenols can be found in high concentrations in products derived from these fruits. A bottle of red wine, a cuppa tea, or a cup of coffee (Kesharwani et al., 2016) contains about 100 mg of polyphenols. Polyphenols can be found in cereals, dried beans, and chocolate. Polyphenols are secondary compounds found in plants that help protect them from UV radiation and illness (Surana et al., 2021b; Tian et al., 2019). Polyphenols can help with bitterness, astringency, color, flavor, odor, and oxidative stability in food. Epidemiological studies and accompanying meta-analyzes near the end of the 20th century suggested that long-term intake of plant polyphenol-rich diets conferred some protection against cancer, cardiovascular disease (CVD), diabetes, osteoporosis, and neurological diseases. Because of their potential health benefits, polyphenols and other dietary phenolics are generating a lot of attention in the scientific world. This study investigates the current state of knowledge due to the biological effects of dietary polyphenolic compounds and their role in human health and disease (Surana et al., 2021a; Belščak-Cvitanović et al., 2018).

1.2 STRUCTURE AND CLASSES OF POLYPHENOLS

In various plant species, over 8,000 polyphenolic compounds have been discovered. All plant polyphenol chemicals are derived from phenylalanine, a common intermediate, or a related precursor, shikimic acid. Although there are direct links between the carbohydrate (polysaccharide or monosaccharide) and an aromatic carbon, conjugated forms with one or more glucose molecules connected to hydroxyl groups are more common. Carboxylic and biological acids, amines, fats (Singh et al., 2017b), and other phenols are some of the compounds that are commonly associated with it. Polyphenols are classified based on the number of phenol rings and the structural elements that link these rings. The key classes include phenolic acids, flavonoids, stilbenes, and lignans (Singla et al., 2019). The chemical structures of the major polyphenol groups are depicted in Figure 1.1.

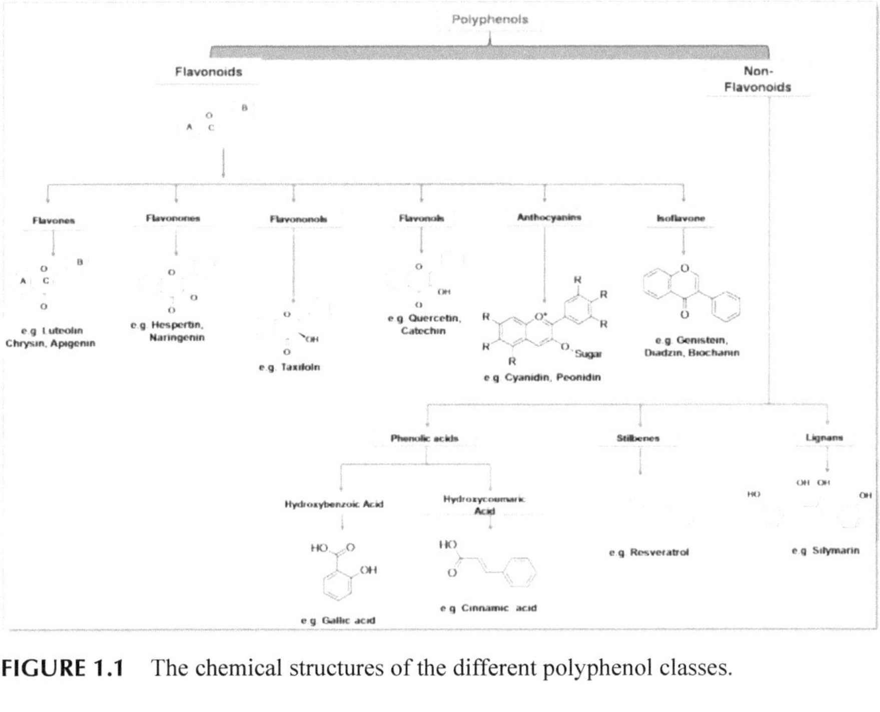

FIGURE 1.1 The chemical structures of the different polyphenol classes.

1.2.1 PHENOLIC ACIDS

Phenolic acids are divided into two categories: benzoic acid derivatives and cinnamic acid derivatives, which are abundant in foods. The hydroxybenzoic acid level in food plants is generally moderate, with the exception of certain red fruits, black radishes, and onions, which can have concentrations of tens of milligrams per kilogram of fresh weight. Hydroxycinnamic acids, which include p-coumaric, caffeic, ferulic, and sinapic, are more common than hydroxybenzoic (Robbins, 2003).

1.2.2 FLAVONOIDS

The polyphenols known as flavonoids have been investigated the most. This group's basic structure is an oxygenated heterocycle made up of two aromatic rings linked by three carbon atoms (Figure 1.2). Flavonoids are made up of approximately 4,000 different types, some of which are willing to take responsibility for the attractive colors of flowers, fruits, and leaves. Flavonoids are classified according to the type of heterocycle involved as flavones, flavanols, flavonols, anthocyanins, flavanones, and isoflavones. Individual variations within each category are caused by differences in the quantity and arrangement of hydroxyl groups, as well as the degree of alkylation and/or glycosylation. Among the most common are quercetin, myricetin, catechins, and other flavonoids (Panche et al., 2016).

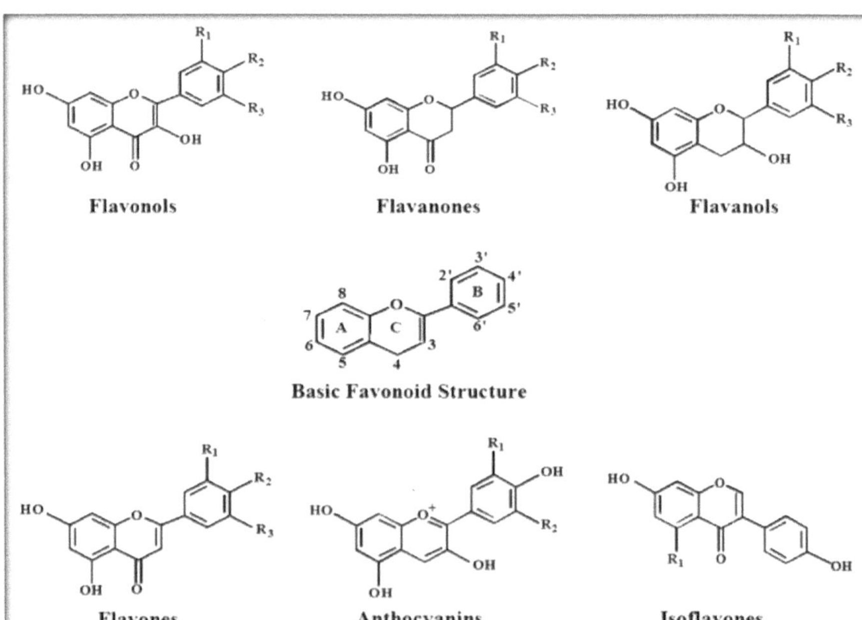

FIGURE 1.2 The chemical structures of flavonoids' subclasses.

1.2.3 STILBENES

A two-carbon methylene bridge connects two phenyl moieties in stilbenes. Stilbenes are extremely rare in human nutrition. The majority of stilbenes in plants are fungicidal phytoalexins, which are compounds produced primarily in response to infection or injury. Resveratrol (3,4,'5-trihydroxystilbene), found primarily in grapes, is among the most well-studied naturally occurring polyphenol stilbenes. Red wine is created from grapes and contains a high amount of antioxidant resveratrol (Shen et al., 2009).

1.2.4 LIGNANS

When two different cinnamic acid residues dimerize, they form lignans, which are diphenolic compounds with a structure of 2,3-dibenzylbutane (Figure 1.1). Several lignans, including secoisolariciresinol, are phytoestrogens. Linseed has the highest concentration of secoisolariciresinol (up to 3.7 g/kg dry weight) and the lowest concentration of matairesinol (Adlercreutz, 2007).

1.3 OCCURRENCE AND CONTENT

Plant phenolic dispersion is inconsistent at the tissue, cellular, and subcellular levels. Plant cell walls contain undissolved phenolics, whereas plant cell vacuoles contain soluble phenolic acids. Certain phenolics, such as quercetin, can be discovered in all plant products, such as fruit, vegetables, cereals, fruit juices, wine, tea, infusions, and so on. In contrast, isoflavones and flavanones can only be found in certain meals. Polyphenols can be found in a variety of foods in various combinations. Plants with higher phenolic content in their outer layers than their inner layers. Several variables influence plant polyphenol concentration, including ripeness during harvest, environmental factors, processing, and storage. Enviro and edaphic factors such as soil type, rainfall, and sun exposure influence the polyphenolic content of foods. The extent of fully ripe significantly influences the amounts and levels of various polyphenols. The quantity of phenolic acid decreases as ripening advances, whereas anthocyanin concentrations rise. Many polyphenols, especially phenolic acids, play a direct role in plants' reactions to a variety of strains. They aid in the repair of damaged areas through lignification, have antibacterial properties, and their concentrations may rise after infection. Another factor that directly impacts the polyphenol concentration of foods in storage. According to studies, the polyphenolic content of foods changes during storage as a result of the simple oxidative stress of these polyphenols. Oxidation reactions produce polymerized chemicals that affect food quality, particularly color and organoleptic properties. Such changes can be beneficial, as in the case of black tea, or harmful, as in the case of fruit browning. Wheat flour loses a substantial proportion of phenolic acids as it is kept. In terms of quality, the flour still had the same polyphenolic compounds after six months of storage, but their concentrations were 70% lower than when it was NewCold storage; on the other hand, it has little effect on the polyphenol content of apples, pears, and onions.

Polyphenol concentrations are significantly affected by cooking. After 15 minutes of boiling, onions and tomatoes lose 75 to 80% of their quercetin content, 65% after cooking in a microwave oven, and 30% after frying (Reddy, 2002; Kuchta et al., 2012). On the other hand, the polyphenol content of apples, pears, and onions is only slightly affected by cold storage.

1.4 BIOAVAILABILITY OF POLYPHENOLS

The proportion of a nutrient that is digested, absorbed, and metabolized normally is referred to as bioavailability. Each polyphenol's bioavailability varies, but there is no correlation between the number of polyphenols in the diet and their bioavailability in the human body. The majority of polyphenols observed in food are esters, glycosides, or polymers that can't be absorbed of their herbal form (Rane et al., 2020a–c). The majority of polyphenols observed in food are esters, glycosides, or polymers that can't be absorbed of their herbal (Sharma et al., 2020) form. Polyphenols (Keservani & Sharma, 2014) are conjugated in intestinal cells and later withinside the liver through methylation, sulfation, and/or glucuronidation, which occurs during absorption. As a result, the forms that reach the bloodstream and tissues differ from those found in food, making it difficult to identify and assess the biological activity of all metabolites. Importantly, rather than their content, the chemical structure of polyphenolic compounds influences the rate and amount of absorption, as well as the environment of the metabolites circulating in the plasma. Phenolic acids biological activities vary greatly because the most lavish polyphenols in our regime do not always have the utmost concentrations of active metabolites in target tissues. The increased plasma antioxidant capacity after polyphenol-rich meals is indirect evidence of absorption through the gastrointestinal barrier. Polyphenols are absorbed at various sites in humans. Some polyphenols are absorbed efficiently in the stomach, while others are absorbed in the intestine or other parts of the digestive tube. All flavonoids (Keservani & Sharma, 2014) in foods, with the exception of flavanols, are glycosylated. Glycosides' fate in the stomach is still unknown. Stomach acid is unlikely to hydrolyze the majority of glycosides. And thus, arrive in the intestine intact; only aglycones and a few glucosides can be absorbed in this environment. According to animal studies, some flavonoids, such as quercetin, can be absorbed in the stomach, but not their glycosides. Furthermore, anthocyanins have recently been discovered to be absorbed from the stomach in rats and mice. One theory holds that glucosides are transported into enterocytes by the sodium-dependent glucose transporter SGLT1 and then degraded by a cytosolic-glucosidase. In contrast to quercetin, isoflavones appear to have less of a glucosylation effect on absorption. Proanthocyanidins vary from maximum different plant polyphenols because of their polymeric shape and excessive molecular weight. Despite the truth that oligomers large than trimers are not going to be absorbed in their local forms in the small bowel, their absorption across the gut barrier should be limited. When hydroxycinnamic acids are consumed in their free form, they are quickly absorbed by the small intestine and conjugated into flavonoids. However, so because intestinal mucosa, liver, and plasma lack esterases capable of hydrolyzing chlorogenic acid to release caffeic acid, and hydrolysis can only be performed by the bacterial flora present in the colon, these compounds are naturally esterified in plant products, reducing absorption. Some polyphenols are not absorbed from the gastrointestinal tract and intestine, despite the fact that the majority are. These polyphenolic compounds reach the large intestine, where bacteria hydrolyze glycosides into aglycones, which are then extensively metabolized into various aromatic acids (D'Archivio et al., 2010). At various points along the heterocycle, aglycones are divided based on their chemical structure, yielding distinct acids that are then metabolized to benzoic acid derivatives. Polyphenols undergo a variety of conjugation mechanisms after absorption. These activities are primarily methylation, sulfation, and glucuronidation, which represent a metabolic detoxication process shared by many xenobiotics that increases hydrophilicity and thus aids biliary and urine excretion. Polyphenol methylation is also very specific; it usually occurs in the polyphenol's C3 position, but it can also occur in the polyphenol's C4' position: in fact, a significant amount of 4′-methyl epigallocatechin (EGC) has been found in human plasma after tea consumption. Sulfotransferases are enzymes that help facilitate the transfer of a sulfate moiety during the sulfonation process. Sulfation occurs primarily in the liver, but the precise location of polyphenol sulfation is unknown. Green tea catechins have high-efficiency complexation mechanisms, and free aglycones are frequently absent or present in low amounts in plasma following nutritional doses. Because the places of the conjugating groups at the polyphenol structure can modify the biological properties of the conjugates, figuring out the circulating metabolites, consisting of the character and placements of the conjugating groups at the phenolic acids structure, is critical (Hu, Wu, & Liu, 2017). Polyphenol metabolites bind to proteins (Jarouliya & Keservani, 2019a) and circulate in the blood system; albumin is the main protein involved in the binding. Albumin is needed for polyphenol bioavailability. Polyphenol bioavailability is dependent on albumin. Polyphenols' affinity for albumin varies depending on their chemical structure. Binding to albumin may affect the rate of metabolite elimination as well as metabolite distribution to cells and tissues. It's possible that the amount of unbound metabolite in the cell influences metabolite absorption. Finally, it is unknown whether polyphenols must be free to be biologically active or whether albumin-bound polyphenols can be biologically active. Polyphenol accumulation in tissues is the utmost significant phase of polyphenol metabolism since this is the concentration at which polyphenols are biologically active and exert their effects. Polyphenols have been shown in studies to permeate tissues, particularly those that process them, such as the intestine and liver. Polyphenols and derivatives are excreted in the bile and urine. Highly conjugated metabolites are more probable to be eliminate waste in bile, whereas tiny conjugates, such as monosulfates, are preferentially excreted in urine. Citrus fruit (Keservani et al., 2010a, b) flavanones have a high urinary excretion percentage, which decreases as you progress from flavonols to isoflavones. As a result, polyphenols' health benefits are dependent on both their intake and bioavailability (Manach et al., 2005; Teng & Chen, 2019).

1.5 HUMAN DISEASES AND POLYPHENOLS

Epidemiologic studies have repeatedly demonstrated a link between the consumption of polyphenolic-rich foods and the threat of chronic human diseases. 1,5-Polyphenolic groups can accept an electron to produce comparatively stable phenoxyl radicals, causing chain oxidation events in biological components to be disrupted. Diets and beverages high in polyphenols have long been thought to increase plasma antioxidant capacity. The presence of decreasing polyphenolic compounds and their metabolites in plasma, their outcomes at the concentrations of different decreasing agents (polyphenolic compounds sparing different endogenous antioxidants), or their impact at the absorption of pro-oxidative food components, such as iron, could all explain the high in plasma antioxidative capacity following polyphenol-rich food consumption. Antioxidant consumption has been linked to lower levels of oxidative damage to lymphocyte DNA. Similar findings with polyphenol-rich foods (PRFs) and beverages have been made, confirming polyphenols' preventive benefits. According to mounting evidence, Polyphenols have the potential to protect cell constituents. From oxidative damage and, as a result, lower the risk of a variety of degenerative diseases linked to oxidative stress (Pandey & Rizvi, 2009; Rasouli et al., 2017).

1.5.1 CARDIO-PROTECTIVE EFFECT

A number of studies have found that consuming polyphenols reduces the risk of coronary heart disease (CHD). Atherosclerosis is an inflammatory disease that affects the vulnerable areas of medium-sized arteries over time. Atherosclerotic lesions can be clinically silent for decades before they become active and cause pathological conditions such as myocardial ischemia, unstable angina, and sudden cardiac arrest. Polyphenols are powerful LDL oxidation inhibitors, which is thought to play an important role in the progression of atherosclerosis. Polyphenols may offer protection against Polyphenols (Keservani et al., 2016a, 2020) may protect against cardiovascular disorders through antioxidant, anti-platelet, and anti-inflammatory properties, in addition to raising HDL and improving endothelial function. Polyphenols also may aid in the retention of the atheroma plaque. Quercetin, an abundant polyphenol found in onions, has been shown to be inversely proportional related to coronary artery disease death by decreasing metalloproteinase 1 (MMP1) expression and atherosclerotic plaque disintegration. Tea catechins have been shown to inhibit smooth muscle cell invasion and proliferation in the artery wall, potentially slowing the progression of atherosclerosis. Because of their ability to prevent platelet aggregation, polyphenols may have antithrombotic properties. Red wine and non-alcoholic wine consumption reduce bleeding time and platelet aggregation. Consumption of red wine and non-alcoholic wine reduces bleeding time and platelet aggregation. Consuming red wine or grape juice helps to prevent thrombosis caused by coronary artery stenosis. Polyphenols have been proposed as a predictive tool for CHD because they can alleviate endothelial dysfunction associated with several atherosclerosis risk factors prior to plaque formation. For 30 days, drinking 450 mL of black tea increased arterial dilation 2 hours after consumption, while drinking 240 mL of red wine prevented endothelial dysfunction caused by a high-fat diet (Ullah et al., 2020). A cross-sectional study of 218 women aged 18 and up was conducted at the age of 70 and discovered that drinking black tea on a regular basis for a long period of time lowers blood pressure (BP). 4-O-methyl gallic acid (4OMGA), a biomarker for tea polyphenols, was measured in the body. Tea consumption, and thus 4OMGA excretion, has been linked to lower BP. Polyphenolic compounds may be the active ingredient in lowering BP. The effect could be attributed to antioxidant activity, improved endothelial function, or estrogen-like activity. Resveratrol, a polyphenol found in wine, inhibits platelet aggregation by inhibiting the activity of cyclooxygenase 1 (COX 1), which produces thromboxane A2, a platelet aggregation inducer and vasoconstrictor. Furthermore, it has been demonstrated that resveratrol relaxes isolated arteries and rat aortic rings. Resveratrol's ability to stimulate Ca++-activated K+ channels and promote nitric oxide (NO) signaling in the endothelium are two other mechanisms by which it exerts a vasorelaxant effect. Currently, there is a well-established link between CVD and LDL oxidation. LDL particle oxidation is linked to an increased risk of coronary artery disease and myocardial infarction. Resveratrol has been shown in studies to reduce LDL particle oxidation by either chelating copper or directly scavenging free radicals. The active ingredient in red wine, resveratrol, is thought to be responsible for the "French Paradox," or the French people's low prevalence of CVD despite their high-fat diet and smoking habits. Several epidemiological studies investigated the relationship between polyphenol intake or consumption of PRFs and the occurrence of CVDs, and it was discovered that consumption of a polyphenol-rich diet was linked to a lower risk of myocardial infarction in both case-control and cohort studies (Luo et al., 2017; Patel et al., 2012).

1.5.2 ANTI-CANCER EFFECT

Polyphenols protect human cancer cell lines, reducing the number of tumors or their growth. These effects have been seen in a variety of places, including the mouth, stomach, duodenum, colon, liver, lung, skin, and mammary gland. Several phenolics have been studied, including quercetin, catechins, isoflavones, lignans, flavanones, ellagic acid, red wine polyphenols, resveratrol, and curcumin; all of these demonstrated barrier protection effects in different models, though their modes of action differed. Carcinogenesis, or cancer development, is a multistage, microevolutionary process. Carcinogenesis is divided into three stages: initiate, promote, and progress. The initiation of a cell is a heritable abnormality. Cells can become cancerous if promotion and advancement are not followed. Promotion, on the contrary, involves the selection and clonal proliferation of newly formed cells and is influenced by variables that do not alter DNA sequences. As mechanisms of action for polyphenols' chemoprevention effect, estrogenic/antiestrogenic activity, antiproliferation, induction of cell cycle arrest or apoptosis, oxidation prevention, induction of detoxification enzymes, host immune (Keservani et al., 2015d) system regulation, anti-inflammatory activity, and changes in cellular signaling have all been identified. Polyphenols have an effect on pro-carcinogen metabolism by modulating the expression of cytochrome P450 enzymes involved in the conversion of pro-carcinogens to carcinogens. They may also aid excretion by boosting the expression of phase II conjugating enzymes. Polyphenol toxicity may be to blame for phase II enzyme stimulation. Polyphenols in the body can produce potentially hazardous quinones, which are also substrates for these enzymes. Consumption of polyphenols may then activate these enzymes for detoxication, resulting in an overall improvement in our defenses against hazardous xenobiotics. In men with high-grade prostatic intraepithelial neoplasia (PIN) lesions, tea catechins in capsule form were shown to have a cancer-preventive effect by preventing the progression of high-grade PIN lesions to cancer. The polyphenols in black tea, theaflavins, and thearubigins, have additionally been proven to have anticancer properties. Black tea polyphenols have been discovered to inhibit boom and growth apoptosis in Du a 145 prostate most cancers cells. A better stage of insulin-like boom factor-1 (IGF-1) has been related to a better threat of growing prostate cancers. IGF-1 binding to its receptor is a part of the signaling cascade that results in mobileular proliferation. Black tea polyphenol has been verified to save you IGF-1-brought on development of prostate most cancers cells into the S section of the mobileular cycle at a dose of 40 mg/ml. When mice have been uncovered to benzo(a) pyrene, quercetin turned into discovered to have anticancer properties, which have been attributed to its loose radical scavenging capacity (Batra & Sharma, 2013; Rawson et al., 2014).

Resveratrol has been verified to be beneficial with inside the prevention of most cancers in any respect stages, inclusive of lung, skin, breast, prostate, stomach, and colorectal most cancers. It has additionally been proven to suppress angiogenesis and metastasis. Resveratrol has been proven to persuade numerous pathways concerned in mobileular proliferation, death, and irritation in human mobileular cultures. Resveratrol has been verified to inhibit cyclooxygenase (COX), hydroperoxidase, protein kinase C, Bcl-2 phosphorylation, Akt, focal adhesion kinase, NFB, matrix metalloprotease-9, and mobileular cycle regulators. These and different *in vitro* and *in vivo* studies indicate that dietary (Keservani et al., 2020; Rane & Keservani, 2020) polyphenols, in combination with chemotherapeutic or cytotoxic agents, can be used in human cancer chemoprevention to effectively treat drug-resistant tumor cells (Ravishankar et al., 2013).

1.5.3 ANTI-DIABETIC EFFECT

Impaired glucose metabolism causes physiological imbalance with the onset of hyperglycemia and, eventually, diabetes mellitus (DM). Diabetes can be classified into two types: type-1 and type-2. Diabetes, according to research, affects a number of physiological indicators in the body. Diabetes causes the progressive development of specific complements such as retinopathy, which affects the eyes and can lead to blindness; nephropathy, which affects renal functions; and neuropathy, which is associated with an increased risk of amputations, foot ulcers, and autonomic disturbances such as sexual dysfunctions. Numerous studies have shown that polyphenols have anti-diabetic properties. The anti-diabetic properties of tea catechins have been studied. Polyphenols can affect glycemia in a number of ways, including inhibiting glucose absorption in the gut and peripheral tissue uptake. Diacetylated anthocyanins had hypoglycemic effects with maltose as a glucose source, but not with sucrose or glucose, at 10 mg/kg food intake. This demonstrates that – glucosidase inhibition in the gastrointestinal mucosa is the cause of these effects. Catechin inhibited amylase and sucrase in rats at doses of 50 mg/kg diet or higher. Polyphenols have been studied for their ability to inhibit glycosidases and glucose transporters in the intestine. Individual polyphenols, such as (+) catechin, (–) epicatechin, (–) epigallocatechin, epicatechin gallate (ECG), isoflavones from soyabeans, tannic acid, glycyrrhizin from licorice root, chlorogenic acid, and saponins, inhibit S-Glut-1-mediated glucose transport in the intestine. Saponins also prevent glucose from being transferred from the stomach to the small intestine. Resveratrol has also been linked to anti-diabetic effects. One of the mechanisms proposed to explain this stilbene's anti-diabetic activity is the modulation of SIRT1, which improves diabetic rats' whole-body glucose homeostasis and insulin sensitivity. Grape seed polyphenols were found to protect LLC-PK1 cells from high glucose-induced cytotoxicity and oxidative stress. Resveratrol suppresses diabetes-induced kidney alterations (diabetic nephropathy) and significantly reduces renal dysfunction and oxidative stress in diabetic rats. Treatment with resveratrol also reduced insulin output and delayed the development of insulin resistance (Babu et al., 2013). A plausible mechanism was suggested to be the blockage of K + ATP and K + V channels in beta cells. Onion polyphenols, particularly quercetin, have been shown to have powerful anti-diabetic properties. A recent study found that quercetin can protect diabetics from oxidative stress-related changes. In diabetics, quercetin significantly reduced lipid peroxidation and inhibited the antioxidant (Singh et al., 2017a; Keservani et al., 2015c) system. Polyphenolic acids, flavonoids, protocatechuic acid, and anthocyanins are all found in Hibiscus sabdariffa extract. Lee and colleagues Polyphenols found in *Hibiscus sabdariffa*. Extracts have been shown to reduce diabetic nephropathy, including pathology, serum lipid profile, and kidney function, and oxidative indicators in the kidney. FA is an effective anti-diabetic drug, according to several lines of evidence. FA reduced blood glucose levels, followed by an increase in plasma insulin and a negative relationship between blood glucose and plasma insulin.

Formalized paraphrase Anti-Aging Property the accumulation of a variety of harmful changes in cells and tissues with increasing age, resulting in an increase in the risk of disease and mortality, is defined as aging (Kawser et al., 2016).

1.5.4 ANTI-AGING EFFECT

Aging is defined as the accumulation of a variety of harmful changes in cells and tissues with increasing age, resulting in an increase in the risk of disease and mortality. The free radical/oxidative strain principle is one of the maximum extensively usual hypotheses for information on the mechanisms of getting older. Some oxidative harm takes place even beneath Neath everyday conditions; however, because the performance of antioxidative and restore mechanisms declines with age, the fee of this harm increases. Plasma antioxidant capability is associated with antioxidant meals consumption; ingesting an antioxidant-wealthy eating regimen has been established to assist lessen the bad consequences of getting older and behavior. Several research have advised that a mixture of antioxidant/anti-inflammatory polyphenolic chemical substances observed in culmination and greens may be powerful anti-aging agents. Anthocyanins, a subset of flavonoids, are ample in brightly colored culmination along with berry culmination, Concord grapes, and grape seeds. Anthocyanins, which can be observed in culmination, were proven to have effective antioxidant and anti-inflammatory properties, in addition to the capacity to suppress lipid peroxidation and the inflammatory mediator's cyclo-oxygenase (COX)-1 and -2. In restoring age-associated deficiencies in mind and behavioral feature in aged rats, in keeping with the findings. According to a brand new study, tea catechins have amazing anti-aging properties, and ingesting in experienced tea wealthy in those catechins might also additionally assist to put off the onset of getting older. Polyphenols also can assist in mitigating the bad consequences of getting older at the anxious device and mind. Dietary polyphenols' ability to cross the blood-brain barrier (BBB), which tightly limits the entry of metabolites, nutrients, and medications into the brain, is critical for their relevance in the protection of the aging brain. Resveratrol has been demonstrated to continuously increase life expectancy; its activity is linked to calorie restriction or partial food deprivation (Li et al., 2019). Resveratrol, a polyphenol found in grapes, is a relatively new anti-aging agent. An early resveratrol target was identified as the sirtuin class of nicotinamide adenine dinucleotide (NAD)-dependent deacetylases. In mammals, seven sirtuins have been discovered, with SIRT-1 thought to mediate the health and longevity benefits of calorie restriction and resveratrol. Resveratrol improved insulin sensitivity, decreased IGF-1 expression, and increased the activity of AMP-activated protein kinase (AMPK) and peroxisome proliferator-activated receptor-c coactivator-1α (PGC-1α). When the mechanism was looked into, it was discovered that it activated forkhead box O (FOXO), which regulates the expression of genes involved in longevity and stress resistance, as well as insulin-like growth factor binding protein 1 (IGFBP-1). Experiments have shown that resveratrol can increase the lifespan of yeast. Candida albicans, Saccharomyces cerevisiae, Drosophila melanogaster (fruit fly), and Candida albicans, *Nothobranchius furzeri* (nematode worm), and *Nothobranchius elegans* (seasonal fish). Freshly, quercetin has been linked to an anti-aging effect (Guo et al., 2019; Imam et al., 2015).

1.5.5 NEURO-PROTECTIVE EFFECTS

Oxidative stress and harm to brain macromolecules are not unusual place in neurodegenerative diseases. Alzheimer's disease (AD) is a not unusual place neurodegenerative disease that influences as much as 18 million human beings worldwide. Polyphenols have an excessive antioxidative capacity, so ingesting them may also shield towards neurological diseases. When compared to individuals who drank much less or did now no longer drink at all, individuals who drank 3 to 4 glasses of wine in keeping with day had an 80% decrease in prevalence of dementia and AD. Resveratrol, that's plentiful in wine, has been related to some of fitness benefits., efficiently scavenges O_2^- and OH• unfastened radicals *in vitro*, in addition to lipid hydroperoxyl free radicals; this effective antioxidant interest is probably to be concerned in mild pink wine consumption's shielding impact towards dementia withinside the elderly. Resveratrol inhibits nuclear issue B signaling and for this reason protects against microglia-dependent-amyloid toxicity in an AD model, and this pastime is related to SIRT-1 activation. Juices from high-polyphenol fruits and vegetables should be consumed at least three times per week, according to research (Keservani et al., 2016b), in order to help prevent Alzheimer's disease. Because in their energy to steer and manipulate numerous cell approaches including signaling, proliferation, apoptosis, redox balance, and differentiation, polyphenols from end result and vegetables seem like treasured ability agents in neuroprotection (Orhan et al., 2015). Polyphenol treatment, according to Aquilano et al. protects against Parkinson's disease (PD), a neurological illness characterized by dopaminergic cell loss in happen in the substantia nigra zona compacta. In nutritional studies, has been associated with a lower risk of PD (Keservani et al., 2015a). Epigallocatechin gallate (EGCG) has been demonstrated in animal models to protect against the neurotoxic MPTP (N-methyl-4-phenyl-1,2,3,6-tetrahydropyridine), a Parkinson's-like disease inducer, by either competitively limiting drug uptake or scavenging MPTP-mediated radical generation. EGCG may also able protect neurons by activating a number of signaling pathways, including those involving MAP kinases, which are essential for cell survival. The ability of catechins to chelate iron contributes to their potential therapeutic role in PD. This property boosts their antioxidant activity by limiting free radical production by redox-active transition metals. Furthermore, the antioxidant function is linked to the activation of antioxidant and detoxifying enzymes, particularly in the brain (Keservani et al., 2015b), which lacks a well-organized antioxidant defense system. Ferulic acid, a polyphenol found in maize bran, has also been shown to be beneficial in the treatment of AD. This effect is due to its antioxidant and anti-inflammatory properties (Oyenihi et al., 2016; Ghaznavi et al., 2015).

1.5.6 OTHERS

Aside from the clinical events mentioned above, polyphenols have a number of additional health benefits. Dietary polyphenols have anti-inflammatory properties that can help with asthma treatment. Asthmatics' airways narrow or block when they become irritated. Because of this, the air has a difficult time moving in and out. Asthma, coughing, shortness of breath, and chest tightness, among other symptoms, can be caused by this narrowing or obstruction. Apple consumption was found to have a negative correlation with asthma prevalence and incidence, as well as a positive correlation with lung function, implying that polyphenols may protect against obstructive lung disease. In asthmatic patients, increased intake of the soy isoflavone genistein was linked to improved lung function.

Polyphenol consumption has also been linked to a lower risk of osteoporosis. Supplementing the weight loss plan with genistein, daidzein, or its glycosides for numerous weeks protects towards the ovariectomy-brought about lower in bone mineral density and trabecular volume. Polyphenols additionally assist to shield the pores and skin from UV rays. Animal research display that polyphenols observed in tea, whilst taken orally or carried out topically, can lessen the bad outcomes of UV publicity at the pores and skin, including pores and skin damage, erythema, and lipid peroxidation. Black tea polyphenols are stated to be useful resource mineral absorption withinside the gut and to have antiviral properties. Theaflavins, which are observed in black tea, had been observed to have anti-HIV-1 activity. These polyphenols inhibited HIV-1 mobileular contamination of goal cells. HIV-1 enters the goal mobileular through fusing the virus's glycoprotein (GP) and envelope with the mobileular membrane of the host mobileular. Heptad repeat devices fuse on the N and C terminals of GP41 (membrane protein) right here on viral membrane to shape the fusion energetic GP41 core, that's a six-helical package deal. The aflavins had been observed to save the virus from forming the specified six-helix package deal for contamination. The aflavins had been determined to save the virus from forming the six-helix package deal required for contamination. It has been proven that the corona virus that causes the severe acute respiratory syndrome (SARS) is inhibited by the aflavin three digallate and three gallate. The inhibition of the chymotrypsin-like protease (3CL Pro), that's worried in posttranslational change during viral multiplication, precipitated the antiviral effect (González et al., 2011; Holiman et al., 1996).

1.6 CONCLUSION

The findings of the studies included in this review provide an up-to-date understanding of polyphenol biological effects and their importance to human health. Polyphenols and polyphenol-rich diets help to prevent the beginning and progression of a variety of chronic pathological illnesses, such as cancer, diabetes, CVD, and aging. Although epidemiological studies can explain some biological effects, the mechanism of action for some polyphenol effects is not fully understood. The design of such studies will be aided by a better understanding of some polyphenol bioavailability characteristics, such as absorption, accumulation, and elimination kinetics.

The role of polyphenols in human health is still being debated. According to current scientific understanding, polyphenols hold great potential for the prevention of chronic human diseases.

ACKNOWLEDGMENT

We, the authors, thank the METs, Institute of Pharmacy, BKC, affiliated under Savitribai Phule Pune University, Adgaon, Nashik, MH, and India. EDA is extremely thankful to the Ministry of Tribal Affairs Government of India (NFST) for providing financial assistance in fellowship (Award No.: 202021-NFST-MAH-01235).

KEYWORDS

- **antioxidants**
- **bioavailability**
- **blood pressure**
- **cardiovascular disease**
- **cyclooxygenase 1**
- **human diseases**
- **polyphenols**

REFERENCES

Adlercreutz, H., (2007). Lignans and human health. *Critical Reviews in Clinical Laboratory Sciences*, *44*(5, 6), 483–525.

Babu, P. V. A., Liu, D., & Gilbert, E. R., (2013). Recent advances in understanding the anti-diabetic actions of dietary flavonoids. *The Journal of Nutritional Biochemistry, 24*, 1777–1789.

Batra, P., & Sharma, A. K., (2013). Anti-cancer potential of flavonoids: Recent trends and future perspectives. *3 Biotech, 3*, 439–459.

Belščak-Cvitanović, A., Durgo, K., Huđek, A., Bačun-Družina, V., & Komes, D., (2018). Overview of polyphenols and their properties. In: *Polyphenols: Properties, Recovery, and Applications* (pp. 3–44). Woodhead Publishing.

D'Archivio, M., Filesi, C., Varì, R., Scazzocchio, B., & Masella, R., (2010). Bioavailability of the polyphenols: Status and controversies. *International Journal of Molecular Sciences, 11*(4), 1321–1342.

Ghaznavi, H., Najafi, R., Mehrzadi, S., Hosseini, A., Tekyemaroof, N., Shakeri-Zadeh, A., Rezayat, M., & Sharifi, A. M., (2015). Neuro-protective effects of cerium and yttrium oxide nanoparticles on high glucose-induced oxidative stress and apoptosis in undifferentiated PC12 cells. *Neurological Research, 37*, 624–632.

González, R., Ballester, I., López-Posadas, R., Suárez, M., Zarzuelo, A., Martínez-Augustin, O., & Medina, F. S. D., (2011). Effects of flavonoids and other polyphenols on inflammation. *Critical Reviews in Food Science and Nutrition, 51*, 331–362.

Guo, C., Zhang, H., Guan, X., & Zhou, Z., (2019). The anti-aging potential of neohesperidin and its synergistic effects with other citrus flavonoids in extending chronological lifespan of saccharomyces cerevisiae BY4742. *Molecules, 24*, 4093.

Holiman, P. C., Hertog, M. G., & Katan, M. B., (1996). Analysis and health effects of flavonoids. *Food Chemistry, 57*, 43–46.

Hu, M., Wu, B., & Liu, Z., (2017). Bioavailability of polyphenols and flavonoids in the era of precision medicine. *Molecular Pharmaceutics, 14*(9), 2861–2863.

Imam, S., Azhar, I., Perveen, S., Hussain, S. G., & Mahmood, Z. A., (2015). Studies on *in vitro* antioxidant activity and total flavonoid contents of a cream formulation to correlate its anti-aging effect. *World Journal of Pharmaceutical Research, 4*, 1646–1655.

Jarouliya, U., & Keservani, R. K., (2019a). Protein function as cell surface and nuclear receptor in human diseases. In: Sanjay, B., & Debarshi, K. M., (eds.), *Medicinal Chemistry with Pharmaceutical Product Development* (pp. 1–32). Apple Academic Press, CRC Press, Taylor & Francis Group. Chapter 1. ISBN: 9781771887106.

Kawser, H. M., Abdal, D. A., Han, J., Yin, Y., Kim, K., Kumar, S. S., Yang, G. M., Choi, H. Y., & Cho, S. G., (2016). Molecular mechanisms of the anti-obesity and anti-diabetic properties of flavonoids. *International Journal of Molecular Sciences, 17*, 569.

Keservani, R. K., & Sharma, A. K., (2014). Flavonoids: Emerging trends and potential health benefits. *Journal of Chinese Pharmaceutical Sciences, 23*(12), 815.

Keservani, R. K., Kesharwani, R. K., & Sharma, A. K., (2015c). Pulmonary and respiratory health: Antioxidants and nutraceuticals. In: Debasis, B., Harry, G. P., & Anand, S., (eds.), *Nutraceutical and Functional Foods in Human Life and Disease Prevention* (pp. 279–295). Boca Raton, Fl. CRC Press, Taylor and Francis, Chapter 17. ISBN: 9781482237214.

Keservani, R. K., Kesharwani, R. K., Sharma, A. K., & Ahmad, M. F., (2015a). Stress, Parkinson's, Alzheimer's disease: Role of dietary supplements. In: Debasis, B., Harry, G. P., & Anand, S., (eds.), *Nutraceutical and Functional Foods in Human Life and Disease Prevention* (pp. 241–253). Boca Raton, Fl. CRC, Press, Taylor and Francis, Chapter 14. ISBN: 9781482237214.

Keservani, R. K., Kesharwani, R. K., Sharma, A. K., & Jarouliya, U., (2015d). Dietary supplements, nutraceutical and functional foods in immune response (immunomodulators). In: Debasis, B., Harry, G. P., & Anand, S., (eds.), *Nutraceutical and Functional Foods in Human Life and Disease Prevention* (pp. 343–358). Boca Raton, Fl. CRC Press, Taylor and Francis, Chapter 20, ISBN: 9781482237214.

Keservani, R. K., Kesharwani, R. K., Sharma, A. K., Vyas, N., & Chadoker, A., (2010b). Nutritional supplements: An overview. *International Journal of cUrrent Pharmaceutical Review and Research, 1*(1), 59–75.

Keservani, R. K., Kesharwani, R. K., Vyas, N., Jain, S., Raghuvanshi, R., & Sharma, A. K., (2010a). Nutraceutical and functional food as future food: A review. *Der Pharmacia Lettre, 2*(1), 106–116.

Keservani, R. K., Sharma, A. K., & Kesharwani, R. K., (2016a). Nutraceutical and functional foods for cardiovascular health. In: *Food Process Engineering* (pp. 257–278, 291–312). Apple Academic Press, CRC Press. ISBN: 9781771884020.

Keservani, R. K., Sharma, A. K., & Kesharwani, R. K., (2016b). Medicinal effect of nutraceutical fruits for the cognition and brain health. *Scientifica*, 10. Article ID 3109254. doi: 10.1155/2016/3109254.

Keservani, R. K., Sharma, A. K., & Kesharwani, R. K., (2020). *Nutraceuticals and Dietary Supplements: Applications in Health Improvement and Disease Management.* CRC Press. ISBN: 9781771888738.

Keservani, R. K., Singh, S., Singh, V., Kesharwani, R. K., & Sharma, A. K., (2015). Nutraceuticals and functional foods in the prevention of mental disorder. In: Debasis, B., Harry, G. P., & Anand, S., (eds.), *Nutraceutical and Functional Foods in Human Life and Disease Prevention* (pp. 255–268). Fl. CRC Press, Taylor and Francis, Chapter 15b. ISBN: 9781482237214.

Kesharwani, R. K., Singh, P., & Keservani, R. K., (2016). Green coffee bean extract and chlorogenic acids: Chemistry and novel antioxidant benefits. In: Bagchi, D., Moriyama, H., & Swaroop, A., (eds.), *Green Coffee Bean Extract in Human Health* (pp. 1–17). Boca Raton, Taylor and Francis, USA. Chapter 1. ISBN: 9781498716376.

Kuchta, K., Ortwein, J., & Rauwald, H. W., (2012). Leonurus japonicus, *Leonurus cardiaca, Leonotis leonurus*: A novel HPLC study on the occurrence and content of the pharmacologically active guanidino derivative leonurine. *Die Pharmazie-An International Journal of Pharmaceutical Sciences, 67*(12), 973–979.

Li, M. R., Zhou, Y. Z., Du, G. H., & Qin, X. M., (2019). Research progress about the anti-aging effect and mechanism of flavonoids from traditional Chinese medicine. *Acta Pharmaceutica Sinica*, 1382–1391.

Luo, Y., Shang, P., & Li, D., (2017). Luteolin: A flavonoid that has multiple cardio-protective effects and its molecular mechanisms. *Frontiers in Pharmacology, 8*, 692.

Manach, C., Williamson, G., Morand, C., Scalbert, A., & Rémésy, C., (2005). Bioavailability and bioefficacy of polyphenols in humans. I. Review of 97 bioavailability studies. *The American Journal of Clinical Nutrition, 81*(1), 230S-242S.

Orhan, I., Daglia, M., Nabavi, S., Loizzo, M., Sobarzo-Sánchez, E., & Nabavi, S., (2015). Flavonoids and dementia: An update. *Current Medicinal Chemistry, 22*, 1004–1015.

Oyenihi, O. R., Afolabi, B. A., Oyenihi, A. B., Ogunmokun, O. J., & Oguntibeju, O. O., (2016). Hepato-and neuro-protective effects of watermelon juice on acute ethanol-induced oxidative stress in rats. *Toxicology Reports, 3*, 288–294.

Panche, A. N., Diwan, A. D., & Chandra, S. R., (2016). Flavonoids: An overview. *Journal of Nutritional Science, 5.*

Pandey, K. B., & Rizvi, S. I., (2009). Plant polyphenols as dietary antioxidants in human health and disease. *Oxidative Medicine and Cellular Longevity, 2*, 270–278.

Patel, D. K., Desai, S. N., Gandhi, H. P., Devkar, R. V., & Ramachandran, A., (2012). Cardio protective effect of *Coriandrum sativum* L. on isoproterenol induced myocardial necrosis in rats. *Food and Chemical Toxicology, 50*, 3120–3125.

Rane, B. R., & Keservani, R., (2020). Introduction to nutraceuticals and dietary supplements. In: *Nutraceuticals and Dietary Supplements* (pp. 35–56). Apple Academic Press. ISBN: 9781771888738.

Rane, B. R., Bharath, M. S., Patil, R. R., Keservani, R. K., & Jain, A. S., (2020b). Novel approaches in nutraceuticals. In: Rajesh, K. K., Raj, K. K., & Anil, K. S., (eds.), *Enhancing the Therapeutic Efficacy of Herbal Formulations Through Novel Drug Delivery Systems* (pp. 241–266). IGI Global International Publisher, Pennsylvania, USA, Chapter 11. ISBN13: 9781799844532.

Rane, B. R., Patil, A. S., Keservani, R. K., & Jain, A. S., (2020a). Novel approaches in herbal formulation. In: Rajesh, K. K., Raj, K. K., & Anil, K. S., (eds.), *Enhancing the Therapeutic Efficacy of Herbal Formulations Through Novel Drug Delivery Systems* (pp. 43–68). IGI Global International Publisher, Pennsylvania, USA, Chapter 2. ISBN13: 9781799844532.

Rane, B. R., Tadavi, S. A., & Keservani, R. K., (2020c). "Naturopathy". In: Anil K. S., Raj, K. K., & Surya, P. G., (eds.), *Herbal Product Development* (pp. 321–347) Apple Academic Press, CRC Press, Taylor & Francis Group, Chapter 12. ISBN: 9781771888776.

Rasouli, H., Farzaei, M. H., & Khodarahmi, R., (2017). Polyphenols and their benefits: A review. *International Journal of Food Properties, 20*, 1700–1741.

Ravishankar, D., Rajora, A. K., Greco, F., & Osborn, H. M., (2013). Flavonoids as prospective compounds for anti-cancer therapy. *The International Journal of Biochemistry & Cell Biology, 45*, 2821–2831.

Rawson, N. E., Ho, C. T., & Li, S., (2014). Efficacious anti-cancer property of flavonoids from citrus peels. *Food Science and Human Wellness, 3*, 104–109.

Reddy, N. R., (2002). Occurrence, distribution, content, and dietary intake of phytate. *Food Phytates, 25*.

Robbins, R. J., (2003). Phenolic acids in foods: An overview of analytical methodology. *Journal of Agricultural and Food Chemistry, 51*(10), 2866–2887.

Sharma, A. K., Keservani, R. K., & Gautam, S. P., (2020). *Herbal Product Development*. Apple Academic Press, CRC Press, Taylor & Francis Group. ISBN: 9781771888776.

Shen, T., Wang, X. N., & Lou, H. X., (2009). Natural stilbenes: An overview. *Natural Product Reports, 26*(7), 916–935.

Singh, P., Kesharwani, R. K., & Keservani, R. K., (2017a). Antioxidants and vitamins: Roles in cellular function and metabolism. In: *Sustained Energy for Enhanced Human Functions and Activity* (pp. 385–407). Academic Press. ISBN:9780128054130.

Singh, P., Kesharwani, R. K., & Keservani, R. K., (2017b). Protein, carbohydrates, and fats: Energy metabolism. In: *Sustained Energy for Enhanced Human Functions and Activity* (pp. 103–115). Academic Press. ISBN:9780128054130.

Singla, R. K., Dubey, A. K., Garg, A., Sharma, R. K., Fiorino, M., Ameen, S. M., Haddad, M. A., & Al-Hiary, M., (2019). Natural polyphenols: Chemical classification, definition of classes, subcategories, and structures. *Journal of AOAC International, 102*(5), 1397–1400.

Surana, K. R., Ahire, E. D., Sonawane, V. N., Talele, S. G., & Talele, G. S., (2021). Informatics and methods in nutrition design and development. *Natural Food Products and Waste Recovery: Healthy Foods, Nutrition Design, and Extraction of Valuable Compounds*, 33.

Surana, K. R., Ahire, E. D., Sonawane, V. N., Talele, S. G., & Talele, G. S., (2021). Molecular modeling: Novel techniques in food and nutrition development. *Natural*

Food Products and Waste Recovery: Healthy Foods, Nutrition Design, and Extraction of Valuable Compounds, 17.

Teng, H., & Chen, L., (2019). Polyphenols and bioavailability: An update. *Critical Reviews in Food Science and Nutrition, 59*(13), 2040–2051.

Tian, S., Sun, Y., Chen, Z., Yang, Y., & Wang, Y., (2019). Functional properties of polyphenols in grains and effects of physicochemical processing on polyphenols. *Journal of Food Quality, 2019*.

Ullah, A., Munir, S., Badshah, S. L., Khan, N., Ghani, L., Poulson, B. G., Emwas, A. H., & Jaremko, M., (2020). Important flavonoids and their role as a therapeutic agent. *Molecules, 25*, 5243.

CHAPTER 2

PHYTONUTRIENTS, ANTHOCYANIDINS, AND ANTHOCYANINS: DIETARY AND MEDICINAL PIGMENTS WITH POSSIBLE HEALTH BENEFITS

SURUCHI SINGH[1], PANKAJ BHATT[2], VIPIN KUMAR[3], and NEHA P. SINGH[4]

[1]*Accurate College of Pharmacy, Knowledge Park III, Greater NOIDA, Uttar Pradesh, India*

[2]*Department of Pharmaceutics, KIET Group of Institutions (KIET School of Pharmacy), Muradnagar, Ghaziabad, Uttar Pradesh, India*

[3]*Department of Pharmaceutical Sciences, Gurukul Kangri (Deemed to be University), Haridwar, Uttarakhand, India*

[4]*School of Pharmacy and Research Dev Bhoomi Uttarakhand University, Dehradun*

ABSTRACT

Anthocyanin is a form of phenolic pigment that dissolves in water. It belongs to the flavonoid group and undergoes glycosylation in fragments. Fruits and vegetables owe their red, purple, or blue coloring to the presence of anthocyanins. Currants and grapes are particularly rich in these compounds, boasting high levels of anthocyanin content. These

Advances in Flavonoids for Human Health and Prevention of Diseases. Nisha Sharma, Deepika Saini, Rajesh K. Kesharwani, Prakash Chandra Gupta, & Raj K. Keservani (Eds.)

natural pigments not only enhance the visual appeal of these fruits and berries but also offer a wide range of nutritional and health benefits. Leafy vegetables, cereals, roots, and tubers with a reddish color to a dark blue have high anthocyanin concentrations. Anthocyanin pigments in plants are primarily composed of cyanidin-3-glucosides (Cy-3-Glu). As a natural food coloring agent, colored anthocyanin pigments have traditionally been used. These parameters affect the pigment's color and stability, such as pH, light, temperature, and structure. When the pH rises, anthocyanins turn blue from red under acidic conditions. Anthocyanins are extensively extracted, separated, and quantified using chromatography. Aside from serving as natural colors, anthocyanidins and anthocyanins are promising medicine components with numerous therapeutic benefits. The antioxidant, antibacterial, and mental- and visual-health-improving properties of anthocyanidins and anthocyanins have been demonstrated in cell cultures, animal models, and human clinical trials. Several studies have shown that anthocyanidins and anthocyanins are potent antioxidants that offer health benefits. The mechanisms and pathways contributing to these compounds ' protective effects are free radical scavenging, the cyclooxygenase (COX) pathway, the mitogen-activated protein kinase pathway, and inflammation-associated cytokines. Here, we will provide an overview of the benefits of natural food dyes anthocyanidins and anthocyanins as nutraceuticals and their functions as food pigments.

2.1 INTRODUCTION

Anthocyanins are a category of plant pigments that are members of the flavonoids or bioflavonoids family of compounds. They are most often seen in higher plants. It is most often found in fruits and flowers, where it imparts colors of red and blue; however, it may also be found in stems, leaves, and root systems. The amount of acidity present determines the color of anthocyanins. Anthocyanins appear red when exposed to acidic circumstances, while blue when exposed to more minor acidic settings. In terms of structure, anthocyanins may be classified into two categories: anthocyanidin aglycons and anthocyanin glycosides. The flavylium ion is the fundamental core structure of anthocyanins, and it has seven distinct side groups to work with. A hydrogen atom, a hydroxide atom, or a methoxy group may be used as the side groups.

Anthocyanidin, a kind of bioflavonoid, is a chemical molecule responsible for the pigmentation of plants and is found in various plants. They are sugar-free counterparts of anthocyanins that are based on the flavylium ion as their active ingredient. The counterion, in this case, is mostly chloride, and it is this positive charge that distinguishes the anthocyanidins from other flavonoids.

It is believed that anthocyanidins, which are antioxidant flavonoid pigments, are responsible for the purple or red color of fruits and vegetables, including grapes, cherries (especially red), raspberries (blue), plums (purple), beets (purple), and purple cabbage (purple). It also enhances the vibrancy of flowers' colors—these aid in attracting a wide range of pollinating agents to the flower's nectar. Because of the pigmentation given by anthocyanidins, plants are also able to retain their developed progeny. Anthocyanidins are phytochemicals that prevent photosynthetic organs in plants from being damaged by direct sunlight.

A category of components found in plant pigments known as bioflavonoids includes anthocyanins and anthocyanidins, both of which are particular kinds of elements. The flavylium ion is found in both substances, and it serves as the fundamental core structure for both. Anthocyanidins are sugar-free counterparts of anthocyanins, which are found in fruits and vegetables. It is possible to produce anthocyanins by attaching various sugars to different aspect groups of the flavylium ion. The primary distinction between Anthocyanin and anthocyanidin is the presence of a flavonoid. Because various kinds of sugars are available, they may be found at distinct side groups, resulting in various anthocyanin types being produced. Both chemicals are pH-dependent and have antioxidant capabilities, making them a good combination. They aid in the pollination process and serve as larval repellants, which prevent larvae from causing damage to the plant's body.

Flowers, fruits, and tubers contain Anthocyanin, a pigment that has been found in many plants. The color of anthocyanin changes from red in an alkaline environment to blue in acidic environments. As a flavonoid, Anthocyanin's oxygen atom is positively ionized in the C-ring. This ion is also known as flavylium (2-phenylchromenylium). In addition to pH, light, heat, and the structure of Anthocyanin, other factors can affect its stability (He et al., 2011). Many plants produce anthocyanins in their flowers and fruits. Anthocyanins were found in abundance in the streaks that were red, purple, and blue. Rhubarb, cherry blossom, pineapple sage, and pink bloom are all examples of flowers with red or pink petals found around the home or garden.

A variety of shades of red are available in these mouthwatering blossoms. Most of the most popular types of culinary flowers for culinary preparation are blue (cornflowers, blue chicory, white rosemary) and purple (rosemary, violets, and violets) (purple mint, purple passionflower, purple sage, standard violets, and lavender). In the past, many of these flowers were employed in traditional Chinese medicine, colors, and even food. These fruits are consumed for their nutritional value as well as their esthetic appeal. Colored foods ranging from red to blue have high levels of the antioxidant anthocyanin pigment. Anthocyanin-rich foods, including purple potatoes, crimson cabbage, and black carrots, have also been consumed to ward against sickness. Plant pigments called anthocyanins have a wide range of applications. As dyes and food colorants, ancient cultures used flowers, fruits, and vegetables. Flowers and fruits high in anthocyanins are used both as medicines to treat various ailments and as natural colorants. The therapeutic potential of plant anthocyanins, on the other hand, has been extensively explored. Among its many beneficial properties, anthocyanins include antidiabetic, anti-inflammatory, antibacterial, and anti-obesity. Because of this, food plant anthocyanins can be used in medicine.

2.2 TYPES OF ANTHOCYANIN IN PLANTS

Anthocyanins are vivid plant pigments that give rise to the red, purple, and blue hues observed in nature. These compounds are part of the phenolic phytochemical group. In contrast to anthocyanins, anthocyanidins are glycosides rather than aglycons. The process of glycosylation enhances the stability and solubility of anthocyanins, making them more easily transported within plants. Anthocyanins are categorized into two classifications in the same way as anthocyanidins, namely acylated anthocyanidins and anthocyanidin glycosides. There are many anthocyanidins, including malvidin, cyanidin, delphinidin, pelargonidin, peonidin, and petunidin. Additionally, to the typical anthocyanins found in plants, acylated anthocyanins have been discovered. According to their acylation, they are classified as acrylate anthocyanins, cumaroylated anthocyanins, caffeoylated Anthocyanins, and malonylated Anthocyanins.

Anthocyanin is created by flavonol, similar to flavylium ions except for the lack of ketone oxygen at position 4. Flavylium ions of Anthocyanin have the empirical formula $C_{15}H_{11}O+$, and their molecular weight is 207.24724 grams per mole. In contrast, glycosylated anthocyanidins are anthocyanins. The conjugated bonds between anthocyanins cause red, blue, or purple hues.

Anthocyanidins can be found in various plants, such as cyanidin, malvidin, peonidin, and petunidin. Half of fruits and vegetables have anthocyanins, while 12% of fruits and vegetables include anthocyanins, while 13% of fruits and vegetables contain anthocyanins, while 13% contain anthocyanins, and 7% contain anthocyanins (Castañeda-Ovando et al., 2009). A pigment of reddish-purple hue, cyanidin, is naturally occurring. Red sweet potatoes and purple maize, and other red fruits and vegetables contain this pigment. The chemical structure of delphinidin is the same as that of most anthocyanidins. When the plant is exposed to sunlight, it develops a blue-reddish or purple pigment. The delphinidin pigment is responsible for the blue hue of flowers (Katsumoto et al., 2007). There are numerous ways in which pelargonidin is different from other anthocyanidins (Bakowska et al., 2005). Flowers become orange when exposed to pelargonidin (Robinson et al., 1936) whereas some fruits and berries turn red (Jaakola et al., 2013). Methylated anthocyanidins, such as peonidin, are also abundant in plants and are a kind of anthocyanidin. The color magenta (Bakowska et al., 2005) shows this well. Berries, grapes, and red wines all contain significant amounts of peonidin. A third anthocyanidin, malvidin, has had its O-methyl group removed. Blue flowers, including the Summer Wave Blue, abound, with purple undertones (Tanaka et al., 1998). Malvidin is also high in red wine (Mazza et al., 1995). When older red wines are poured, this crimson appears darker and dustier (Barnard et al., 2011). Methylated anthocyanidins include petunidin. This pigment is water-soluble and rich in red and purple hues (Bakowska et al., 2005). Petunidin, an antioxidant, is present in blackcurrants (Wolske et al., 2021) and purple flower petals (Yabuya et al., 1997).

2.3 ANTHOCYANIN PIGMENTS: COLOR AND STABILITY

2.3.1 PH-DEPENDENT STABILITY OF ANTHOCYANIN COLOR

Co-pigments, light, heat, pH, metal ions, enzymatic oxygenation and antioxidants all impact anthocyanin stability (Pinheiro et al., 2021). Anthocyanidin may be unstable depending on its B-ring structure and whether or not it contains hydroxyl or methoxyl groups (Castañeda-Ovando et al., 2009). The groups mentioned above reduce the stability of anthocyanidin in the solution. Here, we will talk about how Anthocyanin holds up over time and how its color changes depending on how acidic the environment is.

The color of anthocyanins is affected by the pH. In part, this is because anthocyanins have an ionic molecular structure. When an acidic environment is applied, particular anthocyanins become a vibrant shade of red. At a pH of 8, anthocyanins have a deep purple hue, but when the pH rises, they become blue. Flavylium cations dominate anthocyanin pigments (Bakowska et al., 2005). Since the anthocyanins have a lower pH, they are considerably more stable. Because of the flavylium cation generated at lower pH levels, anthocyanin is highly water-soluble. Reducing the water concentration causes the flavylium cation's deprotonation to occur faster, reducing color stability (Coutinho et al., 2015). As well as influencing pH, anthocyanin-tannin polymerization has the potential to enhance photostability at lower pH levels (Sims et al., 1985).

2.3.2 EFFECT OF CO-PIGMENTATION AND TEMPERATURE ON COLOR CHANGE

The color shift of anthocyanins in a solution may be influenced by copigmentation, temperature, and different pH conditions. When anthocyanidins are copigmented with metallic ions or flavonoids, the anthocyanidins become stronger. A plant's leaves, flowers, and fruits are more stable in color because of copigmentation. Delphinidin is colored blue by including a metal ion in its equilibrium mixture at an appropriate pH value to stabilize carbinol pseudo-base structures (Gürses et al., 2016). Flavonoids co-pigment with floral anthocyanins, intensifying the flower's hue, which changes the color of the bloom (Castañeda-Ovando et al., 2009). The color intensity of Anthocyanin is further enhanced by glycosylation and acylation.

Anthocyanins lose stability as solution temperature rises. The overall anthocyanin concentration of a typical grape was found to be less than half that of control berries grown at 25°C, according to a recent study (Zhang et al., 2018). Despite the solution's low pH, Anthocyanin turns red to orange at temperatures as high as 40°C. Anthocyanin pigment degradation may be prevented by heating an anthocyanin-rich extract solution. Polyphenol oxidase often destroys the extract's phenolic components enzymatically.

In addition, extracts have been found to lose their enzymatic activity when heated to 50°C (Patras et al., 2010). When raw materials are blanched in the food processing sector, polyphenol oxidase cannot oxidize anthocyanins, which would otherwise occur.

2.4 ANTHOCYANIN PIGMENTS AS FOOD COLORANTS AND ADDITIVES

Processed foods and beverages must use natural colorants and additives to get consumer approval. Anthocyanins are colorful plant pigments that exist naturally in plants. Red, blue, and purple pigments produced from plants are known as anthocyanins. Natural dyes are used to make these colors since they are non-toxic. Although synthetic colors are more potent, natural colorants are still safer to consume in higher doses. As natural colorants, anthocyanins provide value-added properties (Bridle et al., 1997). A health benefit of these products includes antioxidants, nutraceuticals, as well as antimicrobial properties.

2.5 ANTHOCYANIN PIGMENTS AND SOLUBILITY

In grapes and wine, there are flavylium ions, which are anthocyanins. Flavylium ions become carbinol pseudo-bases, quinoidal bases, or even chalcones as digested and absorbed into the circulation. Apparently, in acidic conditions, the pigment looks red, whereas, in alkaline conditions, it appears blue. It is not uncommon for these color pigments to be obtained from violet-hued fruits and vegetables, such as blackcurrants and blueberries. Anthocyanin pigments are often extracted using water as the extraction medium.

Anthocyanin hues are often extracted using alcoholic solutions in food preparation. This is due to anthocyanins' solubility in both freshwater and the vast majority of organic solvents. Alternatively, the apolar organic solvent is unable to dissolve Anthocyanin. Similar to acids, this substance is unstable in bases.

Anthocyanidins such as cyanidin readily dissolve in water because they form the flavylium cation at low pH levels or in acidic conditions, which appears red. Flavylium ions remain stable under acidic conditions, and the red color of the anthocyanin pigment is intensified. Since the red anthocyanin pigment remains highly stable in acidic waters because of this characteristic, it is an excellent dye option. The most effective way to use it is as a red colorant for alcoholic beverages. Because quinoidal bases are attracted to alcoholic solutions, flavylium ions deprotonate with increasing pH (Coutinho et al., 2015). When red wine's pH is low, its color can appear purplish-red, while high, its color can be bluish-violet. As the red flavylium cation decomposes in alcohol, it transforms into blue quinoidal species, giving the red wine its bluish-violet color.

2.6 ANTHOCYANINS AND ANTHOCYANIDIN IN PLANTS

Among the many plant parts that contain anthocyanins are flowers, leaves, roots, and grains. Researchers have discovered anthocyanins and anthocyanidins in fruits and vegetables. Anthocyanins are found in fruits and vegetables almost exclusively in the form of cyanidin-3-glucoside (Cy-3-Glu). The pH of a plant's environment can cause them to produce Cy-3-Glu. The acidic character of berries, which all contain cyanidin glycosides, explains this. Berry components including malvidin, peonidin, and petunidin are hard to come by since they are not common. These are anthocyanidins that have had their methyl groups changed.

As a consequence, red berries lack these pigments. Anthocyanidin methylation may have a lower reddening action than unmethylated anthocyanidin. Additionally, these anthocyanidins may be found in a variety of blueberries and other blueberries.

2.7 POTENTIAL USES OF ANTHOCYANIN PIGMENTS

As a result of their separation from plants, anthocyanins have become popular food additives. The food additive E163 is derived from fruit anthocyanins found in grape skin. Use it to dye jams, confectioneries, and beverages purple with a purple tinge! Since recently, the safety and possible damage to human health of synthetic food colors has piqued public interest. This is especially true for the brain and behavioral functioning of children and adolescents. Artificial colors in the diet that contained sodium benzoate (E211) significantly increased hyperactivity in young children aged three and eight to nine years old, causing a significant rise in attention deficit disorders and hyperactivity disorders. Natural food colorants like Anthocyanin, a potentially feasible alternative to synthetic food colors, have generated much attention due to this finding.

Yogurt drinks and certain mixed fruit juices are increasingly using anthocyanin-derived colorants. Certain companies in their products used synthetic colors. Synthetic hues, on the other hand, maybe harmful if consumed in large quantities. Because they are more stable than nonacylated anthocyanins, acylated anthocyanins have recently been used as food coloring agents (McCann et al., 2007).

2.8 NUTRACEUTICAL AND PHARMACEUTICAL EFFECTS OF ANTHOCYANINS

Nutraceuticals and traditional medicine both include Anthocyanin is a phytonutrient. It has long been utilized in phytopharmaceuticals, appetite stimulants, choleretics, and other treatments. Nutraceuticals and medicines use these vivid hues as active ingredients. Anthocyanin's bioavailability as a nutraceutical is critical for staying healthy and preventing disease.

These have a low absorption rate into the circulatory system and high excretion rates in urine and feces due to anthocyanins' poor bioavailability. Anthocyanin's high bioavailability is beneficial to reduce cellular lipid peroxidation and the danger of developing various diseases. Until now, very little research has been done on the bioavailability of essential anthocyanins. A key focus of this review will be on anthocyanidins and anthocyanins' bioavailability in humans.

The two most bioavailable anthocyanins, Cy-3-Glu and malvidin-3-glucoside, are Cy-3-Glu and malvidin-3-glucoside. Previous studies found that peonidin glycosides were the most bioavailable anthocyanins, followed by cyanidin, malvidin, delphinidin, and petunidin glycosides. Despite the low absorption of most red wine anthocyanins, increasing consumption may help increase their efficacy. There is some debate as to whether or not taking too much Anthocyanin is harmful. Some of the health benefits include antioxidative characteristics, antiangiogenesis, cardiovascular disease (CVD) prevention, anticancer activities, antidiabetic activities, improved eye health, weight loss and anti-obesity activities, and neuroprotective qualities. The succeeding subsections explore how anthocyanidin and anthocyanin work to prevent illness.

2.8.1 ANTIOXIDANTS

Anthocyanin's health and therapeutic advantages are mainly due to its antioxidant capabilities (Faria et al., 2013). A recent study showed that anthocyanin chalcones and quinoidal bases containing keto groups could scavenge free radicals effectively. This glycosylated B-ring structure of Anthocyanin, which also contributes to its strength, is greatly enhanced by ortho-hydroxylation and methoxylation, giving it an impressive degree of antioxidant activity.

2.8.2 THE ANGIOGENIC PROCESS AND DISEASE DEVELOPMENT

Angiogenesis is primarily controlled by endothelial cells. The involvement of physiologic angiogenesis in coronary heart disease (CHD), melanoma, and uncontrolled diabetes, primarily diabetic retinopathy and diabetic nephropathy, has been well-established (Wang et al., 1997). For normal angiogenesis, molecules like VEGF, FGF2, TGF, and angiopoietin are necessary, as well as antiangiogenic molecules (angiostatin, endostatin, and thrombospondins) (Xue et al., 2010).

Various studies have repeatedly shown that anthocyanins have antiangiogenic effects. A berry-based preparation significantly inhibits VEGF production in HaCaT cells (human keratinocytes) in response to hydrogen peroxide (H_2O_2) and tumor necrosis factor-alpha (TNF-alpha) (Goodwin et al., 2007). Studies have shown that bilberry anthocyanidins (delphinidin, cyanidin, and malvidin) inhibit VEGF-induced tube formation in the presence of umbilical vein endothelial cells and fibroblasts (Yance et al., 2006).

2.8.3 CARDIOVASCULAR HEALTH

A study in epidemiology has linked anthocyanin-rich foods to CVDs, and anthocyanin intake as a whole has been linked to CVD risk. Testing in a lab setting has also shown that anthocyanins are antithrombotic (Matsunaga et al., 2010). The effects of feeding rats 20% of maize seeds rich in anthocyanins for eight weeks resulted in lower ischemia-reperfusion injury in rats and reduced infarct extent in rats (Rechner et al., 2005).

Chokeberry and bilberry extracts, but not elderberry, contain vasorelaxant anthocyanins, according to Bell and Gochenaur. Furthermore, the response of the coronary arteries to the potent vasodilator nitric oxide (NO) remains unchanged.

2.8.4 ANTICANCER

Anthocyanins have been studied *in vitro* as well as in animal models for their anticancer and antiangiogenesis properties. Angiogenesis plays a significant role in the transition from benign to malignant tumors during cancer development. New blood vessels are needed to provide oxygen to cancerous cells, and antiangiogenesis inhibits this process. Many phytochemicals may act as antiangiogenic agents, including flavonoids and anthocyanins.

In studies on anthocyanins derived from various plant sources, researchers examined their anticancer properties against esophageal, colon, breast, liver, and hematological cancers.

Blueberry anthocyanins and anthocyanin-pyruvic acid adducts (250 mg/ml) were shown to reduce invasion of breast cancer cell lines MDA-MB-231 and MCF7. By inhibiting cancer cell proliferation, these extracts act as chemoinhibitors.

CF-1 mice's colon mucosa contains elevated apoptotic caspase-3, which may explain the presence of abnormal crypt foci in the colon of females with this purple-fleshed sweet potato clone. In addition to causing cell cycle arrest, anti-proliferative mechanisms, and apoptosis, sweet potato P40 high in anthocyanins protects against colorectal cancer. In addition to causing cell cycle arrest, anti-proliferative mechanisms, and apoptosis, sweet potato P40 high in anthocyanins protects against colorectal cancer (Faria et al., 2010). In a study on human hepatoma Hep3B cells, Anthocyanin from the vine, a less common source of Anthocyanin, showed anti-invasive effects.

According to a study, black rice's anthocyanins inhibited the activation of mitogen-activated protein kinases in breast cancer cells. These extracts inhibited the migration and invasion of MDA-MB-453 (HER2+) cells, suppressed the activation of MEK, JNK, and MMP2 and MMP9. Specifically, black rice anthocyanins inhibit the RAS/RAF/MAPK pathways (retrovirus-associated DNA sequences/rapidly-accelerated fibrosarcomas and mitogen-activated protein kinases). As a result, treating cancer patients who are at an advanced stage may be helpful.

2.8.5 ANTIDIABETICS

Many studies have looked at the antidiabetic potential of anthocyanins produced from plants. The high anthocyanin content of Cornu's fruits has been employed in traditional Chinese prescription drugs for diabetic treatment (Chen et al., 2015). Several bioactive components of Cornu's fruits have been identified, such as cyanidin, delphinidin, and pelargonidin glycosides (Chen et al., 2015). By comparing different Anthocyanin and anthocyanidin solutions to chloranthidin-3-glucoside and delphinidin-3-glucoside, Jayaprakasam & colleagues showed that these two compounds enhance insulin secretion (He et al., 2010).

Other studies found that pelargonidin-3-galactoside and pelargonidin-3-galactoside increased insulin secretion by 1.4 times under glucose concentrations indicative of human blood glucose levels of 4 mM. The anthocyanins pelargonidin-3-galactoside, cytosine 3 glucose, and delphinidin-3 glucose have the highest potential to increase insulin production. Findings like these indicate that anthocyanins' insulin-releasing ability depends heavily upon their B-hydroxyl ring's group content. Neither cyanidin, delphinidin, pelargonidin, malvidin, nor petunidin appears to affect insulin secretion substantially.

A 24-week clinical trial using anthocyanins gave 58 people with diabetes two capsules of anthocyanins (160 mg anthocyanins) twice a day (Christison et al., 1993). The placebo group had significantly higher levels of adiponectin and beta-hydroxybutyrate and high fasting plasma glucose and insulin resistance indices. Scientists, on the other hand, are unable to explain how people with diabetes can prevent insulin resistance.

2.8.6 VISUAL HEALTH

Nutraceuticals containing anthocyanin pigments are essential for maintaining good vision. Berries with high anthocyanin content have long been known to improve eyesight and are often associated with improved night vision. Berry varieties with high anthocyanin content include almost all of the berries you will find on the market. Researchers have shown that giving six-week-old C57BL/6 mice bilberry extract (which contains approximately 39% anthocyanins) helps maintain photoreceptor cell activity even when the retina is inflamed (Li et al., 2015). The Humphrey visual field test (anthocyanin supplements) showed that anthocyanin capsules made from bilberries (60 mg anthocyanin per capsule) improved the visual function of 132 patients with normal-tension glaucoma (Miyake et al., 2012).

Berries have been shown to protect the eyes in several studies. Around 19 people with open-angle glaucoma were treated with blackcurrant anthocyanin (50 mg/day) for 24 months and saw an improvement in ocular blood flow. As a result, the intraocular pressure did not differ significantly. Mice with retinal degeneration may be prevented from experiencing retinal degeneration when NMNO (N-methyl-N-nitrosourea) is delivered with anthocyanins from black soybean seed coats (50 mg/kg body weight). Human lens epithelial cells may be protected against H_2O_2-induced oxidative stress by using the extract by 50–200 g/ml (Park et al., 2018). A purple maize seed extract has been proven to decrease lens opacity while decreasing malonaldehyde levels while containing high amounts of anthocyanin (Lu et al., 2018).

2.8.7 THE ANTI-OBESITY EFFECT

Analogous effects of anthocyanin pigment and anthocyanidin on obesity. A recent study found that obese mice on a purple maize Cy-3-Glu enriched diet lost weight, and the weight of white and brown adipose tissue was reduced. The feeding of obese rats with a purple corn diet has been demonstrated to reduce their blood sugar, insulin, and leptin levels and decrease their tumor necrosis factor (TNF-a) mRNA levels. Purple corn reduced sterol regulatory element-binding protein-1 in white adipose tissue and reduced mRNA levels for enzymes involved in the fatty acid synthesis and triacylglycerol synthesis. As a result of these downregulations, white adipose tissue may lack triacylglycerol.

The relationship between obesity and adipocyte dysfunction is strong. In order to prevent weight gain caused by excess body fat, it is essential to regulate adipocyte protein secretion or adipocyte-specific gene expression. Tsuda and his colleagues next determined if anthocyanins, particularly cyanidin and Cy-3-Glu, have anti-obesity effects on isolated rat adipocytes. Anthocyanins were found to increase adiponectin and leptin production in isolated rat adipocytes and expression of gene-specific to adipocytes (Cushnie et al., 2005). Mouse white adipose tissue treated with

Anthocyanin has elevated levels of adiponectin gene expression. Anthocyanin therapy reduces the monophosphate/adenosine triphosphate (ATP) ratio significantly, which may explain the observed changes.

2.8.8 ANTIMICROBIAL

A wide range of pathogens, most notably food-borne illnesses, are suppressed by polyphenolic compounds, including anthocyanins, which include anthocyanins. Anthocyanins' antimicrobial activity is achieved by several mechanisms, including cell wall, membrane, and interstitial matrix destruction-induced cell damage (Soga et al., 2014).

Recent studies have shown that maqui fruit extracts are active against *Aeromonas hydrophila* and *Listeria innocua* (Pojer et al., 2013). As pathogenic microorganisms or degradation microorganisms, these bacteria are often detected in chilled goods (Genskowsky et al., 2016). Cranberry extract was antibacterial against *Enterococcus faecium*, *Pseudomonas aeruginosa*, *Staphylococcus aureus*, and *E. coli*, according to Côté et al. (2011). Cranberry extract's antibacterial activity has not been attributed to anthocyanins and flavonols found in its low pH when adjusted to a pH of 7.

Gram-negative bacteria are inhibited by blueberry, raspberry, blackcurrant, and strawberry anthocyanin-rich extracts, but Gram-positive bacteria are not (Côté et al., 2011). Gram-negative bacteria are protected from hydrophobic substances by the outer membrane, whereas Gram-positive bacteria are not (Puupponen-Pimiä et al., 2001). Gram-positive and Gram-negative bacteria have different structures in their cell walls, which may explain the difference. The anthocyanins found in anthocyanin extracts possess antimicrobial activity through a variety of processes and synergistic effects with other phytochemicals, such as weak organic acids, phenolic acids, or their combination. Therefore, researchers should focus on other chemically complex compounds rather than only studying anthocyanins' antibacterial effects. There is evidence that various microbes are less likely to infect purple, red, and blue fruits and vegetables due to their high levels of anthocyanins.

2.8.9 NEUROPROTECTIVE EFFECT OF ANTHOCYANINS

To prevent nerve cells from damage caused by oxidative stress and neurotoxicity, a strategy known as "neuroprotection" is used. An agent that protects the nervous system from further injury is known as a neuroprotective agent. Anthocyanins have been shown to possess neuroprotective properties in laboratory and animal studies. Cell cultures are used extensively *in vitro* research, while animal models are used extensively *in vivo* research.

This review summarizes the neuroprotective findings of many studies. Human brain cells are protected from oxidative stress by Cy-3-Glu and its aglycone, according to *in vitro* study (Cisowska et al., 2011). Cyanidin pretreatment of SH-SY5Y cells with the antioxidants 100 M cyanidin and 3-glucoside cyanidin appears to have increased the antioxidant activities of the membrane and cytosolic fractions. Cyanidin also increased the proportions of mitochondrial working and inhibited H2O2-induced DNA fragmentation.

Following permanent blockage of the middle cerebral artery in adults, Cy-3-Glu from Prunus cerasus reduced the release of apoptosis-inducing factors from mitochondria under oxidative stress but not that of cytochrome C (Tarozzi et al., 2007). Additionally to what has already been stated (0.9% standard saline treatment), there was a lower level of brain superoxide in the Cy-3-Glu-treated rats.

2.9 DISEASE PREVENTION MECHANISMS

Antioxidants like anthocyanins reduce or prevent diseases. Anthocyanins have been found to reduce the risk of many diseases, both directly and indirectly. Free radical scavengers like colored compounds reduce the risk of acquiring many chronic diseases by reducing oxidative stress. They do this by fighting disease-causing free radicals. Oxidative stress and lipid peroxidation can be reduced indirectly to slow cell development and apoptosis. Historically, anthocyanins have been recognized to be potent anti-free radical scavengers. Anthocyanins may reduce CVD risk by improving lipid profiles and biomarkers in the blood. Reduced levels of specific blood markers have long been linked to a reduced risk of CVD. Many phenolic compounds, including anthocyanins, work to slow cancer cell development. Cyclooxygenase (COX) enzyme inhibition is an established mode of action. They are responsible for synthesizing leukotriene, prostacyclin, prostaglandins (PGs), and thromboxanes (Di Giacomo et al., 2012).

Researchers have demonstrated that inhibiting COX-1 and COX-2 can reduce cancer cell proliferation by inhibiting two COX enzymes.

In addition to inhibiting the mitogen-activated protein kinase pathway, anthocyanins also slowed the spread of cancerous tumors. The cytokine signaling pathway is among the most well-known. A combination of four hydroxylations at positions 5, 7, 31, and 41, a C2–C3 bond, along with a B-ring attached at position C2, is the best way to synthesize monocyte chemoattractant protein 1 (MCP-1) (Pavelecini et al., 2019).

2.9.1 FREE-RADICAL SCAVENGING PATHWAY

A byproduct of oxidative stress is the generation of free radicals inside the biological system. In oxidative stress, free radicals are produced in two forms: reactive oxygen species (ROS) and reactive nitrogen species (RNS). Several biological activities in the human body generate ROS and RNS. The endoplasmic reticulum, peroxisomes, and peroxisomes generate ROS, whereas RNS are synthesized during amino acid metabolism in the cytosol and mitochondria (Garcia-Alonso et al., 2009).

The ROS signaling route involving the respiratory chain is the most well-known. A mechanism of electron transport is used in the mitochondria to produce H_2O_2 by the superoxide dismutase (SOD) enzyme. A Fenton reaction produces hydroxyl radicals (•OH) in mitochondrial Complexes I and III ($H_2O_2 + Fe^{2+}$ •OH + OH + Fe^{3+}). The -ketoglutarate dehydrogenase complex and many oxidoreductases in the mitochondria are additional pathways for producing ROS (O_2•). When it comes to the metabolism of H_2O_2, peroxisomes, and lysosomes, function similarly.

By metabolizing xenobiotics, the endoplasmic reticulum also produces ROS. By reducing an oxygen (O_2) molecule with RH2, lipophilic substrate (such as fatty acids, FH) is reacted to produce RH2 and water (FH + O_2 + RH_2 AOH + R + H_2O). Due to the continual process of microsome formation, ROS are produced. The endoplasmic reticulum and plasma membrane produce NADPH (nicotinamide adenine dinucleotide phosphate), which aids lipid peroxidation. In the presence of NADPH oxidase, O_2 • and H_2O_2 are produced in large quantities. ROS produced in biological systems has therefore been shown to stimulate cells to multiply.

When used in the electron transfer reaction pathway, well-known antioxidant and free radical fighters, anthocyanins may serve as reducing agents. Free radicals may get electrons from antioxidants that have unpaired electrons (Huang et al., 2005). There are two ways by which anthocyanins neutralize free radicals, both of which have been studied extensively. The B-ring is attacked through the first pathway by way of the anthocyanin structure's hydroxyl group(s). The second pathway involves the oxonium ion attacking the C-ring. Both of these attacks are possible. In addition to scavenging free radicals, anthocyanins are potent antioxidants.

Studies have shown that anthocyanin molecules' ability to scavenge radicals is affected by the number of hydroxyl groups on their B rings. When hydroxyl groups are present, scavenging activity increases. However, there is no documented study on the mechanism through which the anthocyanin structure's positive charge on the oxygen atom scavenges free radicals. Another theory holds that the superoxide O_2-radical aids anthocyanin oxonium in its formation (Saint-Cricq de Gaulejac et al., 1999).

2.9.2 MECHANISMS UNDERLYING CARDIOVASCULAR DISEASES (CVDS) AND BIOMARKERS

The damaging effects of ROS and RNS on the cardiovascular system can lead to oxidative stress. Oxidative stress causes vascular inflammation during CVD. Inflammation of the arteries can affect cholesterol levels, lipid profiles, HDL, and VLDL cholesterol, and malondialdehyde and other enzymes in the blood (such as SOD, catalase (CAT), and glutathione peroxidase (GPx)) (Emamat et al., 2020). Plasma lipid profiles and antioxidant biomarkers determine CVD risk. As a result of free radicals, vascular endothelial cells respond by inducing inflammation and increasing the production of cell adhesion molecules and chemokines.

Use of anthocyanins (total intake of 320 mg anthocyanins per day) In a 24-week double-blind clinical trial involving 150 hyperlipidemic subjects aged 40–65 years, significantly reduced serum sensitivity and selective C-reactive protein (22.6%), soluble cellular adhesion molecular-1 (12.3%), and serum interleukin-1 (12.8%) when especially in comparison to the placebo (2.5%, 0.4%, and 12.8%, respectively). Anthocyanins (160 mg twice daily, randomized, double-blind, placebo-controlled study) lowered LDL cholesterol and raised HDL cholesterol by 13.7% when given for 12 weeks compared to placebo (Qin et al., 2009). According to a meta-analysis, anthocyanin supplementation reduced TC, triglyceride, and LDL cholesterol while increasing HDL cholesterol in those with dyslipidemia (Zhu et al., 2013).

2.9.3 CYCLOOXYGENASE (COX) PATHWAY

Thromboxane synthesis in blood platelets and intestinal epithelial integrity both depend on COX-1. This substance may be present in almost all body tissues. As a cell signaling protein, it is essential for maintaining tissue homeostasis and controlling endothelial cell angiogenesis. When arachidonic acid (ARA) is converted into prostaglandin-endoperoxide (PGH2), it is known as prostaglandin-endoperoxide synthase 2 (COX-2) (ARA PGG2 PGH2). Ricciotti & FitzGerald (2011) provided a comprehensive overview of the biosynthesis of PGs.

As ARA is converted to PGH2 by COX-1 and COX-2, various specialized isomerases may be synthesized using PGH2 as a substrate. The four most well studied bioactive PGs are PGE2, PGI2, PGD2, and PGF2. Smooth muscle cells in the blood arteries, platelets, brain, and kidney express PGE2 and PGI2. PGI2 is the major prostanoid in charge of keeping the heart in a healthy state. The expression of PGI2 in vascular cells is increased by oxidative stress. Vasodilation occurs as a result of PGI2 working in combination with PGE2.

Adenocarcinomas and benign polyps both overexpress COX-2. COX-2 plays an essential role in cell proliferation and death, according to an in-depth investigation (Liu et al., 2016). In addition to cell signaling function and proliferation, apoptosis regulation contributes to COX-2's anticancer properties. PGE2 has also been shown to inhibit apoptosis in several *in vitro* conditions, according to the researchers.

2.9.4 A PATHWAY REGULATED BY MAP KINASE

In addition to proliferation and differentiation, MAPKs are also involved in the migration and death of cells (Ricciotti et al., 2011). In addition to ERK 1/2, ERK 5, JNK 1–3, and the many p38 protein isoforms (p38), there is a variety of well-known MAPKs (Williams et al., 1999). Cytokine production is regulated by p38 MAPKs (Sobolewski et al., 2010).

According to a recent study, peonidin-3-glucoside, one of six commonly occurring anthocyanins, inhibits ERK 1/2 expression in the H1299 cell line (Munshi et al., 2013). Furthermore, pomegranate extract with anthocyanins could inhibit UV-B-induced activation of MAPKs in normal human epidermal keratinocytes. Following UV-B illumination, pomegranate extract (20 g/mL) decreased the phosphorylation of MAPK (ERK 1/2, p38 protein, and JNK 1/2) in a time-dependent manner (Cargnello et al., 2011).

Antitumor activity has been found in Anthocyanin extracted from Vitis cognitive Pulliat. The extract (60 g/ml) induces apoptosis in HCT-116 cells by increasing phosphorylation of p38-MAPK and ER. It also includes delphinidin-3,5-diglucoside, cyanidin-3,5-diglucoside, petunidin-3,5-di-glucoside, delphinidin-3-glucoside, malvidin-3,5-glucoside, peonidin-3,5 (Johnson et al., 2002).

2.9.5 SIGNALING BY INFLAMMATION-PROMOTING CYTOKINES

As a consequence of chronic inflammation, one develops a disorder characterized by abnormal signaling pathways, high levels of cytokines, and inflammation cells consuming larger body areas. As the bulk of flavonoids, Anthocyanins work to reduce or eliminate inflammation in several ways. The proliferation and number of the cells in inflammation are influenced by specific cytokines, including IL-1, TNF-, and IL-6. These cytokines activate the hypothalamic-pituitary-adrenal axis during the inflammatory process, resulting in the production of glucocorticoids by the adrenal cortex. IL-1, an inflammatory cytokine, stimulates p38 MAPKs during the inflammatory phase (Yang et al., 2020).

Mice treated with lipopolysaccharide (LPS) and given 10% blueberries had lower levels of TNF- and IL-6 protein and mRNA in their blood than control mice, according to *in vivo* study. According to one randomized controlled trial, hypercholesterolemic people who took a pure anthocyanin mixture for 24 weeks (320 mg/day) had significantly lower plasma IL-1 levels than those who took a placebo. Contrary to expectations (week-24), there are no differences in TNF level reductions between treatment and placebo groups. The use of 10 mg/ml anthocyanin-containing bilberry extract (25% anthocyanin content) on human monocytic THP-1 cells dramatically decreased the production of MCP-1, IL-6, TNF-, intercellular adhesion molecule 1 (ICAM-1) and T-cell specific transcription factor T-bet caused by 100 ng/ml IFN (Afaq et al., 2010).

2.10 CONCLUSIONS

Anthocyanins, which are plant-derived phytochemicals, offer a wide range of health benefits. These natural compounds possess antioxidants and anti-inflammatory properties, support heart health, and enhance cognitive function. Incorporating anthocyanin-rich foods, such as berries and purple sweet potatoes, into one's diet can contribute to overall wellness. These pigments become red when exposed to acid but turn blue when exposed to alkaline. As a result of being acylated or commented, anthocyanidins have more excellent heat stability and maintain their structural integrity across a more comprehensive pH range. In the treatment of CVD, malignancy, diabetes, other metabolic diseases, and microbial infection, anthocyanins may help. In addition to improving vision, these chemicals offer neuroprotective properties. Various preventive mechanisms against various illnesses have been linked to anthocyanidins and anthocyanins. In a nutshell, these pigments function by scavenging free radicals, changing blood biomarkers, activating the COX and MAPK pathways, and signaling inflammatory cytokines, to name just a few of their actions.

KEYWORDS

- **anthocyanins**
- **colorant**
- **cyclooxygenase**
- **disease**
- **flavonoids**
- **health benefits and pigments**
- **monocyte chemoattractant protein 1**
- **necrosis factor-alpha**
- **prostaglandins**

REFERENCES

Afaq, F., Khan, N., Syed, D. N., & Mukhtar, H., (2010). Oral feeding of pomegranate fruit extract inhibits early biomarkers of UVB radiation-induced carcinogenesis in SKH-1 hairless mouse epidermis. *Photochem. Photobiol., 86*(6), 1318–1326.

Bakowska, A. (2005). Acylated anthocyanins as stable, natural food colorants – a review. *Polish Journal of Food and Nutrition Sciences, 14*(2), 107–116.

Barnard, H., Dooley, A. N., Areshian, G., Gasparyan, B., & Faull, K. F., (2011). Chemical evidence for wine production around 4000 BCE in the late chalcolithic near eastern highlands. *J. Archaeol. Sci., 38*(5), 977–984.

Bridle, P., & Timberlake, C. F., (1997). Anthocyanins as natural food colors—Selected aspects. *Food Chem., 58*(1, 2), 103–109.

Cargnello, M., & Roux, P. P., (2011). Activation and function of the MAPKs and their substrates, the MAPK-activated protein kinases. *Microbiol. Mol. Biol. Rev., 75*(1), 50–83.

Castañeda-Ovando, A., Pacheco-Hernández, M. De. L., Páez-Hernández, M. E., Rodríguez, J. A., & Galán-Vidal, C. A., (2009). A. chemical studies of anthocyanins: A review. *Food Chem., 113*(4), 859–871.

Chen, X. Y., Zhou, J., Luo, L. P., Han, B., Li, F., Chen, J. Y., Zhu, Y. F., et al., (2015). Black rice anthocyanins suppress metastasis of breast cancer cells by targeting RAS/RAF/MAPK pathway. *Biomed Res. Int., 2015*, 414250.

Christison, G. B., & MacKenzie, H. A., (1993). Laser photoacoustic determination of physiological glucose concentrations in human whole blood. *Med. Biol. Eng. Comput., 31*(3), 284–290.

Cisowska, A., Wojnicz, D., & Hendrich, A. B., (2011). Anthocyanins as antimicrobial agents of natural plant origin. *Nat. Prod. Commun., 6*(1), 149–156.

Côté, J., Caillet, S., Doyon, G., Dussault, D., Sylvain, J. F., & Lacroix, M., (2011). Antimicrobial effect of cranberry juice and extracts. *Food Control, 22*(8), 1413–1418.

Coutinho, I. B., Freitas, A., Maçanita, A. L., & Lima, J. C., (2015). Effect of water content on the acid-base equilibrium of cyanidin-3-glucoside. *Food Chem., 172*, 476–480.

Cushnie, T. P. T., & Lamb, A. J., (2005). Antimicrobial activity of flavonoids. *Int. J. Antimicrob. Agents, 26*(5), 343–356.

Di Giacomo, C., Acquaviva, R., Santangelo, R., Sorrenti, V., Vanella, L., Li Volti, G., D'Orazio, N., Vanella, A., & Galvano, F., (2012). Effect of treatment with cyanidin-3-O-β-D-glucoside on rat ischemic/reperfusion brain damage. *Evid. Based. Complement. Alternat. Med.*, 285750.

Emamat, H., Zahedmehr, A., Asadian, S., Tangestani, H., & Nasrollahzadeh, J., (2020). Effect of barberry (*Berberis vulgaris*) consumption on blood pressure, plasma lipids, and inflammation in patients with hypertension and other cardiovascular risk factors: Study protocol for a randomized clinical trial. *Trials, 21*(1), 986.

Faria, A., Fernandes, I., Mateus, N., & Calhau, C., (2013). Bioavailability of anthocyanins. In: *Natural Products* (pp. 2465–2487). Springer Berlin Heidelberg: Berlin, Heidelberg.

Faria, A., Pestana, D., Teixeira, D., De Freitas, V., Mateus, N., & Calhau, C., (2010). Blueberry anthocyanins and pyruvic acid adducts: Anticancer properties in breast cancer cell lines. *Phytother. Res., 24*(12), 1862–1869.

Garcia-Alonso, M., Minihane, A. M., Rimbach, G., Rivas-Gonzalo, J. C., & De Pascual-Teresa, S., (2009). Red wine anthocyanins are rapidly absorbed in humans and affect monocyte chemoattractant protein 1 levels and antioxidant capacity of plasma. *J. Nutr. Biochem., 20*(7), 521–529.

Genskowsky, E., Puente, L. A., Pérez-Álvarez, J. A., Fernández-López, J., Muñoz, L. A., & Viuda-Martos, M., (2016). Determination of polyphenolic profile, antioxidant activity and antibacterial properties of maqui [*Aristotelia chilensis* (Molina) Stuntz] a Chilean blackberry: Antioxidant and antibacterial properties of maqui. *J. Sci. Food Agric., 96*(12), 4235–4242.

Goodwin, A. M., (2007). In: Vitro assays of angiogenesis for assessment of angiogenic and anti-angiogenic agents. *Microvasc. Res., 74*(2, 3), 172–183.

Gürses, A., Açıkyıldız, M., Güneş, K., & Gürses, M. S., (2016). Colorants in health and environmental aspects. In: *Springer Briefs in Molecular Science* (pp. 69–83). Springer International Publishing: Cham.

He, J., & Giusti, M. M., (2010). Anthocyanins: Natural colorants with health-promoting properties. *Annu. Rev. Food Sci. Technol., 1*(1), 163–187.

He, K., Li, X., Chen, X., Ye, X., Huang, J., Jin, Y., Li, P., et al., (2011). Evaluation of antidiabetic potential of selected traditional chinese medicines in STZ-induced diabetic mice. *J. Ethnopharmacol., 137*(3), 1135–1142.

Huang, D., Ou, B., & Prior, R. L., (2005). The chemistry behind antioxidant capacity assays. *J. Agric. Food Chem., 53*(6), 1841–1856.

Jaakola, L., (2013). New insights into the regulation of anthocyanin biosynthesis in fruits. *Trends Plant Sci., 18*(9), 477–483.

Johnson, G. L., & Lapadat, R., (2002). Mitogen-activated protein kinase pathways mediated by ERK, JNK, and P38 protein kinases. *Science, 298*(5600), 1911–1912.

Katsumoto, Y., Fukuchi-Mizutani, M., Fukui, Y., Brugliera, F., Holton, T. A., Karan, M., Nakamura, N., et al., (2007). Engineering of the rose flavonoid biosynthetic pathway successfully generated blue-hued flowers accumulating delphinidin. *Plant Cell Physiol., 48*(11), 1589–1600.

Li, D., Zhang, Y., Liu, Y., Sun, R., & Xia, M., (2015). Purified anthocyanin supplementation reduces dyslipidemia, enhances antioxidant capacity, and prevents insulin resistance in diabetic patients. *J. Nutr., 145*(4), 742–748.

Liu, C., Sun, J., Lu, Y., & Bo, Y., (2016). Effects of anthocyanin on serum lipids in dyslipidemia patients: A systematic review and meta-analysis. *PLoS One, 11*(9), e0162089.

Lu, B., Christensen, I. T., Ma, L. W., Wang, X. L., Jiang, L. F., Wang, C. X., Feng, L., et al., (2018). MiR-24-P53 pathway evoked by oxidative stress promotes lens epithelial cell apoptosis in age-related cataracts. *Mol. Med. Rep., 17*(4), 5021–5028.

Matsunaga, N., Tsuruma, K., Shimazawa, M., Yokota, S., & Hara, H., (2010). Inhibitory actions of bilberry anthocyanidins on angiogenesis: Bilberry anthocyanidins and angiogenesis. *Phytother. Res., 24*(S1), S42–7.

Mazza, G., (1995). Anthocyanins in grapes and grape products. *Crit. Rev. Food Sci. Nutr., 35*(4), 341–371.

McCann, D., Barrett, A., Cooper, A., Crumpler, D., Dalen, L., Grimshaw, K., Kitchin, E., et al., (2007). Food additives and hyperactive behavior in 3-year-old and 8/9-year-old children in the community: A randomized, double-blinded, placebo-controlled trial. *Lancet, 370*(9598), 1560–1567.

Miyake, S., Takahashi, N., Sasaki, M., Kobayashi, S., Tsubota, K., & Ozawa, Y., (2012). Vision preservation during retinal inflammation by anthocyanin-rich bilberry extract: Cellular and molecular mechanism. *Lab. Invest., 92*(1), 102–109.

Munshi, A., & Ramesh, R., (2013). Mitogen-activated protein kinases and their role in radiation response. *Genes Cancer, 4*(9, 10), 401–408.

Park, S. W., Choi, J., Kim, J., Jeong, W., Kim, J. S., Jeong, B. K., Shin, S. C., & Kim, J. H., (2018). Anthocyanins from black soybean seed coat prevent radiation-induced skin fibrosis by downregulating TGF-β and Smad3 expression. *Arch. Derm. Res., 310*(5), 401–412.

Patras, A., Brunton, N. P., O'Donnell, C., & Tiwari, B. K., (2010). Effect of thermal processing on anthocyanin stability in foods; mechanisms and kinetics of degradation. *Trends Food Sci. Technol., 21*(1), 3–11.

Pavelecini, M., Zettler, C. G., Fernandes, M. C., & Ely, P. B., (2019). Immunohistochemical expression of cyclooxygenases in hypertrophic scars and keloids. *Plast. Reconstr. Surg. Glob. Open, 7*(2), e2030.

Pinheiro, C. P., Moreira, L. M. K., Alves, S. S., Cadaval, T. R. S. Jr., & Pinto, L. A. A., (2021). Anthocyanins concentration by adsorption onto chitosan and alginate beads: Isotherms, kinetics and thermodynamics parameters. *Int. J. Biol. Macromol., 166*, 934–939.

Pojer, E., Mattivi, F., Johnson, D., & Stockley, C. S., (2013). The case for anthocyanin consumption to promote human health: A review: Anthocyanins and human health. *Compr. Rev. Food Sci. Food Saf., 12*(5), 483–508.

Puupponen-Pimiä, R., Nohynek, L., Meier, C., Kähkönen, M., Heinonen, M., Hopia, A., & Oksman-Caldentey, K. M., (2001). Antimicrobial properties of phenolic compounds from berries. *J. Appl. Microbiol., 90*(4), 494–507.

Qin, Y., Xia, M., Ma, J., Hao, Y., Liu, J., Mou, H., Cao, L., & Ling, W., (2009). Anthocyanin supplementation improves serum LDL- and HDL-cholesterol concentrations associated with the inhibition of cholesteryl ester transfer protein in dyslipidemic subjects. *Am. J. Clin. Nutr., 90*(3), 485–492.

Ramroop, P., & Neetoo, H., (2018). Anti-listerial activity of *Cymbopogon citratus* on crabsticks. *AIMS Microbiol., 4*(1), 67–84.

Rechner, A. R., & Kroner, C., (2005). Anthocyanins and colonic metabolites of dietary polyphenols inhibit platelet function. *Thromb. Res., 116*(4), 327–334.

Ricciotti, E., & FitzGerald, G. A., (2011). Prostaglandins and inflammation. *Arterioscler. Thromb. Vasc. Biol., 31*(5), 986–1000.

Robinson, G. M., & Robinson, R., (1932). A survey of anthocyanins. II. *Biochem. J., 26*(5), 1647–1664.

Saint-Cricq De, G. N., Glories, Y., & Vivas, N., (1999). Free radical scavenging effect of anthocyanins in red wines. *Food Res. Int., 32*(5), 327–333.

Sims, C. A., & Morris, J. R., (1985). A comparison of the color components and color stability of red wine from noble and cabernet sauvignon at various PH levels. *Am. J. Enol. Vitic., 36*(3), 181–184.

Sobolewski, C., Cerella, C., Dicato, M., Ghibelli, L., & Diederich, M., (2010). The role of cyclooxygenase-2 in cell proliferation and cell death in human malignancies. *Int. J. Cell Biol.*, 215158.

Soga, M., Ohashi, A., Taniguchi, M., Matsui, T., & Tsuda, T., (2014). The di-peptide Trp-his activates AMP-activated protein kinase and enhances glucose uptake independently of insulin in L6 myotubes. *FEBS Open Bio., 4*(1), 898–904.

Tanaka, Y., Tsuda, S., & Kusumi, T., (1998). Metabolic engineering to modify flower color. *Plant Cell Physiol., 39*(11), 1119–1126.

Tarozzi, A., Morroni, F., Hrelia, S., Angeloni, C., Marchesi, A., Cantelli-Forti, G., & Hrelia, P., (2007). Neuroprotective effects of anthocyanins and their *in vivo* metabolites in SH-SY5Y cells. *Neurosci. Lett., 424*(1), 36–40.

Wang, H., Cao, G., & Prior, R. L., (1997). Oxygen radical absorbing capacity of anthocyanins. *J. Agric. Food Chem., 45*(2), 304–309.

Williams, C. S., Mann, M., & DuBois, R. N., (1999). The role of cyclooxygenases in inflammation, cancer, and development. *Oncogene, 18*(55), 7908–7916.

Wolske, E., Chatham, L., Juvik, J., & Branham, B., (2021). Berry quality and anthocyanin content of "consort" black currants grown under artificial shade. *Plants, 10*(4). https://doi.org/10.3390/plants10040766.

Xue, Y., Lim, S., Bråkenhielm, E., & Cao, Y., (2010). Adipose angiogenesis: Quantitative methods to study microvessel growth, regression and remodeling *in vivo*. *Nat. Protoc., 5*(5), 912–920.

Yabuya, T., Nakamura, M., Iwashina, T., Yamaguchi, M., & Takehara, T., (1997). *Euphytica, 98*(3), 163–167.

Yance, D. R. Jr., & Sagar, S. M., (2006). Targeting angiogenesis with integrative cancer therapies. *Integr. Cancer Ther., 5*(1), 9–29.

Yang, B., Zhang, D., Qian, J., & Cheng, Y., (2020). Chelerythrine suppresses proliferation and metastasis of human prostate cancer cells via modulating MMP/TIMP/NF-KB system. *Mol. Cell. Biochem., 474*(1, 2), 199–208.

Zhang, Y., Liu, Y., Hu, W., Sun, B., Chen, Q., & Tang, H., (2018). Anthocyanin accumulation and related gene expression affected by low temperature during strawberry coloration. *Acta Physiol. Plant, 40*(11). https://doi.org/10.1007/s11738-018-2767-8.

Zhu, Y., Ling, W., Guo, H., Song, F., Ye, Q., Zou, T., Li, D., et al., (2013). Anti-inflammatory effect of purified dietary anthocyanin in adults with hypercholesterolemia: A randomized controlled trial. *Nutr. Metab. Cardiovasc. Dis., 23*(9), 843–849.

CHAPTER 3

MANAGEMENT AND PREVENTION OF DISEASES BY FLAVONOIDS

SHUBHAM J. KHAIRNAR[1], EKNATH D. AHIRE[2], MOHINI R. JAGTAP[3], KHEMCHAND R. SURANA[4], SANJAY J. KSHIRSAGAR[2], and RAJ K. KESERVANI[5]

[1]*Department of Pharmacology, MET's Institute of Pharmacy, Affiliated to Savitribai Phule Pune University, BKC, Adgaon, Nashik, Maharashtra, India*

[2]*Department of Pharmaceutics, MET's Institute of Pharmacy, Affiliated to Savitribai Phule Pune University, BKC, Adgaon, Nashik, Maharashtra, India*

[3]*JMCTE Institute of Pharmacy, Wadala Road, Nashik, Maharashtra, India*

[4]*Department of Pharmaceutical Chemistry, Shreeshakti Shaikshanik Sanstha, Divine College of Pharmacy, Satana, Nashik, Maharashtra, India*

[5]*Faculty of B.Pharmacy, CSM Group of Institutions, Prayagraj, Uttar Pradesh, India*

ABSTRACT

Noncommunicable diseases (NCDs), often known as chronic diseases, include cardiovascular disease (CVDs), diabetes, neurodegenerative disorders, and cancer, which continue to be a healthcare burden in both

Advances in Flavonoids for Human Health and Prevention of Diseases. Nisha Sharma, Deepika Saini, Rajesh K. Kesharwani, Prakash Chandra Gupta, & Raj K. Keservani (Eds.)

developed and developing countries. NCDs and associated complications are causing an increase in the number of deaths and high medical expenses. Because of these and other considerations, NCD prevention is a top goal for clinical medicine and public health. Flavonoids are a class of bioactive substances found in a wide range of plant-based meals and drinks. High flavonoid consumption appears to play an essential role in the prevention of a number of NCDs, including type two diabetes, CVDs, cataract, neuro-degenerative disorders, and cancer, according to epidemiological research. Furthermore, flavonoids, including quercetin, kaempferol, naringenin, hesperitin, and luteolin have been shown to have key roles in the treatment of cardiovascular and neurological illnesses, as well as cancer, in both *in vitro* and animal investigations. Dietary flavonoids protect the heart via lowering blood pressure (BP), guarding against oxidized low-density lipoprotein (LDL)-induced damage, and acting as an antiplatelet agent. They also have the potential to protect the nervous system by reducing neuroinflammation and axonal damage. Flavonoids' anticancer properties are based on their capacity to inactivate carcinogens and improve DNA repair processes. Flavonoids' free radical scavenging properties provide antioxidative and anti-inflammatory actions, making them significant in the prevention of metabolic disorders such as dyslipidemia, insulin resistance, hypertension, glucose intolerance, inflammation, and oxidative stress. A growing amount of research suggests that flavonoid-rich diets are good for better health and the prevention of NCDs.

3.1 INTRODUCTION

Noncommunicable diseases (NCDs), often known as chronic diseases, are the leading cause of death and disability in both developed and developing countries. NCDs include cardiovascular disease (CVD), obesity, diabetes, cancer, and neurodegenerative disorders. NCDs and its complications are the cause of a large number of deaths as well as high medical costs. As a result, NCD prevention is a top concern for both public health and clinical care. NCDs are mostly caused by unhealthy diets and lifestyle practices. NCD risk can be reduced by eating a plant-based diet. Flavonoids-rich foods play a significant role in preventing NCDs and associated consequences. Flavonoids (Keservani & Sharma, 2014) play an important role in the prevention of metabolic abnormalities such as dyslipidemia, insulin resistance, hypertension, glucose intolerance, inflammation, and oxidative stress, at least in part due to their free radical scavenging properties and enhancement of antioxidant (Singh et al., 2017a) enzymes in their actions in mitigating the onset and progression of various NCDs. Flavonoids are secondary metabolites found in higher plants, with over 10,000 structures known so far. They are commonly used pigments that provide vital activities such as protecting against UV and visible light damage. Flavonoids also serve as visible cues to attract pollinators and seed dispersers. Plants largely produce flavonoids to protect themselves against oxidative stress. Importantly, they retain antioxidant characteristics and contribute to the medicinal and dietary qualities of plant foods. Therefore flavonoids' antioxidant properties are critical for plant and human physiological processes. This chapter discusses the use of different types of flavonoids and their subclasses in the prevention and treatment of NCDs, as well as the current state of knowledge about flavonoids' protective effect against NCDs. The standard clinical trial database (http://clinicaltrials.gov/) and the keyword "flavonoid" were used in our search technique. In April 2016, a search for flavonoids in clinical studies yielded 112 results (Alam et al., 2014; Babu et al., 2013; Panche et al., 2016).

3.2 FLAVONOIDS SUBCLASSES, SOURCES, AND DIETARY INTAKE

Flavonoids are a type of polyphenol that can be split into six subcategories: anthocyanins, flavanols, flavanones, flavones, flavonols, and isoflavones (Table 3.1). Many fruits, vegetables, flowers, cereals, and other plant-derived foods contain anthocyanins, which are responsible for their characteristic red, blue, and purple hue (Garcia et al., 2005). Anthocyanidins are sugar-free aglycones, whereas anthocyanin glycosides are anthocyanin glycosides. Cyaniding, delphinidin, malvidin, peonidin, pelargonidin, and petunidin are the six most frequent anthocyanidins. Flavanols, which are found in tea, fruits, cocoa, and chocolate, are another type of flavonoid. They come as monomers (catechins) and polymers (catechinols). Epicatechin, gallocatechin, and epigallocatechin (EGC) are free catechins, while epicatechin gallate (ECG) and epigallocatechin gallate (EGG) are bound catechins (EGCG) (Beecher, 2003). The flavanones subclass is dominated by hesperidin (HES) and its metabolite hesperetin, as well as naringenin. Another flavonoid subclass is flavones, which contain compounds like sinensetin, isosinensetin, nobiletin, tangeretin, luteolin, apigenin, chrysin, and galangin. The concentration of flavones in various foods. Parsley and oregano are the best suppliers of flavones. The most prevalent flavonoids are flavonols, which are found throughout the plant kingdom. Kaempferol, quercetin, fisetin, isorhamnetin, and myricetin are the most common dietary flavonols. Berries, cowpeas, dock, kale, dark chocolate, parsley, oregano, capers, and green and black tea are the most common sources. It's vital to remember that the presence of specific flavonoids in vegetables and fruits is dependent on the crop variety, region, and type of production, as well as the morphological component of the plant. Flavonoid content differences between plant species are normally minimal, although on a few occasions, extremely high quantities have been found, such as in particular berries and tea made from Quingmao tree leaves. Flavonoids are frequently consumed at a far higher rate in a plant-based diet than other phytochemicals and vitamins (Keservani et al., 2010a, b), such as ascorbic acid (vitamin C), -tocopherol (vitamin E), or carotenoids. There are various databases of flavonoid content in foods that can be used to estimate flavonoid consumption in humans. Databases from the United States Department of Agriculture (USDA) and the online Phenol-Explorer database are the most often utilized (www.phenol-explorer.eu). As a result, comparing nutritional assessment results from diverse data sources may yield different results. Many surveys were conducted in the 1990s and may be underestimated because the database of flavonoid levels in foods at the time was insufficient. The year of release of specific databases should be taken into account for correct interpretation of the results when assessing dietary flavonoids intake (Garcia et al., 2005; Christensen et al., 2012; Garcia-Larsen et al., 2018).

TABLE 3.1 Subclasses of Flavonoids

Subclass	Examples of Compounds
Anthocyanins	Pelargonidin, cyanidin, delphinidin, peonidin, petunidin, malvidin
Flavanols	Epicatechin, catechin, epicatechin gallate (ECG), gallocatechin, epigallocatechin (EGC), epigallocatechin gallate (EGCG)
Flavanones	Naringenin, hesperetin
Flavones	Sinensetin, isosinensetin, nobiletin, tangeretin, luteolin, apigenin, chrysin, galangin
Flavonols	Kaempferol, quercetin, fisetin, isorhamnetin, and myricetin
Isoflavones	Daidzein, genistein

Nutritional assessments obtained on the basis of the various data sources may differ. It should be emphasized that many surveys were performed in the 1990s and may be underestimates, as the database of flavonoid levels in foods was incomplete at that time. For proper interpretation of the results regarding the assessment of dietary flavonoids intake, the year of release of specific databases should be taken into account (Kozłowska & Szostak-Węgierek, 2017a). According to new research, the overall daily intake of flavonoids differs by country and culture. Kim & colleagues calculated the mean daily total flavonoid intake in US adults (NHANES food consumption data 2007–2010, n = 9801) to be 200.1 8.9 mg/day (Kim et al., 2016a). The average total flavonoid intake among European adults (about 30,000 participants) is 428 mg/day, with the Central Region (506 mg/day) having the greatest consumption, followed by the Northern Region (348 mg/day), and the Southern Region (301 mg/day) (Cercato et al., 2021).

According to the Phenol-Explorer database, the total flavonoid consumption among the Polish population (6,666 people from the Polish National Multicenter Health Survey) is 403.5 mg per day. On the other hand, this is equal to 525 mg/day when estimated using USDA databases (Witkowska et al., 2015). These findings contrast with our own studies, which show that Polish students consume an average of 801 mg of flavonoids per day (Kozowska et al., 2015). In turn, a study of 79 older Australians found that they consumed 683 mg of flavonoids per day on average, with flavan-3-ols accounting for 92% of the total, followed by flavonols (4%), flavanones (3%), and flavones (1%). Importantly, total flavonoids consumption varies greatly based on food choices, demographic group, and age. Flavonoid Supplements are also available on the market, in addition to their natural sources. The manufacturers of these supplements advocate daily flavonoid doses that are higher than those found in a flavonoid-rich diet. As a result, potential side effects such as decreased trace element bioavailability, antithyroid, and goitrogenic effects, and flavonoid-drug interactions have been raised. An increasing body of evidence suggests that dietary flavonoids are safe to consume and that dietary flavonoids and their condensed sources can help prevent disease; nonetheless, more clinical and epidemiological research is urgently needed (Kozłowska & Szostak-Węgierek, 2017b).

3.3 FLAVONOIDS AND THE RELATED HEALTH BENEFITS

NCDs have become a severe public health concern in the 21st century due to their rising global prevalence. The prevention of chronic diseases is a top concern, and a flavonoid-rich diet may be a cost-effective and easy-to-implement strategy (Kozlowska & Szostak-Wegierek, 2014). Foods and dietary supplements high in flavonoids have been shown to protect against NCDs such as cardiovascular disease (CVD), type-2 diabetes mellitus (T2DM), obesity, cancer, Alzheimer's disease (AD), and Parkinson's disease (PD) in epidemiological and randomized controlled trials (PD). We tried to summarize the results of epidemiological research and show the major mechanisms of flavonoid action in subsections (Yao et al., 2004; Prasain et al., 2010).

3.3.1 FLAVONOIDS IN CARDIOVASCULAR DISEASE (CVD)

A diet high in fruits (Keservani et al., 2016a, b) and vegetables has long been related to a lower risk of CVD, and multiple studies have found this link. CVD risk was reduced by 12% for every increase in total fruit and vegetable consumption of 477 grams per day, by 16% for 300 grams per day of fruit, and by 18% for 400 grams per day of vegetable consumption (Gan et al., 2015). Similarly, participants in the European Prospective Investigation into Cancer and Nutrition study who ate at least eight parts (80 g each) of fruits and vegetables per day had a 22% lower risk of fatal ischemic heart disease (IHD) than those who ate less than three portions per day (Crowe et al., 2011). In a second study, a linear dose-response relationship revealed that for every daily 200 g increase in fruit and vegetable consumption, the risk of stroke dropped by 32% and 11%, respectively (Hu et al., 2014). The advice of at least five servings of fruits and vegetables per day is backed up by epidemiological studies. This group of foods contains not only a variety of essential nutrients, including fiber, folate, antioxidant vitamins (Keservani et al., 2015c), and potassium, but also a large number of bioactive substances, including flavonoids, which may be beneficial in the prevention of CVD (Gross, 2004). Epidemiological studies that unequivocally show a link between increased flavonoid intake and lower CVD risk are, however, limited. Nonetheless, the findings of a meta-analysis of 15 prospective trials point to such a link (Terao et al., 2008). The combined findings demonstrate that a higher flavonoids intake compared to a lower intake is linked to a lower risk of coronary heart disease (CHD). The researchers also discovered a nonsignificant inverse relationship, with a 5% reduction in CHD risk for every 20 mg/day rise in flavonoids intake (Hodgson, 2008). Despite the fact that numerous research shows an inverse relationship between flavonoid intake and CVD risk, the Framingham Offspring Cohort did not corroborate such findings (Peterson et al., 2012). After multivariable adjustment, the data demonstrate that there is no link between flavonoid intake and CHD incidence (Peluso, 2006). This study, which was based on long-term follow-up of middle-aged and elderly people, implies that the link between flavonoid intake and CVD risk is unclear and needs to be investigated further.

3.3.2 CARDIOPROTECTIVE ACTION OF FLAVONOIDS

Flavonoids benefit the cardiovascular system through a variety of methods, while the direct mechanisms of flavonoids' activities are unknown, it appears that they are complicated and rely on multiple processes. Antioxidant and anti-inflammatory activity, blood pressure (BP) regulation, cholesterol reduction, antiplatelet effects, and protection of low-density lipoprotein (LDL, the predominant atherogenic lipoprotein) against oxidative modification are only a few of the main mechanisms. Flavonoids can reduce inflammation through a variety of methods, including the inactivation of nitric oxide (NO•), the scavenging of ROS, and the reduction of leukocyte entry into inflammatory areas (Cook & Samman, 1996). Consumption of flavanol-rich cocoa lowers plasma LDL cholesterol concentrations, according to a meta-analysis encompassing 215 people from eight clinical trials (Testai, 2015). Furthermore, flavonoids reduce vascular endothelial inflammation and improve vascular function. Flavonoids also inhibit the ROS-producing enzymes xanthine oxidase, nicotinamide adenine dinucleotide phosphate (NADPH) oxidase, and lipoxygenase (LOX). Flavonoids activate the protein kinase 5′ adenosine monophosphate-activated protein kinase (AMPK), which inhibits 3-hydroxy-3-methylglutaryl-coenzyme A (HMG-CoA) reductase, which is the rate-limiting enzyme in cholesterol formation. As a result, higher AMPK activation lowers cholesterol production in the blood (Testai et al., 2013). Flavonoids suppress the synthesis of thromboxane and leukotrienes by inactivating proinflammatory enzymes, including LOX and cyclooxygenase (COX). This results in less vasoconstriction. In males at risk of CVDs, a flavonoid-rich diet may reduce microvascular reactivity and inflammatory state. The males in the study showed increased endothelium-dependent microvascular reactivity after 6 weeks of daily ingestion of two additional servings of high-flavonoid fruits and vegetables. They also had lower plasma concentrations of C-reactive protein, E-selectin, and vascular cell adhesion molecule following 12 weeks of daily eating of an additional four portions of high-flavonoid fruits and vegetables (Khan et al., 2021). Flavonoids' antiplatelet effect is linked to the generation of NO• in the vascular endothelium as well as the prevention of the creation of 12-hydroxyeicosatetraenoic acid (12-HETE), a chemical that degrades endothelial function (Alyane et al., 2008).

3.3.3 ANTIDIABETIC ACTION OF FLAVONOIDS

Flavonoids interact with a variety of metabolic and signaling pathways in pancreatic cells, the liver, adipose tissue, and skeletal muscles to have positive effects on diabetes. Through the translocation of glucose transporter type-4 (GLUT4) vesicles to the plasma membrane, flavonoids improve glucose absorption by peripheral tissues such as skeletal muscle and white adipose tissues. They affect cell mass and function, as well as energy metabolism and insulin sensitivity in peripheral tissues, by activating AMPK and other kinases like extracellular signal-regulated protein kinases 1 and 2 (ERK1/ERK2) and p38 mitogen-activated protein kinase (p38MAPK), which are required for maximal glucose uptake stimulation in response to insulin (Vinayagam & Xu, 2015; Babu et al., 2013).

3.3.4 FLAVONOIDS IN CANCER PREVENTION

Ovarian, breast, and lung cancers can all be reduced by consuming dietary flavonoids. Flavonoid consumption reduces the risk of lung cancer, especially in women who have quit smoking, but there is no evidence that flavonoid consumption reduces the risk of other malignancies (Neuhouser, 2004). According to a meta-analysis of 35 studies (9,88,082 individuals and 8,161 cases), total dietary flavonoids and most flavonoid subclasses are inversely related to the risk of smoking-related cancer. Total dietary flavonoid consumption is inversely linked with aerodigestive tract cancer risk and moderately with lung cancer risk in subgroup analysis by cancer location (Yao et al., 2011). A meta-analysis of 12 research found that postmenopausal women who consumed substantial amounts of flavonoids such flavanols and flavones had a lower risk of breast cancer. The epidemiological data offered above is only for prevention, and the findings of the research are not always clear, and they only apply to certain types of cancers (Romagnolo & Selmin, 2012; Park & Pezzuto, 2012).

3.3.5 FLAVONOIDS IN CANCER: MECHANISMS OF ACTION

Many stages of carcinogenesis are affected by flavonoid, including initiation, promotion, and progression. Carcinogens are inactivated, cell proliferation is inhibited, DNA repair pathways are improved, and oxidative stress is reduced. Flavonoids suppress angiogenesis, induce apoptosis, and have cytotoxic or cytostatic effects on cancer cells during the progression phase. Flavonoids exert their effects in part through interactions with phase I metabolizing enzymes, which are responsible for the metabolism of a variety of endogenous and exogenous substrates, including the activation of carcinogens. Some cytochrome P450 isozymes, such as CYP1A1 and CYP1A2, are inhibited by flavonoids, which protect cells from carcinogens. Another way they work is by inducing phase II enzymes like glutathione S-transferase (GST) and UDP-glucuronyl transferase (UDP-GT), which detoxify and remove carcinogens from the body. Cell cycle inhibition can also explain flavonoids' anticancer properties. CDKs are important regulators of cell cycle progression because they are triggered by mitogenic signals within the cell. CDK activity alteration and dysregulation are pathogenic hallmarks of cancer. Mutations in CDK genes or CDK inhibitor genes have been linked to hyperactivation of CDKs in a variety of malignancies. As a result, a lot of research is being done on drugs that can block or alter CDKs. Flavonoids such genistein, quercetin, daidzein, luteolin, kaempferol, apigenin, and epigallocatechin (EGC) can have these effects (Martinez-Perez et al., 2014).

3.3.6 FLAVONOIDS IN COGNITIVE FUNCTIONS

Dietary flavonoid consumption by the aged was found to support their cognitive functions in the PAQUID study (PersonnesAge'es QUID). After

10 years of monitoring, those who consumed flavonoids in the two highest quartiles had higher cognitive performance than those who consumed less, suggesting that regular use of dietary flavonoids can sustain cognitive function with aging. These findings are in line with those of Samieri et al. who found that a high flavonoid intake in midlife helped women retain good health and well-being as they aged. Higher intake of several flavonoid subclasses (i.e., flavonols, flavones, flavanones, flavan-3-ol polymers, and anthocyanins), as well as total flavonoids in age-adjusted analyzes, was associated with better odds of healthy aging in this large prospective study (13,818 women from the Nurses' Health Study). Healthy aging was defined as living to be at least 70 years old while maintaining four health domains (no major chronic diseases or major impairments in cognitive or physical function or mental health) (Keservani et al., 2015b). Flavonoid subclasses such as flavones and flavanones are positively linked with two of the four domains in the authors' definition of healthy aging in the study of each component of healthy aging, however the connections are lower than for overall healthy aging. Women in the highest quintile of flavone and flavanone intakes, for example, have 32% and 24% better odds of having no mental health limitations, respectively, and 23% and 15% better odds of having no physical function limitations, when compared to women in the lowest quintile. The Mediterranean diet (MD), which is high in flavonoids from vegetables, legumes, fruits, cereals, spices, and wine, has been the subject of some research. MD adherence is linked to a decreased risk of AD and moderate cognitive impairment (MCI). Furthermore, dietary pattern analysis in the prospective Australian Imaging Biomarkers and Lifestyle Study of Ageing (AIBL) suggests that both MCI and AD patients adhere to the MD less than healthy control persons. According to a comprehensive meta-analysis and systematic review, stronger adherence to the MD is linked to a lower risk of MCI and AD, as well as a lower risk of progression from MCI to AD (Ahire et al., 2020; Macready et al., 2009; Schmitt-Schillig et al., 2005).

3.3.7 EFFECTS OF FLAVONOIDS ON THE NERVOUS SYSTEM

Increased oxidative stress and inflammation are important pathogenic processes of neurodegenerative illnesses, including AD. Flavonoids in the diet help to prevent neuronal death caused by oxidative stress (Keservani et al., 2015a). The beneficial effects of flavonoids, which are found in foods and flavonoid-enriched extracts, appear to be linked to a number of other processes, including the regulation of the neuronal signal cascade, which results in the inhibition of cell apoptosis caused by neurotoxic substances such as amyloid (A) and 6-hydroxydopamine (6-OHDA). Flavonoids can also boost levels of brain-derived neurotrophic factor (BDNF), a protein that helps neurons survive and differentiate. Flavonoids can improve blood flow in the peripheral and central nervous systems, which is important for maintaining neuronal and cognitive brain functions. They also induce angiogenesis and the growth of new nerve cells in the hippocampus, which is important for maintaining neuronal and cognitive brain functions (Grosso & Valentão, 2013; Youdim & Shukitt, 2004).

3.4 EFFECTS OF FLAVONOID SUBCLASSES ON HEALTH, PREVENTION, AND TREATMENT OF CHRONIC DISEASES

The efficacy of flavonoids in the prevention and treatment of hyperlipidemia, hypertension, postprandial hyperglycemia, platelet aggregation, cancer, and eye cataracts, as well as in the prevention and treatment of neurodegenerative illnesses, is the subject of much current research. We then go over the findings of some recent studies in order to familiarize the reader with many of the impacts of flavonoids on health and NCDs (Zhang & Gan, 2015; Ahire & Surana, 2020).

3.4.1 ANTHOCYANINS

Anthocyanin intakes of 320 mg twice daily for four weeks do not lower BP in men (n = 31) with a screening BP of 140/90 mm Hg. In a randomized, double-blind, placebo-controlled clinical trial involving 50 hyperlipidemic adults, anthocyanin intake (90 mg of anthocyanins from *Vaccinium arctostaphylos* extract/day/4 weeks) reduced total serum cholesterol by 34.44 mg/dL, LDL-C by 11.44 mg/dL, TG by 69.64 mg/dL, and malondialdehyde (a marker of lipid peroxidation) by Another randomized, double-blind, placebo-controlled clinical trial (n = 80) found that supplementing with anthocyanin-rich fruit extract (whortleberry fruit, 350 mg capsule every 8 h) for two months reduced total cholesterol, triglyceride, and low-density lipoprotein cholesterol (LDL-C) by 27.6%, 19.2%, and 26.3%, respectively, and increased HDL-C by 37.5%.

These data imply that treating hyperlipidemic patients with anthocyanin improves their lipid profile. Furthermore, Zhu et al. show that anthocyanin supplementation (320 mg/day for 24 weeks) reduces proinflammatory markers in hypercholesterolemic patients. When compared to the placebo group, anthocyanin intake lowers serum levels of hs-CRP (21.6% vs. 2.5%), soluble vascular cell adhesion molecule 1 (sVCAM-1) (12.3% vs. 0.4%), and plasma interleukin 1 beta (IL-1, 12.8% vs. 1.3%) (Zhu et al., 2013). Anthocyanins have shown to be effective in the treatment of retinal and eye illnesses. In choroidal neovascularization, vascular endothelial growth factor (VEGF) is a significant participant. VEGF is produced in the retina by retinal pigment epithelial (RPE) cells. Cyanidin lowered the rate of RPE cell necrosis and decreased VEGF release without affecting cell viability in a study employing cultured human RPE cells. Additionally, in human RPE cells, cyanidin-3-glucoside (Cy-3-Glu) isolated from three berries (blueberry, blackberry, and strawberry) shows numerous antioxidant, anti-angiogenic, and anti-aging properties. Under visible light irradiation, Cy-3-glu reduces intracellular ROS levels the most of the four anthocyanins examined (Cy-3-glu, pelargonidin-3-glucoside, delphinidin-3-glucoside, and malvidin-3-glucoside). Furthermore, Cy-3-glu lowers VEGF levels. Anthocyanins can also help prevent cataracts from forming. Apoptosis of the lens epithelial cells is thought to play a part in the pathogenesis of cataract formation, and oxidative stress may be a key cause of lens opacity. A human lens epithelial cell line (HLE-B3) is protected from oxidative stress by anthocyanin from black soybean. Under oxidative stress, this chemical inhibits HLE-B3 cell death. Anthocyanins are potential medicines in the prevention and treatment of hyperlipidemia and several eye disorders when taken together (Wallace & Giusti, 2015; Castañeda-Ovando et al., 2009).

3.4.2 FLAVANOLS

Green tea consumption, which is high in flavanols, is linked to a lower risk of CVD, as well as cardiovascular and overall mortality. Vasodilation is a key mechanism of cardio-protection. The function of NO• synthase (NOS) in endothelial cells and subsequent stimulation of NO• production in arteries is critical for EGCG-induced vasodilation. In the aortic rings of wild-type mice, low doses of EGCG (5–15 M) elicit NO•-dependent vasorelaxation. In rats, black tea (15 mg/kg/day) decreases hypertension-related endothelial dysfunction. Endoplasmic reticulum stress is likely alleviated by EGCG, which inhibits homocysteine metabolic enzymes and reduces ROS generation. Green tea catechins increase energy metabolism at rest and during activity, whether taken in short or long doses (Hodgson et al., 2013a, b). Around 18 patients with relapsing-remitting multiple sclerosis (MS) were given EGCG (600 mg/day) or placebo for 12 weeks in a small randomized, double-blind, placebo-controlled crossover experiment (4-week washout period in between). Because of sex-specific effects on autonomic and endocrine regulation, this supplement enhances muscle metabolism during moderate activity to a larger amount in men than in women. Fasting adipose tissue lactate concentrations are almost twofold greater in men receiving placebo, indicating enhanced lipolytic activity, but corrected with EGCG. Adipose tissue and skeletal muscle perfusion are lower in men but higher in women receiving EGCG. In this study, males with MS benefited from a combination of EGCG and physical activity. As indicated by a decrease in bacterial translocation across swine intestinal epithelial IPEC-J2 cells, EGCG from green tea is an effective antibacterial agent. EGCG appears to improve the epithelial immunological barrier by causing the antimicrobial peptides porcine beta-defensin 1 and 2 to be secreted (pBD-1, pBD-2). This improves the immunological barrier, which protects against bacterial translocation, especially in colitis ulcerosa and Crohn's disease (Wan et al., 2016). In animal and cell culture investigations, flavanols, particularly EGCG, have been shown to have antiviral properties. The most prevalent infectious illnesses in humans are oral herpes simplex virus type-1 (HSV-1) infection and genital herpes simplex virus type-2 (HSV-2). Cantatore et al. studied the effects of adenocarcinomic human alveolar basal epithelial cells and Vero-cultured cells (one of the more extensively used mammalian cell lines in microbiology), finding that black tea extract administration reduced or stopped the formation of infectious HSV-1 virions. These effects are achieved by interference with HSV-1 particle attachment, penetration, and viral DNA replication, which inhibits the virus's infectivity. Furthermore, combining palmitate with the green tea flavanol EGCG improves EGCG's potency as an inhibitor of HSV-1. Autophagy, which is important for viral replication regulation, is also influenced by flavanols. As a result, increasing autophagy may be a viable treatment option for viral infection. EGCG therapy suppressed HBV replication in cultured hepatoma HepG2 cells via increasing lysosomal acidity and autophagy. Flavanols help to protect the cardiovascular system and may also help to prevent viral and bacterial infections, as well as boost muscle metabolism (Heiss et al., 2010; Hollman et al., 2000).

3.4.3 FLAVANONES

HES and its aglycone derivative, hesperetin, are flavonoids from the flavanone subclass that can help with cardiovascular health by lowering blood cholesterol, inflammation, and BP. Human HMG-CoA reductase activity is inhibited by HES, resulting in hypocholesterolemic effects. This enzyme affects plasma cholesterol levels by playing a vital role in the production of cholesterol in humans. When its function is inhibited, intracellular cholesterol levels are reduced, and LDL receptor expression is enhanced. As a result, cellular lipoprotein absorption and cholesterol elimination from the circulation increases. In a randomized, double-blind research, oral treatment of 500 mg of HES daily for three weeks promoted endothelial NO• production and lowered the activity of proinflammatory cytokines in persons with metabolic syndrome (n = 24). HES (0.1% in the diet) and glucosyl hesperidin (GHES) (0.1% in the diet) reduced hypertension and inhibited the mRNA expression of NADPH oxidase subunits and thromboxane A2 synthase in spontaneously hypertensive rats. Bioconverted *Citrus unshiu* (a Mandarin citrus variety) peel extracts inhibited lipid accumulation and suppressed adipocyte-specific transcription factors such as CCAAT-enhancer-binding protein alpha (C/EBP), peroxisome proliferator-activated receptor gamma (PPAR), and sterol regulatory element-binding protein 1c (SREBP1c) in 3T3-L1 preadipocytes culture cells. Supplementing with naringenin may help to treat metabolic syndrome, which includes obesity, diabetes, hypertension, and hyperlipidemia. Furthermore, naringenin decreased postprandial hyperglycemia in streptozotocin-induced diabetic rats fed a high-fat (Singh et al., 2017b) diet. In a rat model of T2DM, oral naringenin (25 mg/kg body) inhibits intestine-glucosidase activity, delaying carbohydrate absorption and lowering postprandial blood glucose levels. Other processes that increase metabolic consequences of obesity are also influenced by naringenin. Macrophage infiltration into adipose tissue is one of these mechanisms, which causes persistent inflammation and leads to obesity-related illnesses like type-2 diabetes and atherosclerosis. Naringenin inhibits macrophage infiltration into adipose tissue in mice during the early stages of obesity caused by a high-fat diet. Treatment with naringenin corrected diet-induced metabolic abnormalities in obese ovariectomized mice (an animal model of menopause). Dietary (Keservani et al., 2015d, 2020; Rane & Keservani, 2020) supplementation with naringenin (3% in the diet) reduces weight growth, hyperglycemia, intra-abdominal obesity, and muscle atrophy (Ke et al., 2016). Furthermore, mice fed a naringenin diet have lower levels of plasma leptin and leptin mRNA in adipose depots. The levels of monocyte chemoattractant protein-1 (MCP1) and interleukin 6 (IL-6) mRNA expression in perigonadal adipose tissue of naringenin-supplemented mice are much lower. Only a few human studies on the role of flavanones in diabetes prevention have been conducted. High flavanone consumption, on the other hand, is linked to a lower risk of diabetes. Over a mean of 5.51 years of follow-up (18,900 person-years), there are 314 new occurrences of diabetes, but after multivariable adjustment, there is a 31% reduction in new-onset diabetes in the lowest tertile of flavanone intake (Graf, 2005). Around 14 metabolites with nominal or substantial changes in concentration between normal-weight, overweight, and obese children were discovered in a pilot investigation of childhood obesity-related features in Mexican-American children. One of these is naringenin, which remained significant when several variables were taken into account. When compared to normal-weight children, obese and overweight children have lower amounts of naringenin. The authors speculated that youngsters of normal weight consume more dietary naringenin sources including grapefruit, orange, and tomato skin. The levels of naringenin in obese children were barely half those in normal-weight youngsters. Naringenin levels were also found to have a negative relationship with systolic BP, diastolic BP, homeostatic model assessment of insulin resistance (HOMA-IR), body mass index, waist circumference, and triglycerides, but a favorable relationship with HDL-C. Risk biomarkers that are available early in life, according to the authors, would allow for early screening and interventions in high-risk children to prevent the development of major health problems later in life. In stomach, colon, breast, lung, and liver malignancies, HES and its derivatives cause apoptosis, which is mediated by nuclear factor-kappa B (NF-κB). In the Swiss albino mouse model of benzo(a)pyrene (B[a]P)-induced lung cancer, HES exerts an antiproliferative effect. By easing lipid peroxidation, modifying antioxidant activities, and diminishing the expression of NF-κB, PCNA, and CYP1A1 (B[a]P metabolic activator), oral treatment of HES (50 mg/kg body weight) suppresses B[a] P-induced lung cancer and its associated preneoplastic symptoms. Future research should concentrate on techniques for enhancing HES bioavailability and absorption, as well as specific molecular processes of HES anticancer benefits, determining the most effective doses for future HES clinical trials, and lastly, evaluating HES anticancer effects in cancer patients. Overall, there is evidence that flavanones have a cardioprotective and anticancer effect. They can also help with obesity metabolic problems (Khan & Dangles, 2014).

3.4.4 FLAVONES

Apigenin and luteolin are the most common dietary flavones. Apigenin and chrysin appear to play essential roles in the prevention of sugar-induced cataractogenesis. The polyol pathway enzyme aldose reductase (AR) is responsible for the reduction of glucose and galactose. Under normoglycemic conditions, the polyol pathway metabolizes less than 3% of glucose. In the case of hyperglycemia, however, more than 30% of glucose enters the polyol pathway, resulting in an excess of sorbitol in the lens, which causes a number of negative pathological alterations, ultimately speeding up the process of cataractogenesis and other diabetes problems. As established in sugar-induced lens cataract model experiments, apigenin and chrysin's anti-cataract activities include suppression of AR. The striatal damage caused by the neurotoxic chemical 6-OHDA leads to damage of the nigrostriatal pathway, which connects the substantia nigra with the striatum, in an animal model of PD. The latter is in charge of, among other things, bodily movement planning. PD is caused by damage to this region. Tangeretin, a flavonoid that also belongs to the flavone subclass, protects the nigrostriatal pathway from the negative effects of 6-OHDA when given to mice. Tangerin's ability to protect striatonigral integrity and function implies that it could be used as a neuroprotective drug (Verma & Pratap, 2010).

Furthermore, nobiletin, a flavone that inhibits the PLC2/PKC cascade and hydroxyl radical (•OH) generation, reduces the activation of Akt and MAPKs by blocking the signaling pathways involved in platelet activation. Ca^{2+} concentrations are reduced, and platelet aggregation is inhibited as a result of these changes. These data imply that nobiletin could be a potential therapeutic agent for thromboembolic diseases prevention or treatment. In animal and human endothelial cell culture studies, luteolin protects against vascular inflammation. In C57BL/6 mice, luteolin supplementation (0.6% luteolin) decreases tumor necrosis factor alpha (TNF-)-induced vascular inflammation. TNF-induced aortic structural alterations in the intimal layer of the artery and breakdown of aortic elastin fibers were avoided by luteolin intake. Furthermore, luteolin (as little as 0.5 M) decreased TNF-induced monocyte adherence to endothelial cells *in vivo*, whereas 20 M of luteolin totally stopped monocyte adhesion. Furthermore, luteolin at a concentration of 0.1 M inhibited TNF-induced gene expression of chemokines monocyte chemotactic protein-1 (MCP-1) and intercellular adhesion molecule 1 (ICAM-1) and vascular cell adhesion molecule 1 (VCAM-1), while 2 M of luteolin almost completely blocked TNF-induced gene expression of these proinflammatory molecules. Importantly, luteolin suppresses TNF-induced activation of the NF-κB signaling pathway, implying that luteolin could be a promising new anti-inflammatory drug. Luteolin has also been suggested as a potential epilepsy therapy. Lutein (LU) (50 or 100 mg/kg/day by oral administration) lowered seizure induction and duration, restored cognitive impairment, and reduced neuronal and oxidative stress damage in pentylenetetrazol-induced epileptic rats. The scientists discovered that luteolin boosted protein kinase A (PKA) and cAMP response element-binding protein (CREB) phosphoactivation, as well as BDNF expression. Flavones isolated from *Alternanthera philoxeroides*, an aquatic plant from South America, luteolin-6-C-d-boivinopyranosyl-3′-O-d-glucopyranoside and chrysoeriol-6-C-d-boivinopyranosyl-4′-O-d-glucopyranoside exhibit substantial anti-hepatitis B virus activity. These compounds stop the HepG2.2.15 cells from secreting HBsAg. As a result, botanical compounds could be an appealing source for a new class of anti-HBV medications. Flavones have an anti-diabetic effect on the pathways that cause diabetes. In a study of women (n = 1997), flavonoid subclass intakes were found to be inversely related to indicators of inflammation, insulin concentration, and the HOMA-IR index (Singh et al., 2014).

When comparing extreme quintiles of intake, participants with high flavone intake had lower HOMA-IR and insulin concentrations and higher adiponectin concentrations, with differences of 0.1 g/ml (P-trend = 0.04), 0.5 g/ml (P-trend = 0.02), and 0.6 g/l (P-trend = 0.01). In the Korean National Health and Nutrition Examination Survey (4,186 men and women), inverse relationships between flavone and flavonol intake, insulin resistance, and insulin level were also found, although only in men. Flavones stimulate adipocyte development and glucose transport in a mouse adipocyte cell line (3T3-L1) via inducing PPAR-mediated adiponectin production and GLUT4 translocation in 3T3-L1 adipocytes. Flavones may have cancer-fighting effects. A meta-analysis of 12 prospective cohort or case-control studies involving 9,513 cases and 1,81,906 controls found that women with high intakes of flavonols and flavones (RR = 0.88, 95% CI 0.80–0.98) and flavones (RR = 0.83, 95% CI 0.76–0.91) have a lower risk of breast cancer than women with low intakes of flavonols and flavones, but not other flavonoid subclasses. In conclusion, flavone intake can help prevent and treat a variety of NCDs, including diabetes, cancer, PD, chronic hepatitis, and epilepsy (Martens & Mithöfer, 2005).

3.4.5 FLAVONOLS

Kaempferol is a flavonoid that has recently been found to have antiproliferative and apoptosis-inducing properties in a variety of cancer cell lines, including human cervical cancer cells (SiHa and Hela), human osteosarcoma cells, cholangiocarcinoma cells, human breast carcinoma (MCF-7) cells, human bladder cancer cells (5637, T24), human stomach carcinoma cells (SGC-7901), and human lung carcinoma. Kaempferol has been suggested as a possible treatment for cancer metastasis. Kaempferol's anti-metastatic activity is due to its ability to inhibit matrix metalloproteinases (MMPs), which aid tumor cells in metastasis by promoting excessive extracellular matrix breakdown and therefore facilitating cancer cell invasion. Kaempferol inhibits lung metastasis in B16F10 murine melanoma cells and inhibits MMP-9 production and activity *in vivo*. Kaempferol also prevents cancer cells from invading by suppressing the activity of the transcription factor activator protein-1 (AP-1) and the MAPK signaling pathway. Kaempferol therapy inhibits enzymatic activity and protein levels of MMP-2, MMP-9, and urokinase plasminogen activator (uPA) in the human osteosarcoma U2OS cell line via attenuation of extracellular signal-regulated kinases (ERK), c-Jun N-terminal kinases (JNK), and p38-MAPKs signaling pathways. This is significant because many structurally related flavonoid compounds have no effect on AP-1, implying that kaempferol could be a potential anticancer drug. To support the potential effects of kaempferol as an anticancer drug for clinical usage, more research is needed. Quercetin possesses anti-inflammatory, antidiabetic, antihypertensive, and antiviral properties, among others. In rats, this chemical has a considerable effect on ritonavir pharmacokinetics and significantly improves its distribution into the brain *in vivo*. A new therapeutic technique that could lead to greater brain distribution, lower toxicity, and enhanced performance of HIV-1 protease inhibitors, which are critical components of highly efficient antiretroviral therapy currently used to treat HIV-1 infection. With a 77% rise in ritonavir accumulation in human brain-microvascular endothelial cells, quercetin is the strongest of the eight flavonoids. Treatment (25 mg/kg body weight/ every 48 h for three consecutive months commencing at 18–21 months of age) reversed-amyloidosis, lowered tauopathy, and reduced astrogliosis and microgliosis in an elderly triple transgenic AD mouse model, protecting cognitive and affective performance. In both a cell-free system and neural cells, natural flavonols (myricetin, quercetin, kaempferol, and morin) and one flavone (apigenin) suppress beta-site APP cleaving enzyme-1 (BACE-1) activity. In neurons, each flavonoid lowers levels of amyloid peptide-40 (A1-40) and A1-42. A is the most common and most harmful component of senile plaques. BACE-1 (-secretase) and -secretase produce A from the amyloid precursor protein (APP). Natural flavonoids may reduce neuronal A levels by inhibiting BACE-1 (Aherne & O'Brien, 2002). Quercetin enhances the conversion of cholesterol to bile acids, resulting in cholesterol efflux. The rate-limiting enzyme in the synthesis of bile acids is cholesterol 7-hydroxylase (CYP7A1). The mRNA and protein expression levels of CYP7A1 in the liver are increased by dietary supplementation with quercetin (0.04% in the diet). In addition, the total bile acid content of feces and bile is higher in the quercetin-supplemented group, although the dry weight of feces and bile excretion is comparable in the treated and control groups. These findings imply that quercetin therapy enhances total bile acid excretion (Zhang et al., 2016). Quercetin has also been shown to be a promising treatment for pulmonary arterial hypertension. Quercetin, at least in part, by modulating the TrkA/AKT signaling pathway, restores the excessive proliferation and apoptotic resistance of pulmonary artery smooth muscle cells exposed to hypoxia. Quercetin has cardioprotective benefits due to its anti-inflammatory, antioxidant, and anti-apoptotic capabilities, which protect rats from myocardial injury. Quercetin reduces TNF-alpha and IL-1-beta mRNA and protein levels, as well as MDA content, and increases superoxide dismutase (SOD) and catalase (CAT) activities in the myocardial tissue of rats in both the low-dose (100 mg quercetin/kg) and high-dose (400 mg quercetin/kg) groups, compared to the control group. Flavonols have been shown to have a positive effect in various epidemiological investigations. A meta-analysis of 13 prospective cohort studies with a total of 3,44,488 participants and 12,445 CVD cases found a negative relationship between flavonol consumption and CVD risk. An investigation of the dose-response relationship found that increasing flavonol consumption by 10 mg per day reduces the risk of CVD by 5%. However, while the majority of studies show a positive link between quercetin consumption and a lower risk of NCDs, a few studies contradict this conclusion. For example, quercetin-3-glucoside supplementation (160 mg/day) showed no effect on flow-mediated dilation, insulin resistance, or other CVD risk variables in a small randomized, double-blind, placebo-controlled crossover experiment. This human investigation, however, had a tiny sample size (n = 37). As a result, more research is needed to determine whether quercetin can be utilized to prevent and cure NCDs. Flavonols appear to be beneficial agents in cardiovascular and anticancer prevention, can improve antiviral therapy efficacy with ritonavir, and may protect against pulmonary arterial hypertension and AD (Perez-Vizcaino & Duarte, 2010).

3.4.6 ISOFLAVONES

The potential benefits of genistein in breast, prostate, and ovarian malignancies have been documented in a variety of human, animal, and cell culture studies. The mechanism of genistein's cancer-fighting effects is complicated and requires further research. It's not always clear whether or not the same molecule has negative or positive effects. Triple negative breast cancer (TNBC, which does not express estrogen, progesterone, or HER2 receptors), HER2, and ER expressing subtypes are the three major subtypes. In this triple scenario, genistein plays a vital role, according to a systematic review. Some studies show that both high and low doses of genistein reduce the growth of ER- and HER2-positive breast cancer cells. High doses, on the other hand, might boost proliferation indicators such fibroblast growth factor receptor 2 (FGFR2), transcription factor (E2F5), budding uninhibited by benzimidazoles 1 (BUB1), cyclin B2 (CCNB2), MYB proto-oncogene like 2 (MYBL2), cell division cycle protein 20 (CDC20), and CDK1 (Ørgaard & Jensen, 2008).

The ATP-binding cassette G2 efflux transporter, commonly known as the breast cancer resistance protein (BCRP), regulates the distribution and excretion of a variety of endogenous and xenobiotic substances (Křížová et al., 2019).

BCRP interacts with genistein and daidzein, as well as their metabolites, as substrates, inhibitors, and/or modulators of gene expression. These isoflavones promote reversion of multidrug resistance and make cancer cells more susceptible to chemotherapy when they function as BCRP inhibitors. However, undesirable consequences can arise, such as an increase in toxicity or a decrease in a drug's efficacy. Breast cancer is less common in Asian countries than it is in North American and European ones. This could be linked to a larger diet of soy, which is high in isoflavones. Soy consumption was linked to a lower incidence of breast cancer in prospective observational studies in Asian populations. A dose-response relationship reveals a considerable reduction in the risk of breast cancer as soy food intake rises, equating to a 16% risk reduction for 10 mg of daily is flavone eaten. In Asian populations, isoflavone intake reduces breast cancer risk (a combined relative risk and odds ratios of 0.68, 95% confidence interval, 0.52–0.89), but not in Western populations (a combined relative risk and odds ratios of 0.98, 95% confidence interval, 0.87, 1.11), according to a meta-analysis of 22 observational studies. In Asian participants, isoflavones had a protective effect on both premenopausal and postmenopausal women. On the other hand, only postmenopausal women benefit from a population-based prospective cohort study in Japan (15,607 women). Postmenopausal breast cancer relative risks are reduced in women who consume more soy (trend $p = 0.023$) and isoflavones (trend $p = 0.046$), but neither soy nor isoflavone intakes are linked to premenopausal breast cancer relative risks. Even in women who consumed moderate amounts of soy (hazard ratio: 0.65 in the second quartile) and isoflavones (hazard ratio: 0.57 in the second quartile), the authors found a lower incidence of postmenopausal breast cancer. These findings imply that soy and isoflavone consumption protects Asian women from postmenopausal breast cancer, however the effect varies depending on the population (Munro et al., 2003).

3.5 CONCLUSIONS

Flavonoids have a favorable influence on health maintenance, according to epidemiological, animal, and *in vitro* research. This holds true for flavonoids found in both natural and supplemented foods. Anti-inflammatory, antioxidant, antiproliferative, anti-angiogenic, antidiabetic, antihypertensive, antiplatelet, and antiviral effects are among the biological actions of these substances. In hyperlipidemic individuals, their ingestion also lowers postprandial hyperglycemia, shields LDL particles from oxidative modification, reduces neuroinflammation, and improves lipid profile. However, a deeper knowledge of the underlying molecular pathways, as well as their metabolism in humans, is critical. Flavonoids have been shown to help with illness prevention and therapy, but less is known about the potential adverse effects of supplements containing supraphysiological amounts of certain flavonoid components. More research, particularly clinical randomized trials, is desperately needed to determine the efficacy and safety of flavonoid use. Our present knowledge backs up the recommendation to consume flavonoids in foods like fruits and vegetables, legumes, nuts, spices, and herbs every day, with a daily fruit and vegetable intake of 600 g. There is also widespread support for suggesting the MD and/or the Dietary Approaches to Stop Hypertension (DASH) diets, both of which are high in flavonoids, which have been proven to promote good health and reduce the risk of NCDs.

ACKNOWLEDGMENT

The authors thank the METs, Institute of Pharmacy, BKC, affiliated under Savitribai Phule Pune University, Adgaon, Nashik, MH, and India. EDA is especially thankful to the Ministry of Tribal Affairs Government of India (NFST) for providing financial assistance in the form of a fellowship (Award No.: 202021-NFST-MAH-01235).

KEYWORDS

- **antioxidants**
- **cardiovascular disease**
- **chronic disease prevention**
- **dietary intake**
- **epicatechin gallate**
- **epigallocatechin**
- **epigallocatechin gallate**
- **flavonoids**

REFERENCES

Ahire, E. D., Surana, K. R., Patil, C. D., Shah, H. S., Sonwane, G. B., & Talele, S. G., (2020). Role of omega-3 fatty acids in different neurodegenerative disorders. *Applied Pharmaceutical Science and Microbiology: Novel Green Chemistry Methods and Natural Products*, 173.

Alam, M. A., Subhan, N., Rahman, M. M., Uddin, S. J., Reza, H. M., & Sarker, S. D., (2014). Effect of citrus flavonoids, naringin and naringenin, on metabolic syndrome and their mechanisms of action. *Advances in Nutrition, 5*, 404–417.

Alyane, M., Kebsa, L. B. W., Boussenane, H., Rouibah, H., & Lahouel, M., (2008). Cardioprotective effects and mechanism of action of polyphenols extracted from propolis against doxorubicin toxicity. *Pakistan Journal of Pharmaceutical Sciences, 21.*

Babu, P. V. A., Liu, D., & Gilbert, E. R., (2013). Recent advances in understanding the anti-diabetic actions of dietary flavonoids. *The Journal of Nutritional Biochemistry, 24*, 1777–1789.

Beecher, G. R., (2003). Overview of dietary flavonoids: Nomenclature, occurrence and intake. *The Journal of Nutrition, 133*, 3248S–3254S.

Cercato, L. M., Oliveira, J. P., Souza, M. T. S., Andrade, N., Martel, F., & Camargo, E. A., (2021). Effect of flavonoids in preclinical models of experimental obesity. *Pharma Nutrition, 16*, 100260.

Christensen, K. Y., Naidu, A., Parent, M. É., Pintos, J., Abrahamowicz, M., Siemiatycki, J., & Koushik, A., (2012). The risk of lung cancer related to dietary intake of flavonoids. *Nutrition and Cancer, 64*, 964–974.

Cook, N. C., & Samman, S., (1996). Flavonoids—Chemistry, metabolism, cardioprotective effects, and dietary sources. *The Journal of Nutritional Biochemistry, 7*, 66–76.

Garcia, V., Arts, I., Sterne, J., Thompson, R., & Shaheen, S. O., (2005). Dietary intake of flavonoids and asthma in adults. *European Respiratory Journal, 26*, 449–452.

Garcia-Larsen, V., Thawer, N., Charles, D., Cassidy, A., Van, Z. T., Thilsing, T., Ahlström, M., et al., (2018). Dietary intake of flavonoids and ventilatory function in European adults: A GA2LEN study. *Nutrients, 10*, 95.

Gross, M., (2004). Flavonoids and cardiovascular disease. *Pharmaceutical Biology, 42*, 21–35.

Hodgson, J. M., (2008). Tea flavonoids and cardiovascular disease. *Asia Pacific Journal of Clinical Nutrition, 17.*

Keservani, R. K., & Sharma, A. K., (2014). Flavonoids: Emerging trends and potential health benefits. *Journal of Chinese Pharmaceutical Sciences, 23*(12), 815.

Keservani, R. K., Kesharwani, R. K., & Sharma, A. K., (2015c). Pulmonary and respiratory health: Antioxidants and nutraceuticals. In: Debasis, B., Harry, G. P., & Anand, S., (eds.), *Nutraceutical and Functional Foods in Human Life and Disease Prevention* (pp. 279–295). Boca Raton, Fl. CRC Press, Taylor and Francis, Chapter 17. ISBN: 9781482237214.

Keservani, R. K., Kesharwani, R. K., Sharma, A. K., & Ahmad, M. F., (2015a). Stress, Parkinson's, Alzheimer's disease: Role of dietary supplements. In: Debasis, B., Harry, G. P., & Anand, S., (eds.), *Nutraceutical and Functional Foods in Human Life and Disease Prevention* (pp. 241–253). Boca Raton, Fl. CRC, Press, Taylor and Francis, Chapter 14. ISBN: 9781482237214.

Keservani, R. K., Kesharwani, R. K., Sharma, A. K., Vyas, N., & Chadoker, A., (2010b). Nutritional supplements: An overview. *International Journal of Current Pharmaceutical Review and Research, 1*(1), 59–75.

Keservani, R. K., Kesharwani, R. K., Vyas, N., Jain, S., Raghuvanshi, R., & Sharma, A. K., (2010a). Nutraceutical and functional food as future food: A review. *Der Pharmacia Lettre, 2*(1), 106–116.

Keservani, R. K., Sharma, A. K., & Kesharwani, R. K., (2016a). Nutraceutical and functional foods for cardiovascular health. In: *Food Process Engineering* (pp. 257–278, 291–312). Apple Academic Press, CRC Press. ISBN: 9781771884020.

Keservani, R. K., Sharma, A. K., & Kesharwani, R. K., (2016b). Medicinal effect of nutraceutical fruits for the cognition and brain health. *Scientifica*, 10. Article ID 3109254. doi: 10.1155/2016/3109254.

Keservani, R. K., Sharma, A. K., & Kesharwani, R. K., (2020). *Nutraceuticals and Dietary Supplements: Applications in Health Improvement and Disease Management*. CRC Press. ISBN: 9781771888738.

Keservani, R. K., Singh, S., Singh, V., Kesharwani, R. K., & Sharma, A. K., (2015). Nutraceuticals and functional foods in the prevention of mental disorder. In: Debasis, B., Harry, G. P., & Anand, S., (eds.), *Nutraceutical and Functional Foods in Human Life and Disease Prevention* (pp. 255–268). Boca Raton, Fl, CRC Press, Taylor and Francis, Chapter 15b. ISBN: 9781482237214.

Khan, J., Deb, P. K., Priya, S., Medina, K. D., Devi, R., Walode, S. G., & Rudrapal, M., (2021). Dietary flavonoids: Cardioprotective potential with antioxidant effects and their pharmacokinetic, toxicological and therapeutic concerns. *Molecules, 26*, 4021.

Kozlowska, A., & Szostak-Wegierek, D., (2014). Flavonoids-food sources and health benefits. *Roczniki Państwowego Zakładu Higieny,* 65.

Kozłowska, A., & Szostak-Węgierek, D., (2017a). Flavonoids–food sources, health benefits, and mechanisms involved. *Bioactive Molecules in Food*, 1–27.

Kozłowska, A., & Szostak-Węgierek, D., (2017b). Plant flavonoids in health, prevention, and treatment of chronic diseases. *Nutritional Antioxidant Therapies: Treatments and Perspectives.* Springer.

Macready, A. L., Kennedy, O. B., Ellis, J. A., Williams, C. M., Spencer, J. P., & Butler, L. T., (2009). Flavonoids and cognitive function: A review of human randomized controlled trial studies and recommendations for future studies. *Genes & Nutrition, 4*, 227–242.

Neuhouser, M. L., (2004). Flavonoids and cancer prevention: What is the evidence in humans? *Pharmaceutical Biology, 42*, 36–45.

Ørgaard, A., & Jensen, L., (2008). The effects of soy isoflavones on obesity. *Experimental Biology and Medicine, 233*, 1066–1080.

Panche, A. N., Diwan, A. D., & Chandra, S. R., (2016). Flavonoids: An overview. *Journal of Nutritional Science, 5*.

Park, E. J., & Pezzuto, J. M., (2012). Flavonoids in cancer prevention. *Anticancer Agents in Medicinal Chemistry (Formerly Current Medicinal Chemistry-Anti-Cancer Agents), 12*, 836–851.

Peluso, M. R., (2006). Flavonoids attenuate cardiovascular disease, inhibit phosphodiesterase, and modulate lipid homeostasis in adipose tissue and liver. *Experimental Biology and Medicine, 231*, 1287–1299.

Peterson, J. J., Dwyer, J. T., Jacques, P. F., & Mccullough, M. L., (2012). Associations between flavonoids and cardiovascular disease incidence or mortality in European and US populations. *Nutrition Reviews, 70*, 491–508.

Prasain, J., Carlson, S., & Wyss, J., (2010). Flavonoids and age-related disease: Risk, benefits and critical windows. *Maturitas, 66*, 163–171.

Rane, B. R., & Keservani, R., (2020). Introduction to nutraceuticals and dietary supplements. In: *Nutraceuticals and Dietary Supplements* (pp. 35–56). Apple Academic Press. ISBN: 9781771888738.

Romagnolo, D. F., & Selmin, O. I., (2012). Flavonoids and cancer prevention: A review of the evidence. *Journal of Nutrition in Gerontology and Geriatrics, 31*, 206–238.

Schmitt-Schillig, S., Schaffer, S., Weber, C., Eckert, G., & Muller, W., (2005). Flavonoids and the aging brain. *Journal of Physiology and Pharmacology Supplement, 56*, 23–36.

Singh, P., Kesharwani, R. K., & Keservani, R. K., (2017a). Antioxidants and vitamins: Roles in cellular function and metabolism. In: *Sustained Energy for Enhanced Human Functions and Activity* (pp. 385–407). Academic Press. ISBN:9780128054130.

Singh, P., Kesharwani, R. K., & Keservani, R. K., (2017b). Protein, carbohydrates, and fats: Energy metabolism. In: *Sustained Energy for Enhanced Human Functions and Activity* (pp. 103–115). Academic Press. ISBN:9780128054130.

Terao, J., Kawai, Y., & Murota, K., (2008). Vegetable flavonoids and cardiovascular disease. *Asia Pacific Journal of Clinical Nutrition, 17*.

Testai, L., (2015). Flavonoids and mitochondrial pharmacology: A new paradigm for cardioprotection. *Life Sciences, 135*, 68–76.

Testai, L., Martelli, A., Cristofaro, M., Breschi, M. C., & Calderone, V., (2013). Cardioprotective effects of different flavonoids against myocardial ischemia/reperfusion injury in Langendorff-perfused rat hearts. *Journal of Pharmacy and Pharmacology, 65*, 750–756.

Vinayagam, R., & Xu, B., (2015). Antidiabetic properties of dietary flavonoids: A cellular mechanism review. *Nutrition & Metabolism, 12*, 1–20.

Yao, H., Xu, W., Shi, X., & Zhang, Z., (2011). Dietary flavonoids as cancer prevention agents. *Journal of Environmental Science and Health, Part C, 29*, 1–31.

Yao, L. H., Jiang, Y. M., Shi, J., Tomas-Barberan, F., Datta, N., Singanusong, R., & Chen, S., (2004). Flavonoids in food and their health benefits. *Plant Foods for Human Nutrition, 59*, 113–122.

CHAPTER 4

FLAVONOIDS AND CARDIOVASCULAR DISEASES

LOVLISH GUPTA[1], MONIKA CHAUHAN[2], AJAY KUMAR[3], DIWAKAR CHAUHAN[4], and PRAGATI SAINI[3]

[1]School of Forensic Sciences, Lok Nayak Jayprakash Narayan National Institute of Criminology and Forensic Sciences, National Forensic Sciences University, Delhi Campus, Ministry of Home Affairs, Government of India

[2]Division of Forensic Science, School of Basic and Applied Sciences, Galgotias University, Greater Noida, Uttar Pradesh, India

[3]Division of Life Science, School of Basic and Applied Sciences, Galgotias University, Greater Noida, Uttar Pradesh, India

[4]Division of Chemistry, School of Basic and Applied Sciences, Galgotias University, Greater Noida, Uttar Pradesh, India

ABSTRACT

Flavonoids are biomolecules found in plants and are also commonly consumed in human diets. They belong to a large class of plant pigments with various sub-classes, including flavones, flavonols, catechism, isoflavones, proanthocyanidines, anthocyanidins, etc. Flavonoids have a generalized structure of 15 carbon skeletons containing 2 phenyl rings and 1 heterocyclic ring. In other words, it includes 2 benzene rings, equal to 12 carbons joined by a linear 3 carbon chain. Flavonoids are found in

Advances in Flavonoids for Human Health and Prevention of Diseases. Nisha Sharma, Deepika Saini, Rajesh K. Kesharwani, Prakash Chandra Gupta, & Raj K. Keservani (Eds.)

various edible plant sources such as spices, various fruits and vegetables, tea, various beverages, etc. However, a diet including flavonoids is rich in nutritional value. It leads to a healthy lifestyle and has been readily in consumption in order to prevent diseases. Flavonoids have been serving as an anti-atherosclerotic agent that helps in the prevention of cardiovascular diseases (CVDs), it regulates various other functions related to CVD, such as prevention of oxidative damage and oxidation of low-density lipoproteins (LDLs), regulation of vasomotor responses and regulates platelet activity. Flavonoids are found to inhibit cholesterol synthesis in order to prevent plaque formation in arteries. Basically, CVDs are multifactorial diseases with two basic types of ischemic heart diseases (IHD) and hemorrhagic heart diseases. Ischemic diseases occur due to blockages in arteries, whereas Hemorrhagic strokes occur due to rupturing of blood vessels. Numerous flavonoids, including those found in tea, vegetables, berries, cocoa, citrus, and other foods, aid in controlling various bodily functions and preventing and treating cardiac diseases.

4.1 INTRODUCTION

4.1.1 FLAVONOIDS

These are natural materials containing polyphenolic groups with low molecular weight, present in various edible items such as fruits, stems, roots, grains, various vegetables, and leaves like tea, wine, etc. They contain favorable biochemical and antioxidants effects associated with various diseases such as cancer, Alzheimer's disease (AD), atherosclerosis, etc. Flavonoids (Keservani et al., 2010a, b; Keservani & Sharma, 2014), have highly variated health-promoting effects. It possesses various antioxidants, anti-inflammatory, and anti-mutagenic properties that regulate many enzymatic functions. Hence, they are called potent inhibitors for several enzymes. Flavonoids are obtained from plants and used as growth regulators. They are also available as flower pigments in angiosperms (Panche et al., 2016). Flavonoids are found to be the most natural products with various physiological properties. Flavonoids consist of C_6-C_3-C_6 carbon framework, and contains phenyl benzopyran. The three main classes of flavonoids are: (i) isoflavonoids; (ii) neoflavonoides; and (iii) flavonoids, which are based on the position of linkage of the aromatic ring of benzopyrans. These different classes possess different functions depending upon their interactions with plants according to the stress created by biotic and abiotic conditions. Metabolites profiling in flavonoids in human and animal organisms is based on glucuronidation, sulfation, or methylation (Grotewold, 2006).

4.1.2 CLASSIFICATION OF FLAVONOIDS

Flavonoids are linked to polyphenolic compounds and are classified into various classes, such as Flavones, Flavanones, Catechins, and anthocyanins. Flavonoids possess various pharmacological and biochemical effects. They are control over various biochemical activities such as Xanthines, lipoxygenase (LOX), etc. Various biological and therapeutic potential have a regulatory role on different hormonal actions, such as androgens, estrogens, and thyroids. Various biological activities include anti-hepatotoxic, anti-inflammatory, anti-ulcer, anti-allergic, antiviral actions, free radical scavenging capacity, cardioprotective, anti-diabetic. A number of Flavonoids are also used in the protection against cardiovascular mortalities and various cardiovascular diseases (CVDs) (Agrawal, 2011; Karak, 2018). Flavonoids prevent the body from getting diseases, as it acts as dietary supplements. Flavonoids are possessed by various plants substances being bioactive polyphenols. They have played and are playing several imperative roles in therapeutics and various medical treatments in ancient and modern times.

4.1.3 DIET AND FLAVONOIDS

Balance Diet have always been an essential part of human beings, for their well-being and mandatories healthy life. Foods such as rice, potato, maize, wheat, are supposed to be traditionally used for delivering various nutrients in the body, such as fibers, lectins, saccharides, amino acids, etc., to the consumers. These diets are responsible for the prevention from various harmful diseases. One of the major diseases are cardiac diseases. Traditional diets are dependent on the foods and beverages which are minimally processed. However, modern diets are found to have lost various nutritive values. They have lost many naturally occurring vitamins, flavonoids, minerals, etc. There are various other constituents that have healthy benefits on the human physiological processes (Meserole, 2002).

4.1.4 *FLAVONOIDS AND CARDIOVASCULAR DISEASES (CVDS)*

Diseases are the multifactorial diseases possessed by the human heart and blood vessels. It includes various CVD such as Coronary artery disease, Heart Arrhythmias, pericardial diseases, cardiomyopathy, etc. CVDs has become the highest factor for risk of illness and death in the world. More than 62 million people are suffering from CVD, and more than 50 million people from hypertension and other cardiac stresses. The data analysis of 2000 has shown an approximation of 9,46,000 deaths attributable to CVDs, i.e., ~39% of deaths in states of the United Kingdom. The firmest factors of CVD are high blood pressure (BP), smoking, high cholesterol levels, diabetes, and family history as a genetic cause. Flavonoids play a greater role in the area of management of CVDs, it possesses specific mechanism for the management of CVD (Keservani et al., 2016a), such as anti-atherosclerotic activities that helps in preventing atherosclerosis. It also includes various anti-inflammatory, antioxidant, anti-proliferative, antiplatelet, etc. Flavonoids contribute to the production of antihyper-tensive and cholesterol-lowering effects. They also inhibit smooth and cardiac muscle cell proliferation, and they also lead to the improvement of the functioning of blood vessels and lymphatic vessels (Gross, 2004; Elizabeth & Nabel, 2003; Tijburg et al., 1997).

4.2 FLAVONOIDS

Flavonoids are a term used that defines a group of naturally occurring substances that are beneficial to human health, which include a C_6-C_3-C_6 backbone of Carbons. It has phenyl benzopyrans functionality. They are majorly categories into isoflavonoids, neoflavnoids, and Flavonoids itself. All these groups are biochemically linked as they share a common group called chalcone (Marais et al., 2006).

4.2.1 *CLASSES OF FLAVONOIDS*

1. **Flavonoids:** Commonly called 2-phenylbenzopyrans with C_6-C_3-C_6 framework of carbon only. Based on the efficiency to get oxidized and tendency towards saturation present in heterocyclic C-ring. Flavonoids are grouped into various sub-classes. Some of them are flavonol, dihydroflavonol, flavan-3-ol, flavan-4-ol, etc.

2. **Isoflavonoids:** They are commonly called 3-phenylchroman skeleton which is being derived from 1,2-aryl delocalization in a 2-phenylchroman precursor. They are low in concentration when their accountability is made accessible in plantae. But based on structural variations they are further subdivided into various subclasses, such as isoflavan, isoflavone, isoflavonone, isoflavonol, rotenoid, etc.
3. **Neoflavnoids:** These are highly related to both flavonoids and isoflavonoids and made up of 4-aryl-2H-1-benzopyran-2-ones and 3,4-dihydro-4-arylcoumarins and neoflavenes.

Flavonoids can further be sub-divided into a number of subclasses such as proanthocyanidins, chalcones, aurones, etc. Anthocyanins are naturally occurring substances in higher plants and absent in 9 caryophylalles families, i.e., Aizocaeae, Amarathaceae, Basellaceae, Cactaceae, Chenopodiaceae, Didiereaceae, Nyctaginaceae, Phytolaccaceae, and Portulacaceae except for Caryophyllaceae and Molluginaceae (Iwashina, 2000). The flavonoid uptake depends on cell type, and the primary reason is internal cellular metabolism—hydrogen donating antioxidant functioning. Flavonoids undergo various intercellular metabolism, such as: (i) conjugation with thiols; (ii) oxidative metabolism; and (iii) P_{450}-related metabolism. However, metabolic changes and modifications in flavonoids tend to produce changes in their chemical antioxidant nature. Flavonoids get interacted to intracellular signaling cascaded and induce selective actions in different components. They produce a variety of inhibitory mechanisms in order to affect cellular functions by alteration of targeted molecules or by producing modifications in the genetic compositions (Williams et al., 2004).

4.2.2 TYPES OF FLAVONOIDS, CHARACTERS, AND EXAMPLES

Flavonoids are potential antioxidants found in various fruits (Keservani et al., 2016b), vegetables, barks, roots, flowers, and certain beverages:

1. **Flavones:** Subclasses involve: catechins and epicatechin.
   - It is antioxidants, anti-atherosclerotic, anti-carcinogenic, anti-mutagenic, anti-microbial, thermogenic, and anti-diabetic.
   - *Example*: In branches, stem, leaves, roots, flowers, and fruits.

2. **Flavonones:** Sub-classes involve: Narigenin and Hesperetin.
   - It is antioxidant, controls carcinogenesis, controls tumor proliferation, controls tumor development inhibition, anti-antherosclerotic.
   - *Example*: Wine, grapes, onions, leafy vegetables, etc.
3. **Flavonols:** Sub-classes involve: Quercetin and Kaempferol.
   - It is anti-osteoporotic, anti-inflammatory, neuroprotective, modulation of cellular signaling, cardioprotective, hepatoprotective.
   - *Example*: Chocolate, beans, apricot, grapes, cherry, kiwi, etc.
4. **Anthocyanidines:** Sub classes: Cyanidin, malvidin, and peoxidin.
   - It is cytotoxic, anti-angiogenic, it improves visual activity.
   - *Example*: Tea, honey, fruits, nuts, olive oil, cocoa, etc.
5. **Isoflavones:** Sub-classes: Daidzein and Getistein.
   - It is anti-angiogenic, immunosuppressive.
   - *Example*: Peas, chickpeas, black beans, Lima beans, etc. (Guven, 2019).

4.2.3 EFFECT OF FLAVONOIDS IN HUMAN BIOLOGY

Flavonoids compounds are consisted of 15-carbon compounds made up of two cyclic (aromatic) rings, i.e., A and B, and a heterocyclic ring C consisting of Oxygen. Flavonoid compounds undergo various biochemical reactions appalling further enzymatic processes such as hydroxylation, methylation, glycosylation, sulfonation, acylation, and prenylation reactions. Due to these processes, the Flavonoids are diversified into more than 500 types (Ibrahim, 2001). Even the 1 mg per day intake of the Flavonoids in human diet plays an important role in mammalian biology, which is responsible for generating immunity that prevents from various infections, inflammation, and carcinogenicity. Flavonoids have anti-inflammatory actions generated through diversified mechanisms such as inhibition of regulatory enzymes and transcription factors. The different characteristics and functions of flavonoids have meaningful impact on cell activation, maturation, signaling transduction, cytokine production, or secretory processes in several immune cells. Other than this flavonoids have various more effects on the physiology of human beings, such as Inhibition of protein kinases and transcription factors, Inhibition of phosphodiesterase that helps inhibiting phosphodiesterase's such as cAMP phosphodiesterase that helps in the regulation of cell functions during inflammation (Malekia et al., 2019).

4.3 CARDIOVASCULAR DISEASES (CVDs)

The chapter focuses on cardiovascular health which is mandatory for maintenance of population and to control of worldwide deaths. Flavonoids are found to be the best agents in order to manage the epidemiology of CVD. The chapter deals with the optimum doses of Flavonoid materials to get disease free. There are various important sources of flavonoids like onions (flavonols); cocoa (proanthocyanidins); tea, apples, and red wine (flavonols and catechins); citrus fruit (flavanones); berries and Cherries (anthocyanidins); and soy (isoflavones) (Geleijnse & Hollman, 2008).

CVD is the primary cause of worldwide deaths, which covers a huge range of diseases of cardiac and smooth muscles, vascular systems supplied to all the organs of the body. The maximum causes of CVD are supposed to be the following:

- Ischemic heart disease (IHD);
- Strokes;
- Congestive heart failure (CHF).

4.3.1 ISCHEMIC HEART DISEASE (IHD)

Ischemic heart disease (IHD) is the major cause of death in developed countries, and this is called a myocardial infarction. The pain that occurs in IHD is called angina. The major reason behind the causation of IHD is atherosclerosis which leads to stenosis of either of the coronary arteries. People suffering with chronic angina have an average annual mortality of 2% or less. Acute myocardial infarction (AMI) is supposed to be the stenosis of the entire coronary artery that causes cardiac muscle necrosis due to the absence of oxygen and nutrients. AMI can be easily visualized by electrocardiogram. AMI is the most severe manifestation of coronary artery disease.

The entire disease leads to several other heart problems, such as Transient ischemic attack that reduces blood flow to the brain and can lead to the production of various strokes like symptoms, endocarditis, causes inflammation of the internal lining of the heart as well as bicuspid, tricuspid valves including semilunar valves which further leads to thickening and hardening of arterial valves, i.e., atherosclerosis.

4.3.2 HEMORRHAGIC STROKES

Strokes are the severance of blood flow in the area of the brain mainly caused due to stenosis in the arteries and veins. It is sometimes caused due to rupture in the blood vessels, specifically known as hemorrhagic stroke.

4.3.3 CONGESTIVE HEART FAILURE (CHF)

The main symptom of this disease is disability in myocardial functioning and nervous and hormonal control causing fatigue, reduced longevity. The main etiology of CHF is hypertension, IHD. In research it has been found that females suffering from hypertension have lesser risk to then hypertensive males. This is true when compared to non-hypertensive people. Also, CHF occurs more commonly in people suffering from AMI than in normal people.

4.4 EPIDEMIOLOGY OF CARDIOVASCULAR DISEASE (CVD)

The Global Burden of Disease Study has found that approx. 30% of all deaths around 2013 was caused by CVD (Bhatnagar et al., 2015). Epidemiology is the scientific study of determinants of health conditions. Epidemiological studies of CVDs are done in order to prevent coronary heart disease (CHD) and various other heart diseases. Various researches have estimated that men are generally more prone to CHD and mortality than women throughout the life cycle. All along with this the risk of CHD being caused increases with the growing age.

One of the major causes of CVD is eating habits, this is another major behavioral contribution to cardiovascular health or disease. People consume fats, calories, and cholesterol is the major cause for getting affected more frequently with CVD. Also, this increases the risk factor for obesity and increasing blood lipids (Jenkins, 1988).

Epidemiological studies state that cigarette smoking is one of the major causes of CVD, found predominantly in males than in females. In case, smoking, nicotine, and carbon monoxide are identified as the major cause of heart fatal. Around 20 Cigarettes a day can multiply the risk of myocardial infarction (Glantz & Parmley, 1991). Other than this physical and behavioral inactivity, alcohol consumption is found to be one of the causes for cardiac inactivity (Malinski et al., 2004).

However, the supplementary cause for almost all CVD is chronically troubling emotional factors such as anxiety, neuroticism, depression, hypochondriasis (somaticizing), life dissatisfaction, interpersonal discord, etc. Many studies have reported that these emotional factors are the cause of CVD, but more predominantly in adults or teenagers (Denollet, 1997).

4.5 FACTORS AFFECTING CARDIOVASCULAR HEALTH

4.5.1 *BLOOD VESSEL FUNCTIONING*

Vessel function is the main lead that causes a sufficient blood supply to all the parts of the body including the brain. The blood is supplied to the body parts where blood vessels, i.e., through veins and arteries. But, when blood is insufficiently supplied to the heart, which leads to IHDs. This imbalance myocardial oxygen supply.

The major cause of IHD is the non-functioning of the blood vessels, various conditions of which is eNo sclerosis, thrombosis/embolus formed in the coronary arteries.

Along with this, aortic stenosis and cardiomyopathy also results into IHD. However, revascularization of coronary artery would repair cardiac conditions (Zhang et al., 2014; Cohn, 2007).

4.5.2 *PLATELET ACTIVITY*

Platelet Activity that affects heart condition is investigated by employing platelet count and serum levels. CHD is a collective condition of lesions of coronary arteries. The most common disease caused due to the reduced activity of platelets is atherosclerosis, which is a disease of major arteries. Due to this least amount of blood is supplied to the myocardium, which causes inflammation of arteries.

There is an atheromatous plaque formed due to injured endothelium. Due to infiltration of macrophages and T-lymphocytes Atherosclerotic lesions are formed. This influences various humoral and cellular factors on blood monocyte interaction. CHD is generally shown due to increased percentage of active platelets in the circulatory system of the patients, which further causes atheromatous disease and ulceration (Assaf, 2017).

4.5.3 VASOMOTOR RESPONSES

Changes in the circulatory system with congestive heart failure (CHF) depend upon the severity of heart disease, and the stress given on the entire cardiovascular system. Vasomotor response in heart failure starts primarily with the constrictions of vessels. Vasoconstrictions results during exercise. The vasoconstrictor activates the coronary and cerebral circulatory system in order to maintain normal BP. This vasoconstriction is part of a total cardiocirculatory response designed to maintain optimal circulatory efficiency.

Increased aortic impedance, vasoconstriction increases burden on the heart already declining in its function increases probability of heart failure (Zelis & Flaim, 1982).

4.5.4 CHOLESTEROL FUNCTIONING

CVD includes CHD, strokes/transient ischemic attack, peripheral vascular diseases, and CVD fatal cases are highly dependent upon the cholesterol Functioning of the body. The main thing to be accessed is low-density lipoprotein cholesterol (LDL-C) responds greatly and invariably in patients increasing the risk factor for CVD. Patients with lesser reduction in LDL-C have increased risk for CVD as compared to the patients with optimum cholesterol response. The benefits of reducing LDL-C are invariantly to reduce CVD outcomes. All other characteristics in the people including age, gender, smoking, emotional stress, and alcohol have found to have modest effects on CVD as compared to LDL-C chronicity (Akyea et al., 2019).

4.6 CORONARY ARTERY DISEASE (CAD)

4.6.1 CORONARY ARTERY DISEASE AND FLAVONOIDS

Coronary artery disease is commonly called CAD as well as Atherosclerotic Heart Disease. This is a serious condition caused by deposition of plaque in the coronary arteries, i.e., the blood vessels that bring oxygenated hemoglobin containing blood to your heart. The plaque is deposited due to cholesterol that causes hardening and narrowing of the blood vessels. It results into least oxygen supply to the heart which causes angina, a severe chest pain or heart attack. In addition to this, heart muscles get weaken that leads to Heart Failure, i.e., heart fails to supply blood insufficiently throughout the body; another condition is called Arrhythmias, which leads to the alter the rhythm of heart beat. CAD or angina may also lead to fatal conditions of heart called cardiac arrest; this is a condition of complete stoppage of heart activity.

Low myocardial oxygen decreases myocardial blood flow, which reduces oxygen requirements and influences factors such as heart rate and the inotropic conditions of the heart (Libby & Theroux, 2005). Flavonoids have a wide range of distribution over in various vegetables, fruits, and they act as various herbal (Sharma et al., 2020; Rane et al., 2020a–c) medicines. Hence, flavonoids are required to be consumed in a regular manner. As it helps in the reduction of various Cardiac factors that leads to heart disease (Guyen et al., 2019).

4.6.2 PATHOGENESIS OF CAD BY FLAVONOIDS

Flavonoids help in curing cardiomyocyte injury deaths that result from ischemia-reperfusion. This reperfusion is caused due to massive reactive oxygen species (ROS) explosion which later causes cellular injuries and deaths. Flavonoids help in preconditioning of heart to prevent this Ischemia-reperfusion injuries and other cardiac injuries. This decrease in the CAD rate is definitely due to the consumption of Flavonoids. Hence, consumption of Flavonoids in a regular manner should be a mandatory action in one's life.

There are two such extracts of the Flavonoids that have special effects in the reduction of various types of CVD. They are *Scutellaria baicalensis* extract (SbE) and grape seed proanthocyanidin extract (GSPE) (Wang et al., 2011).

4.6.2.1 SCUTELLARIA BAICALENSIS

Scutellaria baicalensis is traditionally used in the treatment of disease caused due to inflammation such as hyperlipemia and arteriosclerosis. The major constitutes of this is baicalin, baicalein, wogonoside, and wogonin.

4.6.2.2 GRAPE SEED PROANTHOCYANIDIN EXTRACT (GSPE)

These are produced from grapevines constituting of genus *vitis*. A single *vitis* species and *V. Vinifera.* They have wide therapeutic effect on heart problems. In order to provide cardial benefits, they are given as herbal supplements. They are subdivided into proanthocyanidins and catechins that act as antioxidants (Wang et al., 2011).

The intake of anthocyanidins and proanthocyanidins in our regular diets are inversely complex in terms of CVD. Flavonol-rich foods help in expertization of cardiac protection by controlling platelet reactivity and vessel functioning. Flavonoid consumption results in control of oxidative stress, endothelial dysfunction and inflammation which acts as preventive for atherosclerosis. Acute Coronary events rupture Atherosclerotic plaque and prevent from Myocardial ischemia. Flavonols prevent platelet aggregation and thrombosis that reduces oxidative stress, decreases excitotoxicity that improves in balancing the blood flow rate in the entire brain. Flavonols influence CVD risk, that include on endothelium and vascular smooth muscle, platelet function, BP, inflammation from insulin resistance. Flavonoids are antihypertensive, cholesterol lowering and diabetes medications (Wang et al., 2013; Jacques et al., 2015).

The amount of Flavonoid is regulated greatly due to climate, seasonality, various plant species, manufacturing of the product. Many biomarkers of flavonoids are identified based on epidemiological studies on the effects of Flavonoids. Various research have been studying this from eras that consumption of flavan-3-ols, flavanones, and anthocyanins helps in reducing the risk of CVD. The relationship of Flavonoids to CVD is primarily dependent on the chemical nature of Flavonoids, then their effects on the body, and the amount in which they are consumed (Suen et al., 2016; Felciano et al., 2015).

4.7 FLAVONOIDS PREVENTING HEART DISEASE

4.7.1 DIETARY FLAVONOIDS

Dietary (Keservani et al., 2020) Flavonoids are the major flavonoids that protect from non-communicable diseases and reduces the risk of CVD. Dietary components include components like olive oil, nuts, whole grains, cocoa, tea, etc. That benefits cardiovascular health (Micek et al., 2021; Keservani et al., 2016a).

These Flavonoids enhances cure from hypertension, CVD mortality, and severe diseases like cancer, etc. Compounds have diverse chemical structures, a variety of biological activities that influence Cardiovascular Health.

They also found to be increasing the bioavailability and bioactivity of nitric oxide (NO) that regulates vascular relaxing factors. With their Antioxidants and anti-inflammatory activity, it regulates endothelial functioning, BP, and composition of blood lipids in a controlled amount (Mateos et al., 2014).

4.7.2 VEGETABLE FLAVONOIDS

Like other flavonoids vegetable flavonoids also have antioxidants, anti-inflammatory, anti-thrombotic activities. They help in improving blood lipid profiles that reduces or maintains BP, and low-density lipoproteins (LDLs).

Adults are found to be more in impact with the vegetable flavonoids, it has been found in various studies that vegetable flavonoids also help in reducing low-density levels of cholesterol (LDL-C). Vegetable flavonoids are energy-based Flavonoids which are standardized by different age group and sex. However, the males consume more Flavonoids from vegetables that leads to an increase in higher HDL-C levels, whereas it is inversely lowered in females, i.e., lower LDL-C. Along with this, these Flavonoids are found to maintain serum-lipid profiles that maintain the equipotential cardiac stresses (Penczynski et al., 2018; Klinder et al., 2016).

4.7.3 TEA FLAVONOIDS

Drinking tea has been the greatest thing in the regulation of CVH (Cardiovascular Health) for around 2000 years. There are numerous benefits of tea in CVD. Most of the tea flavonoids are polyphenols and are constituted of benzene rings to which hydroxyl group is attached to each benzene ring. Tea and tea flavonoids are linked with reduced risk for CVDs. This reduces oxidative damage, inflammation, excessive platelet activation, excessively high BP, that promotes good cardiac health (Hodgson et al., 2010; Kesharwani et al., 2016).

From several years, tea is being known for the good source of flavonoids that decreases the risk of various CVDs such as atherosclerotic lesions, smooth muscle cell proliferation, reduction in the NO content that causes hypertension. Tea flavonoids are supposed to be maintaining the systolic BP by causing vasodilation. Tea flavonoids reduces oxidative stress, cholesterol levels and prevent from vascular diseases by acting anti-inflammatory to it (Grassi et al., 2008).

Excessive deposition or activation of platelet induces clotting in aggregation of blood in the arteries that might result in the blockage of the artery, which further cause heart attack. This factor is immensely controlled by the regular intake of tea. Also, along with this tea flavonoids help in controlling diabetes type-2, which indirectly supports good cardiac health of the person (Tijuburg et al., 1997).

4.7.3.1 GREEN TEA FLAVONOIDS

Green tea is a beverage for the prevention of CVDs. It acts as dietary supplements, produced from the leaves of *Camellia sinensis.* It contains oxidative enzymes, caffeine, theophylline, carbohydrates, lipids, proteins, amino acids, sterols, minerals, and trace elements, vitamins (B, C, E), pigments, etc., which helps in maintaining vascular homeostasis, anti-inflammatory effects, antioxidant effects, antiproliferative effect, and also regulate antithrombotic activity, balances leopard metabolism. All these factors are collectively helpful in preventing coronary artery disease, arrhythmia, peripheral arterial disease, strokes, etc.

Green tea also prevents atherosclerosis by controlling hyperlipidemia and hypertensive blood (Islam, 2012; Hodgson et al., 1999).

4.7.3.2 BLACK TEA FLAVONOIDS

Black tea flavonoids are the major source of flavonoid intake. Mainly 30–40% of total flavonoid is weighed by both black and green tea. Black tea contains more Flavonoid oxidation products, this under polymerization which produces various therapeutic and physiological effects that enhance heart health. Both black and green tea polyphenols were found to attenuate increased BP.

The beneficiary effects of black tea are long term and is increases on both endothelial functioning and BP (Hodgson, 2006).

4.7.4 COCOA FLAVONOIDS

Cocoa products have special performance on CVD, there are reliable medical proves which indicates that the Cocoa products reduce BP to normal levels. Its consumption is moreover suggested to the people suffering from hypertension. On the other hand, these Flavonoids improves vascular endothelial functioning, increases insulin sensitivity. However, they are insufficient in putting beneficiary effects towards lipids and body weight. Cocoa products are the most abundant proud found to have flavonoids as nutrients (Bauer et al., 2011).

Chocolate is the major cocoa products which is balance calorie and weight of the body as it is energy rich in turns reduces risk factors of CVD. Cocoa products are the most abundant proud found to have flavonoids as nutrients. The production process of chocolate involves fermentation of the cocoa beans from the *Theobroma cacao* followed by drying and later separated into fat-free cocoa powder and cocoa butter, which are in addition added to different cocoa products in different ratios. Mainly chocolate contains catechin, epicatechin, and procyanidin. There are numerous evidences to show that cocoa products also provide fortification to treat multiple CVDs.

One such factor is the energy center of each cell, which is considered as mitochondria, which generates energy in the form of ATP, cocoa flavonoids help in maintaining the proper functioning of mitochondria in order to ensure the production of oxygen during ATP generation. This is a serious condition that needs to be maintained by such flavonoids because even a minute oxidative stress or deficiency of oxygen causes myocardial ischemia that overload cardiac muscles with calcium ions triggering to the swelling in mitochondria. This affects optimum cellular bioenergetics, causes cell death or loss of organelles structure or its dysfunctioning. Hence, the consumption of an organized number of cocoa products, highly rich in flavonoids of flavonoids are meant to reduce cardio metabolic risk, and also the risk of various other diseases such as AD, diabetes type-2, etc. (Blumberg et al., 2014).

This has been found in research that mode consumption of cocoa products lowers both systole and diastolic BP.

Chocolate induces vasodilation by acting as a supplement for endothelial function primarily during hypertensive nature. It has no effect on low density lipoprotein, and low-density cholesterol levels (Matsumoto, 2018).

4.7.5 *CITRUS FLAVONOIDS*

CVD is becoming the major source for comorbidity in case of diabetes, which is becoming a vast reason of early deaths. Citrus flavonoids produce several biological and therapeutic effects in order to prevent various CVDs such as maintenance of glucose tolerance level and insulin sensitivity, on the other hand it regulates leopard and adipocyte differentiation prevents from inflammation of arteries (Mahmoud et al., 2019).

Citrus flavonoids commonly act against a group of CVDs including oxidative stress, endothelial dysfunction, atherosclerosis, etc. Diabetes mellitus (DM) is one of the diseases that cause significant morbidity and mortality and hyperglycemia in the patient suffering from DM get associated with a high risk of CVDs as a diabetic heart has to suffer from various physiological issues such as oxidative stress that leads to diabetic cardiomyopathy. Citrus flavonoids have a great control over this condition; it is found that in the people consuming citrus flavonoids are generally free from this oxidative stress.

Citrus flavonoids are anti-inflammatory, cytoprotective, cures obesity, diabetes, CVDs. Their chemical composition indicates them to be oxygen quenchers, and hydrogen donating antioxidants that prevent cell damage directly through free radical scavenging.

The major cause of CVDs mainly atherosclerosis is age, hypertension, diabetes, etc. Citrus flavonoids maintain vascular endothelial, that prevents atherogenic lipoproteins to increase aggressiveness and permeability of arterial intima. This flavonoids module lipid homeostasis which further regulates the efflux of cholesterol with respect to Sterol content in the cell. Therefore, intake of flavonoid rich diets evaluates a better lifestyle and prevents cardiovascular mortality (Mulvihill et al., 2012).

This flavonoid is present in various citrus fruits such as limes, lemons, oranges, pomelos, bergamots, etc. The major example of such flavonoids is hesperidin and aglycone. Hesperetin, these are two flavonoids present in oranges and lemons are known to be beneficiary as they are hepato-protective, antidiabetic, antihypertensive, anticarcinogenic, and antiviral. Therefore, citrus flavonoids are known to be e having a huge therapeutic effect over diabetes and are found to be cardio metabolically protective in nature (Taher et al., 2020).

4.7.6 BERRY FLAVONOIDS

Berries are extremely rich sources of flavonoids, various findings have indicated that these flavonoids help in the reduction of arterial stiffness, maintains BP, endothelial functioning, regulate blood lipid levels. The major source of berries are blueberries, strawberries, blackberries, cranberries (Mateos et al., 2014).

The major flavonoid found in berries are anthocyanin, and different type of anthocyanins are found in different berries, for example pelargonidin conjugates are present in strawberries, on the other hand malvidin, petunidin, peonidin conjugates are present in blueberries. Whereas a significant number of other flavonoids are also present such as chlorogenic acids, phenolic acids, hydroxycinnamates, etc. (Chong et at., 2010).

The majority of CVDs occur due to hypertension which develops atherosclerosis and various other diseases. However, the blackberries and found to be the most preventive from hypertension and an alternative to it is raspberry. On the other hand, blueberries are found to have a great impact on the treatment of arterial stiffness, an alternative to blueberry is cranberry juice containing polyphenols (400 mg polyphenols in 21 mg of anthocyanins). Bilberry and blackcurrant anthocyanins reduce the blood lipids levels and prevents hypercholesterolemia, which itself is the major source of atherosclerosis (Wightman & Heuberger, 2015).

The other major impact of these flavonoids is to inhibit the mechanism meant for platelet aggregation and activation, as physiological excessive deposits of platelets main cause coronary thrombosis which can be inhibited by the actions of flavan-3-ols (Rio et al., 2010).

4.8 CONCLUSION

In the 21st century, people are living a fast life in which they are leading an irregular lifestyle. They are suffering from various emotional stress, financial stress, and hence, diseases are becoming friends in one's body. One such chronic disease that has led to the destruction of millions of lives and families is CVDs. CVD is the leading source of mortality and morbidity in human beings after cancer. Therefore, a numerous dietary priority has to be associated, and various strategies has to be made to provide nutrient supplements in order to prevent CVD risk. One such specific favorable nutrient are flavonoids. Flavonoids are polyphenols derived from plants have several medical benefits. Flavonoids nowadays are a big source of herbs for the treatment of innumerable diseases. These flavonoids are bioactive compounds present as dietary sources in various vegetables, nuts, seeds, and spices. Based on the chemical structure, flavonoids are divided into six major classes, these are: flavanones, flavones, flavonols, flavan-3-ols, anthocyanins, and isoflavones. Over 10,000 compounds in the plant kingdom are found to have flavonoids. Flavonoid lowers the risk of CVDs as they are antihypertensive, antioxidant, anti-atherosclerotic, because of its antioxidant nature flavonoids in activate oxygen species to prevent the counter act of plasma that causes inflammation of blood vessel endothelium. They have anti-arteriosclerotic action that reduces inflammation of blood vessels by influx of leucocytes. Other than CVDs flavonoids are found to you have treatments in other diseases such as neurodegenerative diseases that causes oxidative neuronal damage, it also has concern towards Parkinson's and AD. Along with this flavonoid also produces anti-carcinogenic effects. To conclude, flavonoids are rich source available in plant products, various plant species such as tea, berries, various fruits, and vegetables have an adequate number of flavonoids with them.

KEYWORDS

- **acute myocardial infarction**
- **cardiovascular diseases**
- **congestive heart failure**
- **flavonoids**
- **hemorrhagic heart diseases**
- **ischemic heart diseases**
- **Preventory Flavonoids**

REFERENCES

Agrawal, A. D., (2011). International journal of pharmaceutical sciences and nanotechnology. *Pharmacological Activities of Flavonoids: A Review, 4*, 1394–1398.

Akyea, R. K., Kai, J., Qureshi, N., Iyen, B., & Weng, S. F., (2019). Sub-optimal cholesterol response to initiation of statins and future risk of cardiovascular disease. *Heart, 105*(13), 975–981.

Assaf, A. I., (2017). Platelet activity and some immunological findings in atherosclerosis and coronary heart diseases. *Haitham Journal for Pure and Applied Science, 19*(2), 1–10.

Bauer, S. R., Ding, E. L., & Smit, L. A., (2011). Cocoa consumption, cocoa flavonoids, and effects on cardiovascular risk factors: An evidence-based review. *Current Cardiovasc Risk Reports, 51,* 120–127.

Bhatnagar, P., Wickramasinghe, K., Williams, J., Rayner, M., & Townsend, N., (2015). The epidemiology of cardiovascular disease in the UK. *Heart, 101,* 1182–1189.

Blumberg, J. B., Ding, E. L., Dixon, R., Pasinetti, G. M., & Villarreal, G. F., (2014). The science of cocoa flavanols: Bioavailability, emerging evidence, and proposed mechanism. *Advance Nutrition, 5,* 547–549.

Boersma, E., Mercado, N., Poldermans, D., Gardien, M., Vos, J., & Simoons, M. L., (2003). Acute myocardial infarction. *The Lancet, 361,* 847–848.

Chong, M. F. F., Macdonald, R., & Lovegrove, J. A., (2010). Fruit polyphenols and CVD risk: A review of human intervention studies. *British Journal of Nutrition, 104*, 28–39.

Cohn, J. N., (2007). Is it the blood or the blood vessel? *Journal of American Society of Hypertension, 1*(1), 5–16.

Denollet, J., (1997). Personality, emotional distress and coronary heart diseases. *European Journal of Personality, 11*(5), 343–357.

Elizabeth, G., & Nabel, M. D., (2003). *The New England Journal of Medicine, Cardiovascular Diseases, 349,* 60–72.

Felciano, R. P., Pritzel, S., Heiss, C., & Mateos, R., (2015). Flavonoid intake and cardiovascular disease risk. *Current Opinion in Food Science, 2,* 92–99.

Gaziano, T., Reddy, K. S., Paccaud, F., Horton, S., & Chaturvedi, V. (2006). *Disease Control Priorities in Developing Countries, Cardiovascular Disease*, 645–662.

Geleijnse, J. M., & Hollman, P. C. H., (2008). Flavonoids and cardiovascular health: Which compounds, what mechanisms? *The American Journal of Clinical Nutrition, 88*(1), 12, 13.

Glantz, S. A., & Parmley, W. W., (1991). Passive smoking and heart disease: Epidemiology, physiology, and biochemistry. *Circulation, 83*(1), 1–12.

Gofman, J. W., Young, W., Tandy, R., & Shah, A., (1996). Atherosclerosis, and longevity. *Journal, Ischemic Heart Disease, 34,* 679–697.

Grassi, D., Aggio, A., Onori, L., Croce, G., Tiberti, S., Ferri, C., Ferri, L., & Desideri, G., (2008). Tea, flavonoids, and nitric oxide-mediated vascular reactivity. *The Journal of Nutrition, 138*(8), 1554S–1560S.

Gross, M., (2004). *Pharmaceutical Biology, Flavonoids and Cardiovascular Disease, 42*, 21–35.

Grotewold, E., (2006). Springer. *The Science of Flavonoids*, 1–47.

Guven, H., Arici, A., & Simsek, O., (2019). Flavonoids in our foods. *J. Basic Clin. Health Sci., 3*, 96–106.

Hodgson, J. M., & Croft, K. D., (2010). Tea flavonoids and cardiovascular health. *Molecular Aspects of Medicines, 31*(6), 495–502.

Hodgson, J. M., (2006). Effects of tea and tea flavonoids on endothelial functioning and blood pressure: A brief review. *Clinical and Experimental Pharmacology and Physiology, 33,* 838–841.

Hodgson, J. M., Puddey, I. B., Burke, V., Beilin, L., & Jordan, N., (1999). Effects on blood pressure of drinking green and black tea. *Journal of Hypertension, 17*(4), 457–463.

Ibrahim, R. K., (2001). *Encyclopedia of Life Sciences, Flavonoids*, 1–6.

Islam, M. A., (2012). Cardiovascular effects of green tea catechins: Progress and promise. *Recent Patent on Cardiovascular Drug Discovery, 2*(7), 88, 89.

Iwashina, T., (2000). The structure and distribution of the flavonoids in plants. *The Journal of Plant Research, 113*, 287–299.

Jacques, P. F., Cassidy, A., Rogers, G., Peterson, J. J., & Dwyer, J. T., (2015). Dietary flavonoid intake and cardiovascular disease incidence in the Framingham offspring cohort. *British Journal of Nutrition, 114,* 1496–1503.

Jenkins, C. D., (1988). Epidemiology of cardiovascular diseases. *Journal of Consulting and Clinical Psychology, 56*(3), 324–332.

Karak, P., (2018). Biological activities of flavonoids: An overview. *International Journals of Pharmaceutical Sciences and Research, 10*(4), 1567–1574.

Keservani, R. K., & Sharma, A. K., (2014). Flavonoids: Emerging trends and potential health benefits. *Journal of Chinese Pharmaceutical Sciences, 23*(12), 815.

Keservani, R. K., Kesharwani, R. K., Sharma, A. K., Vyas, N., & Chadoker, A., (2010b). Nutritional supplements: An overview. *International Journal of Current Pharmaceutical Review and Research, 1*(1), 59–75.

Keservani, R. K., Kesharwani, R. K., Vyas, N., Jain, S., Raghuvanshi, R., & Sharma, A. K., (2010a). Nutraceutical and functional food as future food: A review. *Der Pharmacia Lettre, 2*(1), 106–116.

Keservani, R. K., Sharma, A. K., & Kesharwani, R. K., (2016a). Nutraceutical and functional foods for cardiovascular health. In: *Food Process Engineering* (pp. 257–278, 291–312). Apple Academic Press, CRC Press. ISBN: 9781771884020.

Keservani, R. K., Sharma, A. K., & Kesharwani, R. K., (2016b). Medicinal effect of nutraceutical fruits for the cognition and brain health. *Scientifica*, 10. Article ID 3109254. doi: 10.1155/2016/3109254.

Keservani, R. K., Sharma, A. K., & Kesharwani, R. K., (2020). *Nutraceuticals and Dietary Supplements: Applications in Health Improvement and Disease Management*. CRC Press. ISBN: 9781771888738.

Kesharwani, R. K., Singh, P., & Keservani, R. K., (2016). Green coffee bean extract and chlorogenic acids: Chemistry and novel antioxidant benefits. In: Bagchi, D., Moriyama, H., & Swaroop, A., (eds.), *Green Coffee Bean Extract in Human Health* (pp. 1–17). Boca Raton, Taylor and Francis, USA. Chapter 1. ISBN: 9781498716376.

Klinder, A., Shen, Q., Susanne, H., Lovegrove, J. A., Rowland, I., & Tuohy, R. M., (2016). Impact of increasing fruit and vegetables and flavonoid intake on the human gut microbiota. *Food and Function, 7*(4), 1788–1796.

Levy, D., Larson, M. G., Vasan, R. S., Kannel, W. B., & Ho, K. K. L., (1996). The progression from hypertension to congestive heart failure. *JAMA, 275*(20), 1557–1562.

Libby, P., & Theroux, P., (2005). Pathophysiology of coronary artery disease. *Circulation, 111*, 3481–3488.

Lovegrove, J. A., Stainer, A., & Hobbs, D. A., (2017). Role of flavonoids and nitrates in cardiovascular health. *Proceedings of the Nutrition Society, 76,* 83–95.

Mahmoud, A. M., Bautista, R. J., Sandhu, M. A., & Hussein, O. E., (2019). Beneficial effects of citrus flavonoids on cardiovascular and metabolic health. *Oxidative Medicine and Cellular Longevity*, 1–19.

Malekia, S. J., Crespob, J. F., & Cabanillas, B., (2019). Anti-inflammatory effects of flavonoids. *Food Chemistry, 299*(125), 1–11.

Malinski, M. K., Sesso, H. D., Jimenez, F. L., Buring, E. J., & Gaziano, J. M., (2004). Alcohol consumption and cardiovascular disease mortality in hypertensive men. *Archives of Internal Medicine, 164*(6), 623–628.

Marais, P. J., Deavours, B., Dixon, R. A., & Ferreira, D., (2006). *The Stereochemistry of Flavonoids,* 1–47. Springer.

Mateos, A. R., Heiss, C., Borges, G., & Crozier, A., (2014). Berry (poly)phenols and cardiovascular health. *Journal of Agriculture and Food Chemistry, 62*(18), 3842–3851.

Mateos, A. R., Vauzour, D., Krueger, C. G., Shanmuganayagam, D., Reed, J., Calani, L., Mena, P., Rio, D. D., & Creozoer, A., (2014). Bioavailability, bioactivity and impact on health of dietary flavonoids and related compounds: An update. *Archives of Toxicity, 88*(10), 1803–1853.

Matsumoto, C., (2018). Cocoa polyphenols: Evidence from epidemiological studies. *Current Pharmaceutical Design, 24*(2), 140–145.

Meserole, L., (2002). Ethnomedicine and drug discovers, Elsevier science. *Health Foods in Anti-Aging Therapy: Reducers of Physiological Decline and Degenerative Diseases*, 173–180.

Micek, A., Godos, J., Rio, D. D., Galvano, F., & Grosso, G., (2021). Dietary flavonoids and cardiovascular diseases: A comprehensive dose-response meta-analysis. *Molecular Nutrition Food Research, 65*, 1–11.

Mulvihill, E. E., & Huff, M. W., (2012). Drug targets (formerly current drug targets-cardiovascular & hematological disorders), citrus flavonoids and the prevention of atherosclerosis. *Cardiovascular & Haematological Disorders, 12*(2), 84–91.

Panche, A. N., Diwan, A. D., & Chandra, S. R., (2016). Flavonoids: An overview. *Journal of Nutritional Science, 5*(47), 1–15.

Penczynski, K. J., Remer, T., Herder, C., Kalhoff, H., Reinks, J., Markgraff, D. F., Roden, M., & Buyken, A. E., (2018). Habitual flavonoid intake from fruit and vegetables during adolescence and serum lipid levels in early adulthood: A prospective analysis. *Nutrients, 10*(4), 1–13.

Rane, B. R., Bharath, M. S., Patil, R. R., Keservani, R. K., & Jain, A. S., (2020b). Novel approaches in nutraceuticals. In: Rajesh, K. K., Raj, K. K., & Anil, K. S., (eds.), *Enhancing the Therapeutic Efficacy of Herbal Formulations Through Novel Drug Delivery Systems* (pp. 241–266). IGI Global International Publisher, Pennsylvania, USA, Chapter 11. ISBN13: 9781799844532.

Rane, B. R., Patil, A. S., Keservani, R. K., & Jain, A. S., (2020a). Novel approaches in herbal formulation. In: Rajesh, K. K., Raj, K. K., & Anil, K. S., (eds.), *Enhancing the*

Therapeutic Efficacy of Herbal Formulations Through Novel Drug Delivery Systems (pp. 43–68). IGI Global International Publisher, Pennsylvania, USA, Chapter 2. ISBN13: 9781799844532.

Rane, B. R., Tadavi, S. A., & Keservani, R. K., (2020c). Naturopathy. In: Anil, K. S., Raj, K. K., & Surya, P. G., (eds.), *Herbal Product Development* (pp. 321–347). Apple Academic Press, CRC Press, Taylor & Francis Group, Chapter 12. ISBN: 9781771888776.

Rio, D. D., Borges, G., & Crozier, A., (2010). Berry flavonoids and phenolics: Bioavailability and evidence of protective effects. *British Journal of Nutrition, 104*, S67–S90.

Sharma, A. K., Keservani, R. K., & Gautam, S. P., (2020). *Herbal Product Development*. Apple Academic Press, CRC Press, Taylor & Francis Group. ISBN: 9781771888776.

Suen, J., Thomas, J., Kranz, A., Vun, S., & Miller, M., (2016). Effect of flavonoids on oxidative stress and inflammation in adults at risk of cardiovascular disease: A systematic review. *Healthcare, 4*, 1–23.

Taher, M., Shukry, N. A. S. B., Susanty, D., Saleh, W. N. H. W., & Syukri, Y., (2020). Plant-derived bioactives. *Citrus Flavonoids in Preventing Cardiovascular Diseases*, 495–508.

Tijburg, L. M. B., Mattern, T., Folts, J. D., Weisgerber, U. N., & Katan, M. B., (1997). Critical review in food science and nutrition. *Tea Flavonoids and Cardiovascular Diseases, 37*(8), 771–785.

Tijuburg, L. B. M., Mattern, T., Folts, J. D., Weisgerber, U. M., & Katan, M. B., (1997). Tea flavonoids and cardiovascular disease: A review. *Critical Reviews in Food Science and Nutrition, 37*(8), 771–785.

Wang, C. Z., Mehendale, S., Calway, T., & Yuan, C. S., (2011). Botanicals flavonoids on coronary heart disease. *American Journal of Chinese Medicine, 39*(4), 661–671.

Wang, X., Ouyang, Y. Y., Jiu, J., & Zhao, G., (2013). Flavonoid intake and risk of cardiovascular disease: A systemic review and meta-analysis of prospective cohort studies. *British Journal of Nutrition, 111,* 1–11.

Wightman, J. D., & Heuberger, R. A., (2015). Effect of grape and other berries on cardiovascular health. *Journal of the Science of Food and Agriculture, 95*(8), 1584–1597.

Williams, R. J., Spencer, J. P. E., & Evans, C. R., (2004). Free radical biology and medicine. *Flavonoids: Antioxidants or Signaling Molecules?, 36*(7), 838–849.

Zelis, R., & Flaim, S. F., (1982). Alterations in vasomotor tone in congestive heart failure. *Progress in Cardiovascular Diseases, 34*(6), 437–459.

Zhang, H., Olden, C. V., Sweeney, D., & Random, E. M., (2014). Blood vessel repair and regeneration in the ischaemic heart. *Open Heart, 1*(1), 1–14.

CHAPTER 5

NEUROPROTECTIVE EFFECTS OF FLAVONOIDS

VIPIN KESHARWANI[1], SAGARIKA KABRA[1], BHUPESH CHANDRA SEMWAL[2], and DEEPIKA SAINI[3]

[1]*School of Pharmaceutical Sciences, Maharishi University of Information Technology, Lucknow, Uttar Pradesh, India*

[2]*Institute of Pharmaceutical Research, GLA University, Mathura, Uttar Pradesh, India*

[3]*Department of Zoology, Chaman Lal Mahavidhyalya, Landhaura, Haridwar, Uttarakhand, India*

ABSTRACT

The damage to the neurons is a combination of various outcomes cranked by inflammation in neurons, as seen in neurodegenerative illnesses such as Parkinson's disease (PD) and Alzheimer's disease (AD). As day by day, the incidences of neuro-degenerative illness are increasing in old age populations, various new technologies are helpful in preventing or treating dysfunctions in the brain as well as its related cognitive loss is needed. So, it was found that the use of flavonoids is very helpful in treating various health ailments. We learned that they have a wide range of neuroprotective effects on the brain, including protection of the brain against many neuro-toxins and suppression of the process of neuron inflammation, as this will enhance memory, learning, and various comprehensible functions. The compounds such as berries, vegetables, roots, stems, flowers, and other

Advances in Flavonoids for Human Health and Prevention of Diseases. Nisha Sharma, Deepika Saini, Rajesh K. Kesharwani, Prakash Chandra Gupta, & Raj K. Keservani (Eds.)

parts of the plant contain flavonoids in the form of a collection of natural compounds which have varying phenolic structures. Due to this nature, these natural compounds are majorly known to have health benefits, and many researchers are working on extracting the components known as flavonoids. They are presently used in a major range of pharmacological, medical, and other nutraceutical areas and many more. All of this is due to the capability to control crucial cell enzyme functioning and also in activities such as anti-oxidative, anti-inflammatory, etc. The function-based mechanism of flavonoids is still not completely known. On the other hand, compounds that are derived from plants have been seen for a long time as they contain a wide zone of biological activities. The current trends in the area of research and development on the activities of flavonoids on health benefits are due to the isolation, identification, characterization, and functions of flavonoids. The purpose of the study presented is to label the current trends in research of flavonoids as well as their development, mechanism of action of flavonoids, functions of flavonoids as well as prediction on the medications used to treat various chronic diseases.

5.1 INTRODUCTION

Flavonoids come in a major category of products derived from nature, mainly the metabolites are secondary of plant origin which have a design like polyphenols as it is detected in vegetables, berries, and some liquid refreshments. It is seen that they are diverse agreeable effects such as they show chemical processes and protects cells from damage by free radicals having connection in much illness for example can abnormal mass of tissues, atherosclerosis, etc. (Burka et al., 1999; Ovando et al., 2009; Lee et al., 2009). The flavonoids are having connection with a broad-spectrum well-being result and found to be an essential element in an area of pharmaceuticals, medicals, and various face-related demulcents. It takes place because of integrity with the capacity in modulating key cellular enzymatic functions with a combination of anti-oxidative, anti-inflammatory, anti-mutagenic, and anti-carcinogenic activities. Flavonoids have one more property as active inhibitors in a majority of enzymes Txanthine oxidase, cyclooxygenase (COX), lipoxygenase (LOX), and phosphoinositide 3-kinase (Metodiewa et al., 1997; Hayashi et al., 1998; Walker et al., 2000). In the environment, these components have outcomes pulled out from plants and as well as present in various areas in the growing plants. These compounds are extensively dispersed in the plant kingdom and come under low molecular weight phenolic compounds. In higher plants, they are expanded and found to be the most characteristic class of compounds. In many angiosperm families these flavonoids are simply identified in the form of flower pigments. In combination with nature, these flavonoids are also found in food and liquid containing drinks such as tea, wine due to this, they are known as dietary flavonoids. They are divided into many sub-groups that is chalcones, flavones, flavonols, and iso-flavones. These compounds frolic in a range of organic living functioning in herbs and creatures move on land, water, and micro-organisms. The flowers, its texture and fragrance come due to flavonoids as they are synthesized at a particular site. In addition to it, the dispersion of fruits in seed germination and spore germination is due to these compounds as in future it leads to further growth and development of seedlings (Havsteen, 2002). They also play a role in protection from ultraviolet filters, stresses of both types biotic and abiotic and many more (Griesbach, 2005). The different functions of these flavonoids are signal molecules, allopathic molecules, detoxification, and defend compound which have anti-microbial properties. Some other characteristics of this compound are action against frost hardiness, resistance in drought, freezing tolerance, etc. (Jorgenesen, 1995) about early progress of floral genetics is occur of mutation technique which makes an image upon those flavonoids which are derived from colors of flowers, and displayed silencing of functional gene in plants which is associated with biosynthesis of flavonoids. These compounds also have a plus action on the well-being of living beings, and currently it is gaining interest in disease therapy and chemo-prevention. Presently, it is found that 600 flavonoids that furnish the colored pigments of berries, vegetables, and plants that have medical origin. Dixon & Pasinetti (Kumar et al., 2013) evaluated plant flavonoids and iso-flavonoids in detail and explained their application in areas of agriculture and neuroscience in human beings. In Alzheimer's disease (AD), its current therapeutic methods the uses of flavonoids in the form of plant secondary metabolites are used in the treatment of this disease with its mechanism involvement. In this review, various ventures are made in discussing today's trends in areas of research for flavonoids, its use in the form of dietary forms, and its benefits on the health of individual comes with a broader classification and its future research directions.

5.2 NEURODEGENERATIVE DISORDERS AND FLAVONOIDS METABOLISM: INFLUENCE OF FLAVONOIDS

They come under a unique category of therapeutic molecules due to its therapeutically active proportion of flavonoids. Certain flavonoids for example, Querecetin, Astilbin, Catechin, Epicatechin, Rutin, Apigenin, Epigallactechin they all act in curing of various neuro-degenerative disorders in protection of brain against these disorders. Certain flavonoids have the property to squash neuro-inflammation due to the presence of medicinal plants as well as they have the ability to protect the head (CNS) in contrary to various neuronal injuries and promote thinking, learning, etc. (Figure 5.1).

5.3 FLAVONOID MECHANISM

The potential to function as an antioxidant is almost found in each and every group of flavonoids. It has been released that flavones and catechins are the most powerful acting flavonoid which protect the human body against the action of flavonoids various active O_2 molecules. In normal metabolism of oxygen this cellular damage results in alternate net charge of cell, which alters the osmotic pressure, causing inflammation and finally cell death. Free radicals can catch a variety of mediators of inflammation, triggering an action caused by inflammatory mediators and causing tissue loss. Those living creatures have initiated a number of effective methods in defending themselves from various reactive O_2 molecules (Halliwell, 1995). Not just enzymes like superoxide dismutase (SOD), catalase (CAT), and glutathione peroxidase (GPx), but as well as non-enzymatic analogs like glutathione, ascorbic acid, and -tocopherol, are part in body's antioxidant defense mechanisms. The increased generation of reactive oxygen species (ROS) during damage causes the endogenous scavenging chemicals to be consumed and depleted. Flavonoids may work in conjunction with endogenous scavenging chemicals (Nijveldt, 1995; Codormin-Hernandez, 2007) Docking experiments were conducted out to better Flavonoid–protein interactions should be properly understood. The findings revealed that hydrophilic amino acid residues were more prevalent interact with flavonoid molecules with high affinity, as expected in the way of theoretical order of affinity. The docking methods of catechin molecules and 4 proteins further corroborate this theory (human serum albumin, elastase, renin, etc.). The summative affinity in association with between flavonoids and amino acid residues looks a lot of promise in terms of theoretical predictions of flavonoid–protein interactions as a high-quality technique of understanding flavonoid biological activity.

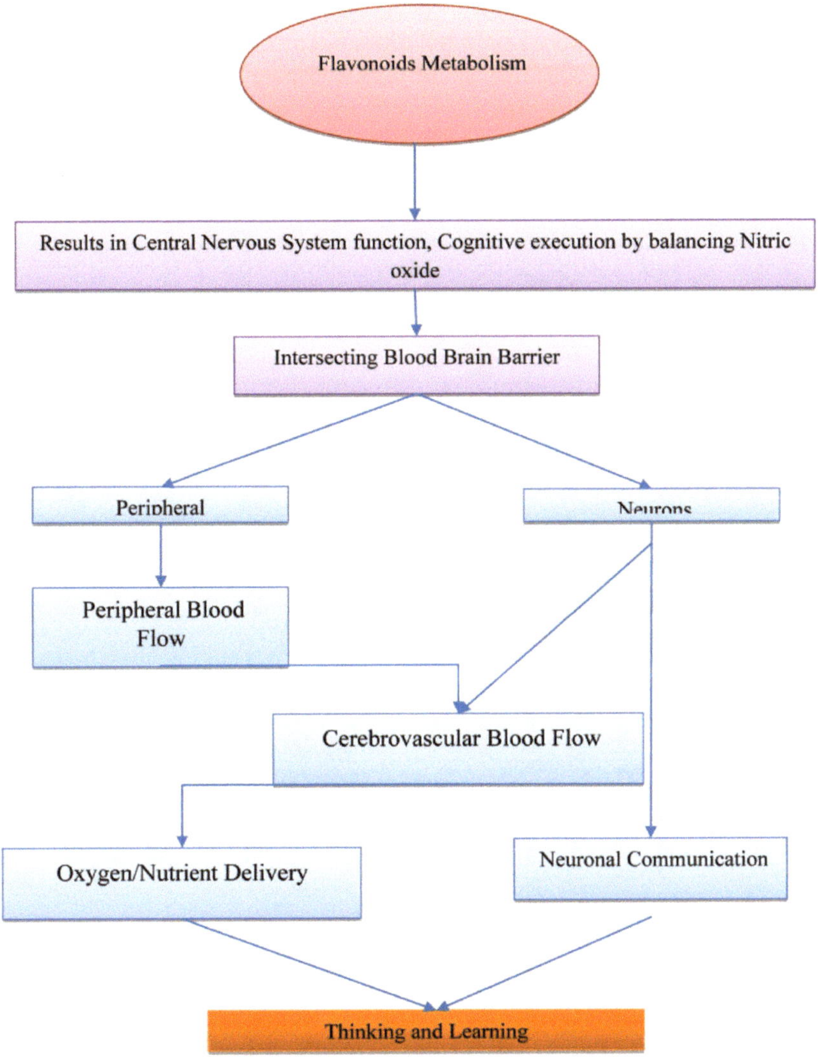

FIGURE 5.1 Mechanism: Effect of dietary flavonoids in memory and learning with functions affecting the regions of the brain.

5.4 CLASSES OF FLAVONOIDS

By seeing upon the Carbon of C ring on which B ring is attached and oxidation of C ring with degree of unsaturation, on these points the subdivision of Flavonoids into different sub-groups is done. The name of iso-flavones is given to those flavonoids in which the B ring attached to the C ring in number 3rd. In continuation to it there is one more category called neoflavonoids as B ring is linked on 4 position and at last on the structural feature of C ring subgroups are divided in which B ring is linked on position 2nd. Then the corresponding flavones are Flavonols, flavanones, flavanonols, anthocyanins, and chalcones (Figures 5.2 and 5.3).

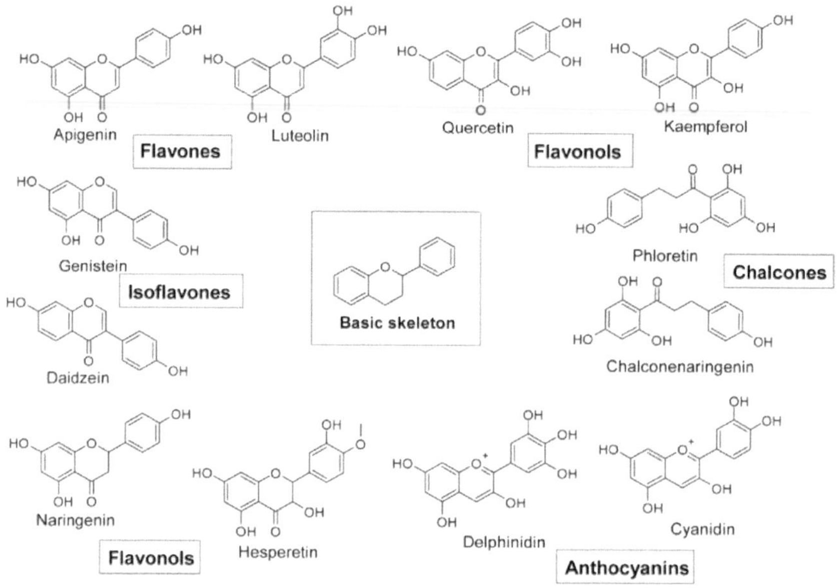

FIGURE 5.2 Flavonoids' skeletal structure and classification.

5.4.1 FLAVONES

They are major sub-groups of flavonoids. Majorly they are found in flowers, fruits, and leaves in the form of glucosides. The main sources are celery, parsley, chamomile, mint, etc. The subclass of these flavones is luteolin, apigenin, and tangeritin.

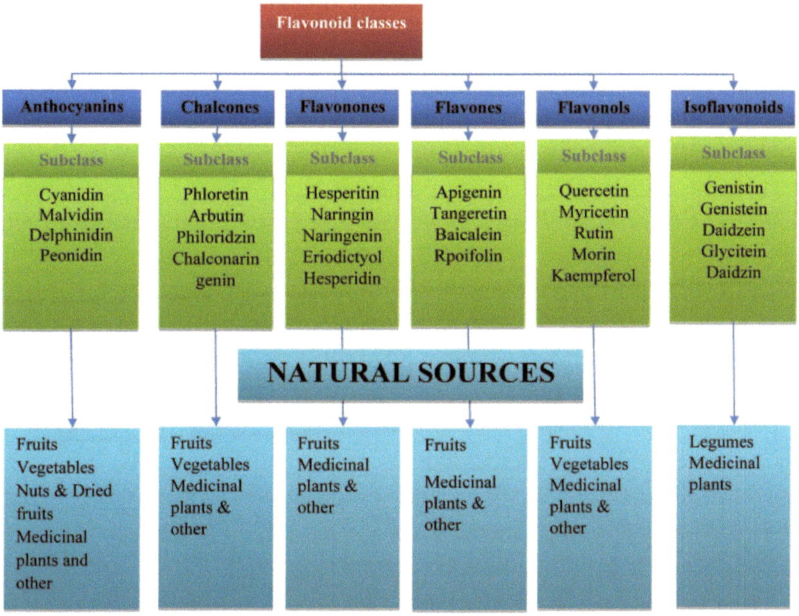

FIGURE 5.3 Flavonoids classes, subclasses, and natural sources.

5.4.2 FLAVONOLS

These are those flavonoids which contains a ketone GP and as well as also essential components of proanthocyanins. They are found profusely in a number of fruits and vegetables. As majorly studied flavonols are kaempferol, quercetin, fistin. The prosperous origins of flavonols example onions, kale, tomatoes, apples, etc., and in certain sources like tea, red wine, vegetables also show some origins of flavonols. There are many health benefits associated with flavonols, e.g., antioxidant potential and decreased risk of vascular diseases. These compounds have a high diversity in methylation and hydroxylation patterns. Due to this, they are one of the major common and largest sub-groups found in fruits as well as leafy vegetables.

5.4.3 FLAVANONES

They are an additional major class seen all citrus fruits for ex. Grapes, lemons, and oranges for example eriodictyol, naringenin, Hesperitin (Keservani et al., 2010a). All of them have a free radical scavenging property which is associated with a majority of health benefits. The bitterness of taste in juice and peels of citrus fruits take place in the presence of this compound. They also have many pharmacological effects for ex. Antioxidant, anti-inflammatory, blood lipid lowering agents. The only structural difference between these two groups is the presence of saturated C ring and on other side there is double bond present between position 2 and 3 is saturated. The increment in the number of flavanones has severely increased in the past 15 years (Iwashina, 2013).

5.4.4 ISOFLAVONOIDS

It is a large and very particular subgroup of flavonoids. There major contribution is found in the plant kingdom and mainly in soyabeans and leguminous plants. There are also present in microbes (Matthies et al., 2008). During plant microbe interaction they shows a major role as precursor in the development of phytoalexins (Aoki et al., 2000; Dixon et al., 2002). They have an extreme potential to reduce a number of diseases. It is found that in certain animal models they have estrogenic activity as they contain some isoflavones (genistein, daidzein) in major distribution.

5.4.5 NEOFLAVONOIDS

It is a class of poly-phenolic compounds. The difference between flavonoids and neoflavonoids is backbone structure in one it is 2-phenylchromen-4-one and other 4-phenylchromen with number hydroxyl group at position second substitution. The 1st neoflavone detected from environmental source in the year 1951 was calophyllolide from *Calophyllum inophyllum* seeds.

5.4.6 FLAVANOLS, FLAVON-3-OLS, OR CATECHINS

Flavanonols (Keservani et al., 2010b; Keservani & Sharma, 2014) are 3-hydroxy derivatives of flavanones, frequently named as dihydroflavonols or catechins. They are multi-substituted and extremely diverse subclass due to this hydroxyl GP is always attached to post. 3 of the C ring, flavanols are also known as flavan-3-ols. As well as there not any kind of double bond between positions second and third. Bananas, apples, blueberries, etc., contain more amount of flavonols.

5.4.7 ANTHOCYANINS

They are pigments that provide color to plants, fruits, and flowers in nature. The highly studied anthocyanins are cyanidin, delphinidin, malvidin, pelargonidin, and peonidin. They are majorly seen in cell external layer of red grapes, raspberries, and many more. Their stability as well as its health benefits make it useful in a high majority of areas in the areas of food distribution (Giusti et al., 2003). The color is affected by pH and methylation or acylation of the hydroxyl groups on the A and B rings.

5.4.8 CHALCONES

Chalcones are a flavonoid subclass. The absence of 'ring C' of the basic flavonoid skeleton structure depicted in distinguishes them. Due to this nature, they come under open-chain flavonoids. Some of its examples are Phloridzin, arbutin, phloretin, and chalconaringenin. Tomatoes, pears some products of wheat all contain substantial levels of chalcones. Due to their multiple nutritional and biological benefits, chalcones, and their derivatives are known to gather major concern and attention. Table 5.1 lists the food sources for all of the dietary flavonoids covered in this chapter in terms of its bioactivity and trends in research field (Hertog et al., 1993; Lopez-Lazaro et al., 2009). The simplest and safest strategy to prevent diseases and adjust activities may be to consume flavonoids through food sources.

5.5 SUBCLASSES OF FLAVONOID

5.5.1 RUTIN

It comes under one of the vital constituent present in leaves of tea, apple, etc. It contains a variety of pharmacological effects in human body. This compound is also known by another name called vitamin P/rutoside and

TABLE 5.1 Botanical Sources of Quercetin

SL. No.	Botanical Name	Family	General Name	Main Contents	References
1.	*Punica granatum*	Lythraceae	Pomegranate	Fruits	Choudhary et al. (2015)
2.	*Ruta graveolens*	Rutaceae	Rue	Leaves	–
3.	*Camellia sinensis*	Theaceae	Green tea	Leaves	Zhou et al. (2016); Thakur et al. (2011)
4.	*Allium cepa*	Amaryllidaceae	Red onion	Fruits	Henagan et al. (2015); Oliveria et al. (2015)
5.	*Mangifera indica*	Anacardiaceae	Mangu	Fruits	Gondi et al. (2015)
6.	*Moringa oleifera*	Moringaceae	Drumstick tree	Leaves	Leone et al. (2015); Saini et al. (2016)
7.	*Cydonia oblonga*	Rosaceae	Quince	Fruits and leaves	Oliveira et al. (2012)
8.	*Solidago candensis* L.	Compositae/Asteraceae	Goldenrod	Flowering parts	Apati et al. (2002)
9.	*Ginkgo biloba*	Ginkgoaceae	Maidenhair tree	Leaves	Shepperd et al. (2008); Chan et al. (2007)
10.	*Achillea millefolium* L.	Compositae/Asteraceae	Yarrow	Flowering tops	Benedek et al. (2007)

it contains many nutraceutical effects also. The other pharmacological activities of this compound are anti-diabetic, analgesic, stroke, anti-Alzheimer, etc. Apart from this the other effects are also present in human body such as hypertension, coagulation of blood, anti-platelet aggregatory effects, cardio-protective effects in CVS (Keservani et al., 2016a), anti-ulcer effects in GIT System, effect in bones and other effects which are associated with respiratory system. It also diuretic effects in excretory system and ophthalmic effects in regions of the eye. In the reproductive system, it helps in increasing sperm quality and enhancement of male reproductive organs. This compound has many protective functions such as anti-cancer, chemotherapeutic action, anti-bacterial action, anti-fungal action, anti-mycobacterial, larvicidal, anti-malarial, anti-retroviral, anti-viral, neuroprotective, hepatoprotective, blood vasculature protection, relieving of wounds, modulation of radio effect, etc.

5.5.2 APIGENIN

The activities reported in this compound are anti-inflammatory, anti-carcinogenic, and other pharmacological properties. At the cellular and molecular level, this compound shows a progress while studying its biological effects. Some studies show that the cancer chemo-protective effects of this compound are in organ specification format which evaluated its restrictions and has a probability of developing as an cancer chemoprotective agent (Patel et al., 2007).

5.5.3 HESPERIDIN

This compound has remarkable properties such as anti-cancer, anti-viral, anti-fungal, anti-inflammatory, etc. (Gaur et al., 2010). The noticeable contributions of this compound such as reduced number of neurons death and intracellular oxidant defense system. In addition to it, other actions such as hydroxyl radicals, singlet oxygen and action against superoxide are also noticeable (Galleano et al., 2010). On the basis of its pharmacological activity it is shown to have acting on the central nervous system in the form of sedatives and sleep inducing compounds (Marder et al., 2003). In dopamine induced neurotoxicity it shows neuroprotective property (Obregon et al., 2006).

5.5.4 *KAEMPFEROL*

The various plant species and some food materials such as red wine, onions naturally contain a flavonoid named quercetin, which contains anti-apoptotic, anti-oxidant, pharmacological, and some cardio-protective properties. Certain neuro-degenerative diseases such as Parkinson's disease (PD) or Alzheimer's disease (AD) have neuroprotective action which is found in the prevention of neurodegenerative disorders (Spagnuolo et al., 2017).

5.5.5 *NARINGENIN*

It is a flavonoid which is obtained from milk bramble of species Silybum marianum is Silibinin, (2R, 3R)-3, 5, 7-trihydroxy-2-[(2R, 3R)-3-(4-hydroxy-3-methoxyphenyl)-2-(hydroxymethyl)-2, 3-dihydro-1, 4-benzodioxin-6-yl]-2, 3-dihydro-4H-chromen-4-one (Švagera et al., 2003). The various actions of Silibinin inflammatory process, oxidative disturbances in balance between ROS, apoptosis (Wang et al., 2016) and in treatment of locomotion activity was recently investigated by Song. The response of free radicals in higher amount is due to concentration of proinflammatory cytokines IL-1 beta, Inter Leukin-4 and level of antioxidant enzyme glutathione and malondialdehyde which is liquid peroxidase marker. The $NF\text{-}_{K}B$ which is a product of glial cells which provides swelling and pain in the response of the brain. Two components named as P-53 (component of acute cell response) is a tumor suppressor agent and p-p53 is its phosphorylated component. P-53 is a major object of acute cell stress response. The further results observed are decreased in anxiety behavior, flexibility in memory and recognization of new objects. The major action is in neuroprotective potential which acts on inflammatory responses and death of neurons (Song et al., 2017).

5.5.6 *ANTHOCYANINS*

These compounds are mostly found in vegetables, fruits, flowers, grains having poly-phenolic flavonoids (Wang et al., 2016). The properties they have are antioxidant, anti-apoptosis, and improvement in memory as well as cognition (Winter et al., 2017). The main characteristic is that it contains a flavylium core which has 2 aromatic rings conjugated with 3 C-unit and compressed by an O_2 molecule. This mol. contains 2 or 3 portions they are an aglycone and may contain a composition of organic acids (Smeriglio et al., 2016). It is found that many evidences are there which proves about the role of anthocyanins in apprehension and thinking which a main factor of neuro-protectives is. It is found that in improvement of memory loss, synaptic defects these anthocyanin particles (ethylene gycol) gold particles named PEG-AuNPs are very helpful in curing of this disease. The increase in behavior alteration is due to anthocyanins and anthocyanin-loaded PEG-AuNPs. In the manufacturing of beta-amyloid protein the flavonoid action potential was found to be decreased in Anthocyanin-loaded PEGAuNPs, the level A beta (Beta-amyloid protein), BACE-1 which is a major enzyme found during development of beta-amyloid protein. In organization of free anthocyanins and anthocyanin stacked with PEG-AuNPs mitigated the effect, a beta-1-42 and escalated the expression level of synaptophysin, PSD95 and SNAP23; as all these molecules shows their relation with the synapse process which occurs between the neurons. The process of memory can be improved by two compounds that is Anthocyanins and anthocyanins loaded with PEG-AuNPs which further increase the phosphorylation of GluR1 at position Ser845 and this increased the expression level of p-CREB (Serum 133). When the intake of free anthocyanins and anthocyanins which are loaded with PEG-AuNPs increase phosphorylation elevation level of p-PI3K and p-AKtat Ser 473, increase level p-GSK3 beta at position Serum 9, decreased level of p-tau at 413 and Ser404 this will reduce the level of formation of fibrillar components. Further an increase in number of surviving neurons is due to reduced level of A-beta-1-42-induced degenerated neuronal cells present in cortex and hippocampus regions of brain. Another study was also done on mouse model A-beta-1-42-induced AD to see the treatment efficacy only anthocyanins and those which are loaded with PEG-AuNPs. After that it was found that the movement of anthocyanins-loaded PEG-AuNPs across the BBB and collection in the A-beta injected mice. Some of the other results were also seen, such as reduced expression of beta-amyloid, BACE-1 expression and halt of tau hyperphosphorylation GSK-3 beta/ CDK 5 pathway by anthocyanin loaded PEG-AuNPs. One more report showed some evidences, such as inhibition of activated astrocytes and many inflammatory markers (p-NF-κB), tumor necrosis factor alpha by anthocyanins in the regions of hippocampus and cortex of D-gal treated rats (Rehman et al., 2017).

5.5.7 NARINGIN

All the citrus fruits, such as grape fruit have a flavonoid 4′,5,7-trihydroxy-flavanone 7-rhamnoglucoside of family C_6-C_3-C_6 polyphenol compounds having efficacy to reduce inflammation and anti-oxidative properties (Qin et al., 2016). The effect of neuroprotective was demonstrated by modulating endogenous biomarkers and down regulation of free radicals as well as cytokines which includes TNF-alpha in streptozotocin persuaded peripheral neuropathy (Kandhare et al., 2012). As action in this compound against deltamethrin induced neurotoxicity in male Wistar rats, when their cure is done, it leads to revival of oxidative status this further conforms the safeguarding effect of naringin. This compound has no effect on the locomotion activity while improving the potential of episodic memory as a behavioral analysis the effect of naringin on memory deficiency in pharmacological animal models has no influence. The flavonoids reversed the time induced episodic memory deficiency with enhanced in standard object exploitation time when contrasted with an object that is own nature improvised the recognition and discriminative inducts. As a result, the compound reversed the episodic memory with is of very short duration its deficits induced by scopolamine and enhanced the discrimination and recognition (Ramalingayya et al., 2016). The death of neurons is due to a neurotoxin KA and excitotoxicity. The compound naringin used as an antiepileptic agent showed that there was a significant decrease in the frequency of chronic spontaneous seizures in KA treated mice in contrary with mice KA alone. Further, the loss of hippocampal neurons in KA treated the CA 1 region was impaired by treating with this compound which indicates autophagic tension lowering effects and also seen to be involved in death of neurons. In addition to it, increase in Tumor Necrosis Factor-alpha in its Ibal +ve microglia in the KA treated hippocampus shows the characteristics of many diseases such as Progressive neurological disease and mental deterioration (Jeong et al., 2015).

5.5.8 BAICALEIN

This in starting secluded from the roots of *Scutellaria baicalensis* and *Scutellaria lateriflorais* Baicalein (Makino et al., 2008), shows anti-oxidant, reduces the chances of inflammation activity as well as it is neuro-protective against PD (Li et al., 2005). The noticed enfeeblement of the motor deficits after giving cure of baicalein with increase in striatal neurotransmitters are dopamine, 3-4-dihydroxyphenylacetic acid and homovanillic acid. The rise in a number of TH + ve neurons and decrease in the amount of alpha-syn in the enteric nervous system as seen at the time of the analysis of intensity of fluorescence by microscopy. It was seen that the decrease of alpha-synoligomers which are not monomers as found in ileum, spinal cord, midbrain, this compound has no effect on alpha-syn mRNA expression, as seen in PD it works by reducing the development of alpha-synoligomers (Hu et al., 2016). Rotenone-induced PD rats were used for investigating the therapeutic effects of baicalein. When the study was completed then the results showed that the motor dysfunction was partially ameliorated by baicalein and TH+ cells in the substance nigra increased in number. As it was found that the neurons in SN were also against rotenone-induced apoptosis. The distortion of mitochondrial complex-I in the ventral midbrain damaged by rotenone was enhanced by baicalein. In addition to it the protein level of PGC-1 alpha, NRF-1 and TFAM in ventral midbrain their affinity increased when intake of baicalein provide enhanced in response of brain against oxidative stress and neuronal loss as seen in PD (Zhang et al., 2017). The compound baicalein is very effective against PD as it shows protections of neuronal cells on mitochondrial activity and in activation of glial cells as well as increase the chances due to neurodegenerative diseases.

5.5.9 CATECHIN

From the previous decades, it is found that the clinical uses of herbal (Rane et al., 2020a–c) medicines originated from China are helpful in treating AD with a high effectiveness (Su et al., 2014). They belong to a group of polyphenols. The main poly-phenols present in green tea such as epicatechin, epicatechin-3-gallate, epigallocatechin (EGC), epigallocatechin-3-gallate, etc. (Gaur & Kumar, 2010; Esselen et al., 2014) There are many fruits in which this compound is present such as kiwi, blueberries, apples, strawberries, etc. There are many molecular processes in the human body which are affected by catechin are angiogenesis, degradation of extracellular matrix, regulating cell death, multiple drug resistance in cancer patients and many more. In contrary to it there are some beneficial effects also due to its antioxidant property as well as it helps in treating hypertension, reduces the chances of swelling as well as pain, anti-thrombogenic, and excessive secretion of lipids (Anand et al., 2014).

5.5.10 QUERCETIN

A flavonoid origination of plants and potentially anti-oxidant shows its predominance in the category of flavonoids in edible plants (Brull et al., 2015). The major class of poly-phenols represented by flavonols of class flavonoids. The person's diet intake of complete flavonoids is found to be approx. 200 to 350 mg per day with quercetin intake of 10 to 16 mg per day. The dose which is prescribed of quercetin aglycone in the form of dietary supplement is 1 gram per day (Kim et al., 2019; Harwood et al., 2007). The various effect which are beneficial to health of human beings such as anti-carcinogenic, anti-inflammatory, anti-infective is found in it. It also has the potential in inhibition of lipid-peroxidation and aggregation of platelets which further stimulates the biogenesis of mitochondria (Li et al., 2016). Various *in-vitro* as well as *in-vivo* models of neurodegeneration such as impairment of dementia, ischemia, traumatic injuries, PDs, etc., in all of them the neuro-protective effects of quercetin are observed and studied.

5.6 PRESENT DAY RESEARCH AND TRENDS ON FLAVONOIDS

5.6.1 ACTIVITY OF ANTI-CHOLINESTERASE

In AD (Perry et al., 1978), the symptomatic relief from mild to moderate condition is the therapy in which Acetyl-cholinesterase is a major enzyme in CNS whose inhibition leads to an increase in neural Ach level. To fight against AD the drug development focuses upon the inhibition of cholinesterase centrally in the whole nervous system. The concentration dependent inhibitory action against AChE and BChE (Khan et al., 2009) is done by quercetin and macluraxanthone in which flavonoids like quercetin, rutin, macluraxanthone as shown by *in-vitro* inhibitory studies. When the docking study was done it was found that macluraxanthonen shows more tightened binding as seen in both enzymes as compared to quercetin. Sheng et al. (2009) while drafting, synthesizing, and evaluating derivatives of flavonoids which act as potent AChE inhibitors in this it was observed that many of flavonoid derivatives have the potency of inhibiting AChE. A potent inhibitor, isoflavone derivative 10d, inhibits AChE with an IC50 of 4 nM, showing a high BChE: AChE inhibition ratio (4,575-fold), superior to donepezil (IC50 = 12 nM, 389-fold). Many molecular docking studies are performed in exploring complete interaction with Acetylcholinesterase.

5.6.2 INFLAMMATION REDUCING ACTIVITIES

Cyclo-oxygenase an naturally occurring enzyme that converts arachidonic acid (ARA) to prostaglandins (PGs) and thromboxanes (Smith et al., 2000). COX-1 and COX-2 are the two isoforms of the enzyme. Cyclooxygenase-1 is a constitutive enzyme that produces PGs that sustain the probity of stomach mucosa and maintain proper vascular homeostasis as compared to COX-2 is an inducible enzyme that only expresses in response to stimulators that cause inflammation (Kurumbali et al., 1996). Cycloxygenase-2 is a protein that produces PGs, which are used to cause inflammation and pain (Kujubu et al., 1991). As the binding methods of flavonoids with cyclo-oxygenase 2 were studied using *in silico* methods, and it was discovered some flavonols and flavones with a 2,3-double bond may act as preferential inhibitors of COX-2 (D'Mello et al., 2011). The flavonol, flavone, flavanone or isoflavone classes all had these findings. This discovery paved the way for the development of selective COX-2 inhibitors, a class of anti-inflammatory drugs with fewer gastrointestinal side effects. Flavonoids such as silbinin, galangin, scopoletin, hesperitin, genistein, celecoxib, etc., which are commercially available, were also tested for COX inhibitory action (Madeswaran et al., 2012). When compared to the standard (–3,473 kJ/mol), the selected flavonoids had higher binding energy ranging from 877 to 624 kcal/mol (–3,669 to –2,611 kJ/mol), leading to the creation of effective COX inhibitors used in the treatment of inflammation and colleagues *In silico* docking studies were used to assess flavonoids' COX inhibitory efficacy. Flavonoids such as farobin-A, gericudranin-B, glaziovianin-A, rutin, and xanthotoxin were used in this study. As a result, more research could lead to the development of effective aldose reductase (AR) inhibitors for the treatment of diabetes. *In silico* docking investigations of commercially available flavonoids' lipoxygenase (LOX) inhibitory activities were also described. Flavonoids such as aromadedrin, eriodictyol, fisetin, homoeriodictyol, pachypodol, and azelastine were investigated in this light. Taiwan-homo-flavone C and eight known lignans justicidin B, justicidin C, justicidin D, chinensinaphthol methyl ether, procumphthalide A, procumbenoside A, and ciliatosides A and B from medicinal herbal plants *Cephalotaxus wilsoniana* and Justicia species, respectively, were the flavonoids studied. Ginkgetin, Taiwan-homo-flavone C, justicidin B, and justicidin D were found to have platelet reducing effects among these flavonoids.

5.6.3 MODIFICATIONS OF STEROID GENESIS

Abyssinone with other linked flavonoids can be employed as possible steroid-genesis modulators against 3 steroid-genesis pathway enzymes: 3-hydroxysteroid dehydrogenase (HSD), 17-hydroxysteroid dehydrogenase (HSD) and aromatase (Hatti et al., 2009). As virtual screening experiment revealed that flavonones have a stronger affinity for their respective chalcones. Flavonones have a constant combining affinity for all three enzymes studied and are more effective steroidogenesis modulators in hormone depending cancers.

5.6.4 MODULATORS OF XANTHINE OXIDES

It is a catalytic enzyme that converts hypoxanthine into xanthine and then xanthine is converted into uric acid. Hyperuricemia an increment in uric acid levels in the blood serum that leads to serious consequences like gout, kidney stones (Borges et al., 2002). In focused on naturally acting flavonoids in the search for a possible XO inhibitor The most effective inhibitor of XO was Licoisoflavone-A, which was isolated from the roots of Glycyrrhiza glabra. *In silico* docking studies were used by (Umamaheswari et al., 2011) to assess flavonoids' XO-inhibitory action. The flavonoids butein, fisetin, diosmetin, tricetin, genistein, tricin, vitexycarpin, herbacetin, okanin, etc., were investigated, and all flavonoids were found to have inhibitory activity. Its XO-inhibitory activity would have been aided contains a benzopyran ring in its basic nucleus. This molecular docking study could pave the way for the creation of effective Xenthine oxidase inhibitors used in preventing and treating gout and many other inflammation causing diseases. A fresh strategy has been examined highlighting the importance of nature derived compounds as a primary solution to unresolved concerns such as the treatment of the 'silent killer' polycystic kidney disease (PKD) (Shoba et al., 2010). The main protein, cystic fibrosis transmembrane conductance regulator (which causes PKD), and its altered 3-D structure were exposed to molecular docking and *in silico* toxicity experiments using flavonoids derived through plants. The findings suggested that flavonoids derived from plant sources could be used as natural medicinal agents to fight PKD.

5.6.5 ANTIBIOTIC RESISTANCE SHOULD BE TREATED

Beta-Ketoacyl acyl carrier protein synthase III, a critical target enzyme for combating antibiotic resistance, begins fatty acid synthesis in bacteria (Lee et al., 2011), flavonoids such as 5,7,4′-trihydroxyflavanone and 5,7,3′,4′-tetrahydroxyflavanone are strong anti-microbial inhibitors of Staphylococcus aureus KAS III, discovered during working on flavonoid inhibitors of -KAS III against the methicillin-resistant bacteria Staphylococcus aureus (Ganugapati et al., 2011) investigated the *in silico* modeling and docking of a superbug enzyme called New Delhi metallo—lactamase-1 which is present in *E. coli*. This enzyme belongs to the B1 subclass of metallo-lactamases and is known to cause resistance to regular intravenous antibiotics. Using the *in silico* molecular docking technique, researchers investigated the suppression of NDM-1 in superbugs by flavonoids (Padmavathi et al., 2012). As many currently non-effective antibiotics against the NDM-1-positive pathogen, so this research will help researchers better understand the molecular basis of NDM-1's extended antibiotic resistance and, as a result, speed up the development of new antibiotics against the NDM-1-positive strain in clinical trials.

5.7 FLAVONOIDS HAVE A VARIETY OF FUNCTIONS AND ACTIVITIES

As we know the leafy plants contains a broad span of organic chemicals, in which a large number of them do not appear to take part in direct form in development and growth. So chemicals that is flavonoids, often found in variable amounts in different taxonomic groups in plant kingdom (Croteau et al., 200). Further divided into three groups: alkaloids, terpenoids, and phenolics. In human healthy body they provide a large number of protective roles. Many flavonoids have evolved as bioactive chemicals with antibacterial, insecticidal, and pharmacological activities that interfere with nucleic acids. As a result, they are of interest in medicines and in agriculture as pesticides (Wink et al., 2004).

5.7.1 IN-VITRO TECHNOLOGY IS A TYPE OF IN-VITRO TESTING

The efficiency of plant cell tissue culture in creating the carbon copy one valued chemical compound as similar to parent plant has been revealed by *in vitro* technology (Anand, 2010). These procedures in development of flavonoids have advanced well far away presumptions. An aseptic technology that allows for the production of favorable quality and quantity of plants and metabolites by manipulating nutrients, culture conditions, and phyto-hormone supply. It is feasible to acquire manufacturing of the desired chemicals at levels comparable to those found in plants using differentiated cell culture. They are connected to an extensive range of properties which help in the production of health factors. Then they perform an major action in various scope of food processing nutritional values, pharmacological, medicaments, and other beauty products usage compounds. This is due to their ability to alter important cellular enzyme functions as well as their protect cell from damage caused by free radicals (Srivastava et al., 2015), reduces the chances of inflammation, interferes with the mutagenicity of substance (Snijman et al., 2007), and reduces the development of cancer (Lejeune et al., 2015) capabilities. Flavonoids are antioxidants, antimicrobials, photoreceptors, visual attractants, feeding repellents, and light screeners in plants. Numerous studies have revealed that they have living human beings properties for example anti-allergenic, antiviral, anti-inflammatory, and vasodilator properties. Several research have been conducted in upcoming years on the ability of flavonoids to serve as antioxidants *in vitro*, and key structure-activity connections of antioxidant activity have been identified. Described the primary molecular mechanisms of action in various conditions in their research on flavonoids and anticancer drugs. They indicated that flavonoids have an inhibitory effect on cytochrome P450, which helps to prevent carcinogens. Activities of particular P450 isozymes, which are involved in the synthesis of a variety of pro-carcinogens Flavonoids also aid in the creation of metabolizing enzymes such as glutathione-S-transferase, quinone reductase, etc., which help cancer reducing agents to be detoxified and therefore removed from human body. In further, it would also laid in preventing flavonoids from in contrary as chemotherapeutic agents against cancer causing agents. Structural features and *in vitro* antioxidant capacity of most prevalent flavonoids, as well as *in vitro* antioxidant activity and impacts on endogenous antioxidants. They have been discovered to be particularly helpful in avoiding lipid peroxidation, which is linked to diseases including atherosclerosis, high blood sugar level, hepatotoxicity, swelling, pain, as well as the aging process. On the basis of research, quercetin helps in the prevention of lipid peroxidation. Other that aid to suppress the generation of superoxide radicals include myricetin, quercetin, and rutin, in addition to quercetin.

5.7.2 ANTIMICROBIAL PROPERTIES

Flavonoids are also known as it have many anti-microbial properties as so many scientists have isolated and done identification about structures of flavonoids with its anti-fungal, anti-viral, and anti-bacterial properties. As a result of this feature, many of them are presently useful in areas of nutrition, food safety, and health (Wang et al., 1998). Flavonoids have antiviral properties, especially in the treatment of many airborne infections, it has been demonstrated. Antiviral activity varies amongst flavonoids such as quercetin, naringin, and many more. Many ribo-nucleic and deoxy-ribonucleic acid viruses' reproduction and infectivity are affected. Quercetin and apigenin are two of the highly studied flavonoids, both of which are known to have antibacterial properties. Quercetin derived from lotus leaves could be a promising antibacterial agent for many infectious diseases (Li & Xu, 2008).

5.7.3 HORMONES ACTIVITY

From a large quantity, there are few flavonoids that have hormone-like properties and look a lot like steroid hormones mainly estrogens. Fruits, vegetables, and grains all contain flavonoids. Hormone as steroids, particularly estrogen, which has advancing effects in protecting neurons in regions of the brain, are well known in the prevention of numerous chronic disorders. Several flavonoids, including genistein, daidzein, and equal as examined in clinical studies to determine their estrogenic activity. The investigations looked into their potential for treating a variety of chronic conditions like cancer, heart disease, and osteoporosis (Wiseman et al., 200). The flavonoid genistein has the best chance of reducing postmenopausal bone loss in women. A number of dietary flavonoids have been demonstrated to have a positive impact on the disease of the arteries due to deposition of fatty material related indicators such as lipoprotein oxidation, blood platelet aggregation, and many more (Comalada et al., 2005).

5.7.4 ANTI-INFLAMMATORY PROPERTIES

In this, the up marks of flavonoids notably quercetin, in a number of procedures causing inflammation and immunological actions have been studied and it is seen discovered that some flavonoids can assist to reduce inflammation and strengthen the immune system. Flavonoids and tannins from the leaves of the plant *Spilanthes paniculata* have been shown to have anti-inflammatory properties (Hossain et al., 2012).

5.7.5 ANTI-CANCER ACTIVITY

Various studies have been conducted on reducing the cancerous effects of flavonoids such as tangeritin, 3-hydroxyflavone, 3′,4′-dihydroxyflavone, daidzein, and genistein. A large number of flavonoids, for example, tannins, stilbenes, curcuminoids and other flavonoids have chemo-preventive actions. It provides in inducing the death of cells by arresting the cell cycle, regulating carcinogen metabolism (Ren et al., 2003; Huang et al., 2010). They went on to say that these flavonoids contain supporting and protruding mechanisms in cancer prevention, which includes anti-oxidant activity and scavenging free radicals, modulation of carcinogen metabolism, regulation of gene expression on oncogenes, and many other activities.

5.7.6 NEURODEGENERATIVE ACTIVITY

Flavonoids and their influence on central nervous system (Keservani et al., 2016b) protection are of particular interest, especially in the case of neurodegenerative illness induced by a combination of oxidative stress, swelling with pain and transition metal deposition; this is wealth of material accessible. AD and associated dementias are two of the most common neurodegenerative diseases (Keservani et al., 2015a). Flavonoids, such as flavonols, have been linked to lower dementia rates in the general population. Citrus flavanones crosses the blood brain barrier and may be useful for the treatment of neurodegenerative illnesses (Hwang & Yen, 2008). They are also been linked to anti-diabetic and anti-aging properties.

5.8 CONCLUSION

As the data of presented work it is summarized that the plants have originated of Indian herbal are habitately owned in the form of medicines in the field of Ayurveda. In neuro-degeneration disease products derived from nature such as alkaloids, poly-phenolic compounds extracted from plants have the ability to reduce the chances of neurodegeneration as further it will improve memory as well cognitive functions. These compounds having residence in many herbal plants as they have power in protecting the brain against many injuries caused due to neurons. In addition to it, they also have ability in reducing the inflammation of neuron as well as its potential in promoting memory, thinking, and learning functions. Therefore, flavonoids come under powerful genera of being a major ancestor molecule for progression of advanced generation compounds of drugs that are having brain intensifying activities. Reactive O_2 species comes in reversed limit as they can stabilize through the action of antioxidants which is called oxidative stress neuro-degeneration. As we know neuro-degeneration causes disease, such as Alzheimer's, Parkinson's, and migraines. Certain flavonoids, such as Rutin, Apigenin, Hesperidin, Kaempferol, Naringenin, etc., show a wide range of neuro-protective effects. So future aspects show that the majority of flavonoids are helpful in treating diseases related to the brain and its parts.

KEYWORDS

- **Alzheimer's disease**
- **anthocyanins**
- **antioxidant**
- **chalcones**
- **flavonoids**
- **inflammation**
- **Parkinson's disease**
- **polycystic kidney disease**

REFERENCES

Alim, T., Kim, M. J., Rehman, S. U., Ahmad, A., & Kim, M. O., (2016). Anthocyanin-loaded PEG-gold nanoparticles enhanced the neuroprotection of anthocyanins in an A β1-42 mouse model of Alzheimer's disease. *Molecular Neurobiology, 54*(8), 6490–6506.

Anand, A. Z., & Subhash, L. B., (2014). *Polyphenols in Chronic Diseases and Their Mechanisms of Action in Polyphenols in Human Health and Disease, b9780*, 123.

Anand, S., (2010). Various approaches for secondary metabolite production through plant tissue culture. *Pharmacia., 1*, 1–7.

Aoki, T., Akashi, T., & Ayabe, S., (2000). Flavonoids of leguminous plants: Structure, biological activity, and biosynthesis. *J. Plant Res., 113*, 475–488.

Apati, P., Szentmihályi, K., Balázs, A., Baumann, D., Hamburger, M., Kristó, T. S., Sz″oke, É., & Kéry, Á., (2002). HPLC analysis of the flavonoids in pharmaceutical preparations from Canadian goldenrod (*Solidago canadensis*). *Chromatographia, 56*, S65–S68.

Benedek, B., & Kopp, B., (2007). *Achillea millefolium* L. sl revisited: Recent findings confirm the traditional use. *Wien. Med. Wochenschr., 157*, 312–314.

Borges, F., Fernandes, E., & Roleira, F., (2002). Progress towards the discovery of xanthine oxidase inhibitors. *Curr. Med. Chem., 9*, 195–217.

Brüll, V., Burak, C., Stoffel-Wagner, B., Wolffram, S., Nickenig, G., Müller, C., Langguth, P., et al., (2015). Effects of a quercetin-rich onion skin extract on 24 h ambulatory blood pressure and endothelial function in overweight-to-obese patients with (pre-) hypertension: A randomized double-blinded placebo-controlled cross-over trial. *Br. J. Nutr., 114*, 1263–1277.

Burak, M., & Imen, Y., (1999). Flavonoids and their antioxidant properties. *Turkiye Klin Tip Bil Derg., 19*, 296–304.

Chan, P. C., Xia, Q., & Fu, P. P., (2007). Ginkgo biloba leave extract: Biological, medicinal, and toxicological effects. *J. Environ. Sci. Health C Environ. Carcinog. Ecotoxicol. Rev., 25*, 211–244.

Choudhary, M., Kumar, V., Malhotra, H., & Singh, S., (2015). Medicinal plants with potential anti-arthritic activity. *J. Intercult. Ethnopharmacol., 4*, 147.

Codorniu-Hernández, E., Rolo-Naranjoa, A., & Montero-Cabrera, L. A., (2007). Theoretical affinity order among flavonoids and amino acid residues: An approach to understand flavonoid–protein interactions. *J. Mol. Struc: Theochem., 819*, 121–129.

Comalada, M., Camuesco, D., Sierra, S., et al., (2005). *In vivo* quercitrin anti-inflammatory effect involves the release of quercetin, which inhibits inflammation through down-regulation of the NF-κB pathway. *Eur. J. Immunol., 35*, 584–592.

Croteau, R., Kutchan, T. M., & Lewis, N. G., (2000). Natural products (secondary metabolites). In: Buchanan, B., Gruissem, W., & Jones, R., (eds.), *Biochemistry and Molecular Biology of Plants* (pp. 1250–1318). Rockville, MD: American Society of Plant Biologists.

Dixon, R., & Ferreira, D., (2002). Molecules of interest: Genistein. *Phytochemistry, 60*, 205–211.

Dixon, R., & Pasinetti, G., (2010). Flavonoids and isoflavonoids: From plant biology to agriculture and neuroscience. *Plant Physiol., 154*, 453–457.

DMello, P., Gadhwal, M., Joshi, U., et al., (2011). Modeling of COX-2 inhibitory activity of flavonoids. *Int. J. Pharm. Pharm. Sci., 3*, 33–40.

Esselen, M., & Stephan, W., (2014). *Barth Food-Borne Topoisomerase Inhibitors in Advances in Molecular Toxicology B9780, 8,* 123–171.

Galleano, M., Pechanova, O., & Fraga, C. G., (2010). Hypertension, nitric oxide, oxidants, and dietary plant polyphenols. *Curr. Pharm. Biotechnol., 11*, 837–848.

Ganugapati, J., Sirisha, V., Mukkavalli, S., et al., (2011). Insilico modeling and docking studies of New Delhi metallo β lactamase-1 (super bug). *Int. J. Eng. Sci. Tech., 3*, 2427–2434.

Gaur, V., & Kumar, A., (2010). Hesperidin pre-treatment attenuates NO-mediated cerebral ischemic reperfusion injury and memory dysfunction. *Pharmacol. Rep., 62*, 635–648.

Giusti, M., & Wrolstad, R., (2003). Acylated anthocyanins from edible sources and their applications in food systems. *Biochem. Eng. J., 14*, 217–225.

Gondi, M., & Rao, U. P., (2015). Ethanol extract of mango (*Mangifera indica* L.) peel inhibits α-amylase and α-glucosidase activities, and ameliorates diabetes related biochemical parameters in streptozotocin (STZ)-induced diabetic rats. *J. Food Sci. Technol., 52*, 7883–7893.

Griesbach, R., (2005). Biochemistry and genetics of flower color. *Plant Breed Rev., 25*, 89–114.

Halliwell, B., (1995). How to characterize an antioxidant: An update. *Biochem. Soc. Symp., 61*, 73–101.

Harwood, M., Danielewska-Nikiel, B., Borzelleca, J., Flamm, G., Williams, G., & Lines, T., (2007). A critical review of the data related to the safety of quercetin and lack of evidence of *in vivo* toxicity, including lack of genotoxic/carcinogenic properties. *Food Chem. Toxicol., 45*, 2179–2205.

Hatti, K., Diwakar, L., Rao, G., et al., (2009). Abyssinones and related flavonoids as potential steroidogenesis modulators. *Bioinformation, 3*, 399–402.

Havsteen, B., (2002). The biochemistry and medical significance of the flavonoids. *Pharmacol. Ther., 96*, 67–202.

Hayashi, T., Sawa, K., Kawasaki, M., et al., (1988). Inhibition of cow's milk xanthine oxidase by flavonoids. *J. Nat. Prod., 51*, 345–348.

Henagan, T., Cefalu, W., Ribnicky, D., Noland, R., Dunville, K., Campbell, W., Stewart, L., et al., (2015). *In vivo* effects of dietary quercetin and quercetin-rich red onion extract on skeletal muscle mitochondria, metabolism, and insulin sensitivity. *Genes Nutr., 10*, 2.

Hertog, M. G., Hollman, P. C., & Van De, P. B., (1993). Content of potentially anticarcinogenic flavonoids of tea infusions, wines, and fruit juices. *J. Agric. Food Chem., 41*, 1242–1246.

Hossain, H., Shahid-Ud-Daula, A., Jahan, I., et al., (2012). Evaluation of antinociceptive and antioxidant potential from the leaves of *Spilanthes paniculata* growing in Bangladesh. *Int. J. Pharm. Phytopharmacol. Res, 1*, 178–186.

Hu, Q., Uversky, V. N., & Huang, M., (2016). Baicalein inhibits α-synuclein oligomer formation and prevents progression of α-synuclein accumulation in a rotenone mouse model of Parkinson's disease. *Biochimica et Biophysica Acta (BBA) – Molecular Basis of Disease, 1862*(10), 1883–1890.

Huang, J. H., Huang, C. C., Fang, J. Y., et al., (2010). Protective effects of myricetin against ultraviolet-B-induced damage in human keratinocytes. *Toxicol. in Vitro, 24*, 21–28.

Hwang, S., & Yen, G., (2008). Neuroprotective effects of the citrus flavanones against H_2O_2-induced cytotoxicity in PC12 cells. *J. Agric. Food Chem., 56*, 859–864.

Iwashina, T., (2013). Flavonoid properties of five families newly incorporated into the order caryophyllales (review). *Bull. Natl. Mus. Nat. Sci., 39*, 25–51.

Jeong, K. H., Jung, U. J., & Kim, S. R., (2015). Naringin attenuates autophagic stress and neuroinflammation in kainic acid treated hippocampus *in vivo*. *Evidence-based Complementary and Alternative.* Article ID 354326, 9.

Jorgensen, R., (1995). Co-suppression, flower color patterns, and metastable gene expression states. *Science, 268*, 686–691.

Kandhare, D., Raygude, K. S., Ghosh, P., Ghule, A. E., & Bodhankar, S. L., (2012). Neuro-protective effect of naringin by modulation of endogenous biomarkers in streptozotocin induced painful diabetic neuropathy, *Fitoterapia, 834*, 650–659.

Keservani, R. K., & Sharma, A. K., (2014). Flavonoids: Emerging trends and potential health benefits. *Journal of Chinese Pharmaceutical Sciences, 23*(12), 815.

Keservani, R. K., Kesharwani, R. K., Sharma, A. K., & Ahmad, M. F., (2015a). Stress, Parkinson's, Alzheimer's disease: Role of dietary supplements. In: Debasis, B., Harry, G. P., & Anand, S., (eds.), *Nutraceutical and Functional Foods in Human Life and Disease Prevention* (pp. 241–253). Boca Raton, Fl. CRC, Press, Taylor and Francis, Chapter 14. ISBN: 9781482237214.

Keservani, R. K., Kesharwani, R. K., Sharma, A. K., Vyas, N., & Chadoker, A., (2010b). Nutritional supplements: An overview. *International Journal of Current Pharmaceutical Review and Research, 1*(1), 59–75.

Keservani, R. K., Kesharwani, R. K., Vyas, N., Jain, S., Raghuvanshi, R., & Sharma, A. K., (2010a). Nutraceutical and functional food as future food: A review. *Der Pharmacia Lettre, 2*(1), 106–116.

Keservani, R. K., Sharma, A. K., & Kesharwani, R. K., (2016a). Nutraceutical and functional foods for cardiovascular health. In: *Food Process Engineering* (pp. 257–278, 291–312). Apple Academic Press, CRC Press. ISBN: 9781771884020.

Keservani, R. K., Sharma, A. K., & Kesharwani, R. K., (2016b). Medicinal effect of nutraceutical fruits for the cognition and brain health. *Scientifica*, 10. Article ID 3109254. doi: 10.1155/2016/3109254.

Kim, D. H., Khan, H., Ullah, H., Hassan, S. T., Šmejkal, K., Efferth, T., Mahamoodally, M. F., et al., (2019). MicroRNA targeting by quercetin in cancer treatment and chemoprotection. *Pharmacol. Res.*, 104346.

Kujubu, D., Fletcher, B., Varnum, B., et al., (1991). TIS10, a phorbol ester tumor promoter-inducible mRNA from Swiss 3T3 cells, encodes a novel prostaglandin synthase/cyclooxygenase homologue. *J. Biol. Chem., 266*, 12866–12872.

Kumar, S., & Pandey, A. K., (2013). Chemistry and biological activities of flavonoids: An overview. *Scientific World Journal*, 162750.

Kurumbail, R., Stevens, A., Gierse, J., et al., (1996). Structural basis for selective inhibition of cyclooxygenase 2 by anti-inflammatory agents. *Nature, 384*, 644–648.

Lee, J., Jeong, K., Shin, S., et al., (2011). Antimicrobial natural products as β-ketoacyl-acyl carrier protein synthase III inhibitors. *Bioorg. Med. Chem., 17*, 5408–5413.

Lee, Y., Yuk, D., Lee, J., et al., (2009). Epigallocatechin-3-gallate prevents lipopolysaccharide-induced elevation of beta amyloid generation and memory deficiency. *Brain Res., 1250*, 164–174.

LeJeune, T. M., Tsui, H. Y., Parsons, L. B., et al., (2015). Mechanism of action of two flavone isomers targeting cancer cells with varying cell differentiation status. *PLOS One, 10*, e0142928.
Leone, A., Spada, A., Battezzati, A., Schiraldi, A., Aristil, J., & Bertoli, S., (2015). Cultivation, genetic, ethnopharmacology, phytochemistry and pharmacology of *Moringa oleifera* leaves: An overview. *Int. J. Mol. Sci., 16*, 12791–12835.
Li, M., & Xu, Z., (2008). Quercetin in a lotus leaves extract may be responsible for antibacterial activity. *Arch. Pharm. Res., 31*, 640–644.
Li, Q., Wang, T., Pei, Z., Liu, B., & Hong, J. S., (2005). Inhibition of microglial activation by the herbal flavonoid baicalein 20 oxidative medicine and cellular longevity attenuates inflammation-mediated degeneration of dopaminergic neurons. *Journal of Neural Transmission, 112*(3), 331–347.
Li, Y., Yao, J., Han, C., Yang, J., Chaudhry, M. T., Wang, S., Liu, H., & Yin, Y., (2016). Quercetin, inflammation and immunity. *Nutrients, 8*, 167.
López-Lázaro, M., (2009). Distribution and biological activities of the flavonoid luteolin. *Mini Rev. Med. Chem., 9*, 31–59.
Madeswaran, A., Umamaheswari, M., Asokkumar, K., et al., (2012). In-silico docking studies of cyclooxygenase inhibitory activity of commercially available flavonoids. *Asian J. Pharm Life Sci., 2,* 174–181.
Makino, T., Hishida, A., Goda, Y., & Mizukami, H., (2008). Comparison of the major flavonoid content of S. Baicalensis, S. Lateriflora, and their commercial products. *Journal of Natural Medicines, 62*(3), 294–299.
Marder, M., Viola, H., & Wasowski, C., (2003). 6-Methylapigenin and hesperidin: New valeriana flavonoids with activity on the CNS. *Pharmacol. Biochem. Behav., 75*, 537–745.
Matthies, A., Clavel, T., Gütschow, M., et al., (2008). Conversion of daidzein and genistein by an anaerobic bacterium newly isolated from the mouse intestine. *Appl. Envrion. Microbiol., 74*, 4847–4852.
Metodiewa, D., Kochman, A., & Karolczak, S., (1997). Evidence for antiradical and antioxidant properties of four biologically active N, N, diethylaminoethyl ethers of flavanone oximes: A comparison with natural polyphenolic flavonoid (rutin) action. *Biochem. Mol. Biol. Int., 41*, 1067–1075.
Nijveldt, R., Nood, E., Hoorn, D., et al., (2001). Flavonoids: A review of probable mechanisms of action and potential applications. *Am. J. Clin. Nutr., 74*, 418–425.
Obregon, D. F., Rezai-Zadeh, K., & Bai, Y., (2006). ADAM10 activation is required for green tea (-) epigallocatechin-3 gallate induced a-secretase cleavage of amyloid precursor protein. *J. Biol. Chem., 281*, 16419–16427.
Oliveira, A. P., Costa, R. M., Magalhães, A. S., Pereira, J. A., Carvalho, M., Valentão, P., Andrade, P. B., & Silva, B. M., (2012). Targeted metabolites and biological activities of *Cydonia oblonga* miller leaves. *Food Res. Int., 46*, 496–504. [CrossRef].
Oliveira, T. T., Campos, K. M., Cerqueira-Lima, A. T., Carneiro, T. C. B., Da Silva, V. E., Melo, I. C. A. R., Figueiredo, E. A., et al., (2015). Potential therapeutic effect of *Allium cepa* L. and quercetin in a murine model of *Blomia* tropicalis induced asthma. *Daru J. Pharm. Sci., 23*, 18.
Ovando, C., Hernandez, D., Hernandez, E., et al., (2009). Chemical studies of anthocyanins: A review. *Food Chem., 113,* 859–871.

Padmavathi, M., Prasanth, R. V., & Rao, R., (2012). Inhibition of NDM-1 in superbugs by flavonoids – an *in silico* approach. *J. Adv. Bioinfo. App. Res, 3*, 328–332.

Patel, D., Shukla, S., & Gupta, S., (2007). Apigenin and cancer chemoprevention: Progress, potential and promise (review). *Int. J. Oncol., 30*(1), 233–245.

Perry, E. K., Tomlinson, B. E., Blessed, G., et al., (1978). Correlation of cholinergic abnormalities with senile plaques and mental test scores in senile dementia. *Br. Med. J., 2*, 1457–1459.

Qin, S., Chen, Q., & Wu, H., (2016). Effects of naringin on learning and memory dysfunction induced by gp120 in rats. *Brain Research Bulletin, 124*, 164–171.

Ramalingayya, V., Nampoothiri, M., & Nayak, P. G., (2016). Naringin and rutin alleviates episodic memory deficits in two differentially challenged object recognition tasks. *Pharmacognosy Magazine., 12*(45), 63–70.

Rane, B. R., Bharath, M. S., Patil, R. R., Keservani, R. K., & Jain, A. S., (2020b). Novel approaches in nutraceuticals. In: Rajesh, K. K., Raj, K. K., & Anil, K. S., (eds.), *Enhancing the Therapeutic Efficacy of Herbal Formulations Through Novel Drug Delivery Systems* (pp. 241–266). IGI Global International Publisher, Pennsylvania, USA, Chapter 11. ISBN13: 9781799844532.

Rane, B. R., Patil, A. S., Keservani, R. K., & Jain, A. S., (2020a). Novel approaches in herbal formulation. In: Rajesh, K. K., Raj, K. K., & Anil, K. S., (eds.), *Enhancing the Therapeutic Efficacy of Herbal Formulations Through Novel Drug Delivery Systems* (pp. 43–68). IGI Global International Publisher, Pennsylvania, USA, Chapter 2. ISBN13: 9781799844532.

Rane, B. R., Tadavi, S. A., & Keservani, R. K., (2020c). Naturopathy. In: Anil, K. S., Raj, K. K., & Surya, P. G., (eds.), *Herbal Product Development* (pp. 321–347). Apple Academic Press, CRC Press, Taylor & Francis Group, Chapter 12. ISBN: 9781771888776.

Rehman, S. U., Shah, S. A., Ali, T., Chung, J. I., & Kim, M. O., (2017). Anthocyanins reversed D-galactose-induced oxidative stress and neuroinflammation mediated cognitive impairment in adult rats. *Molecular Neurobiology., 54*(1), 255–271.

Ren, W., Qiao, Z., Wang, H., et al., (2003). Flavonoids: Promising anticancer agents. *Med. Res. Rev., 23*, 519–534.

Saini, R. K., Sivanesan, I., & Keum, Y. S., (2016). Phytochemicals of *Moringa oleifera*: A review of their nutritional, therapeutic and industrial significance. *3 Biotech, 6*, 203.

Sheng, R., Lin, X., Zhang, J., et al., (2009). Design, synthesis and evaluation of flavonoid derivatives as potent AChE inhibitors. *Bioorg. Med. Chem., 17*, 6692–6698.

Shepperd, W. D., (2008). *Ginkgo biloba*, Ginkgo. L. In: Bonner, F. T., & Karrfalt, R. P., (eds.), *The Woody Plant Seed Manual; Agricultural Handbook No. 727* (pp. 559–561). Department of Agriculture, Forest Service: Washington, DC, USA.

Smeriglio, D., Barreca, D., Bellocco, E., & Trombetta, D., (2016). Chemistry, pharmacology and health benefits of anthocyanins. *Phytotherapy Research, 30*(8), 1265–1286.

Smith, R., DeWitt, D., & Garavito, R., (2000). Cyclooxygenases: Structural, cellular and molecular biology. *Ann. Rev. Biochem., 69*, 145–182.

Snijman, P., Swanevelder, S., Joubert, S., et al., (2007). The antimutagenic activity of the major flavonoids of rooibos (*Aspalathus linearis*): Some dose–response effects on mutagen activation – flavonoid interactions. *Mutat. Res., 631*, 111–123.

Song, X., Zhou, B., & Cui, L., (2017). Silibinin ameliorates A*β*25-35-induced memory deficits in rats by modulating autophagy and attenuating neuroinflammation as well as oxidative stress. *Neurochemical Research, 42*(4), 1073–1083.

Spagnuolo, C., Moccia, S., & Russo, G. L., (2017). Anti-inflammatory effects of flavonoids in neurodegenerative disorders. *European Journal of Medicinal Chemistry, 17*, 30683–30689.

Srivastava, N., & Bezwada, R., (2015). *Flavonoids: The Health Boosters*. White Paper. Hillsborough, NJ: Indofine Chemical Company.

Su, Y., Wang, Q., Wang, C., Chan, K., Sun, Y., & Kuang, H., (2014). The treatment of Alzheimer's disease using Chinese medicinal plants: From disease models to potential clinical applications. *Journal of Ethnopharmacology, 152*(3), 403–423.

Švagera, N., & Škottová, V. P., (2003). Plasma lipoproteins in transport of silibinin, an antioxidant flavonolignan from *Silybum marianum*. *Phytotherapy Research, 17*(5), 524–530.

Thakur, D., Das, S., Sabhapondit, S., Tamuly, P., & Deka, D., (2011). Antimicrobial activities of tocklai vegetative tea clones. *Indian J. Microbiol., 51*, 450–455.

Walker, E., Pacold, M., Perisic, O., et al., (2000). Structural determinations of phosphoinositide 3-kinase inhibition by wortmannin, LY294002, quercetin, myricetin, and staurosporine. *Mol Cell, 6*, 909–919.

Wang, H., Xia, Y., Yang, Z., et al., (1998). Recent advances in the discovery and development of flavonoids and their analogues as antitumor and anti-HIV agents. *Adv. Exp. Med. Biol., 439*, 191–225.

Wang, M., Li, Y. J., & Ding, Y., (2016). Silibinin prevents autophagic cell death upon oxidative stress in cortical neurons and cerebral ischemia-reperfusion injury. *Molecular Neurobiology, 53*(2), 932–943.

Wink, M., (2004). Phytochemical diversity of secondary metabolites. In: *Encyclopedia of Plant and Crop Science* (pp. 915–919). New York: Marcel Dekker, Inc.

Winter, N., Ross, E. K., & Wilkins, H. M., (2017). An anthocyanin-enriched extract from strawberries delays disease onset and extends survival in the hSOD1G93A mouse model of amyotrophic lateral sclerosis. *Nutritional Neuroscience, 9*, 1–13.

Wiseman, H., (2000). The therapeutic potential of phytoestrogens. *Exp. Opin. Investig. Drugs, 9*, 1829–1840.

Zhang, X., Du, L., Zhang, W., Yang, Y., Zhou, Q., & Du, G., (2017). Therapeutic effects of baicalein on rotenone-induced Parkinson's disease through protecting mitochondrial function and biogenesis. *Scientific Reports, 7*, 1. Article 9968.

Zhou, T. S., Zhou, R., Yu, Y. B., Xiao, Y., Li, D. H., Xiao, B., Yu, O., & Yang, Y. J., (2016). Cloning and characterization of a flavonoid 30-hydroxylase gene from tea plant (*Camellia sinensis*). *Int. J. Mol. Sci., 17*, 261.

CHAPTER 6

OVERVIEW AND POTENTIAL BENEFITS OF FLAVONOIDS IN DIABETES

SAGARIKA KABRA[1], AKHAND PRATAP SINGH[1], VIPIN KESHARWANI[1], and DEEPIKA SAINI[2]

[1]*School of Pharmaceutical Sciences, Maharishi University of Information Technology, Lucknow, Uttar Pradesh, India*

[2]*Department of Zoology, Chaman Lal Mahavidhyalya, Landhaura, Haridwar, Uttarakhand, India*

ABSTRACT

The use of herbal medicines has been advantageous since ancient times in treating a variety of diseases. There is an expanding number of diseases in India and worldwide in which diabetes mellitus (DM) is the third leading cause, accompanied by cancer, cardiovascular diseases (CVDs), etc. Therefore, in the treatment of these diseases, a large variety of organic and inorganic compounds are synthesized. In addition to it, flavonoids come in the category of curing this disease as they are a group of naturally occurring polyphenolic chemicals found in plants. As we all know, DM has been categorized into two types: (i) Type I diabetes, and (ii) Type II diabetes. Dietary flavonoids are secondary plant metabolites found in a variety of fruits, vegetables, and seeds. To understand the function of flavonoids, their mechanism of action is a significant concern, which we study about PARP inhibitors, glucose transporting proteins, glucose transporters, etc.

Advances in Flavonoids for Human Health and Prevention of Diseases. Nisha Sharma, Deepika Saini, Rajesh K. Kesharwani, Prakash Chandra Gupta, & Raj K. Keservani (Eds.)

Various enzymes in our bowel, liver, and colon are hydrolyzed and conjugated by flavonoids and further converted into monomeric units that act as metabolites in flavonoids. Many antidiabetic properties are associated with flavonoids, such as Flavonols, Quercetin, Rutin, Kaempferol, Fiseten, Morin, Flavanones, and Hesperidin, etc. So, going through the complete data, it is concluded that flavonoids contain many essential compounds which are found in fruits and vegetables, which are a daily intake of human individuals and have a positive impact on disease outcomes. However, more research will be helpful in further treating the diseases with the help of flavonoids.

6.1 INTRODUCTION

The medicines having herbal (Sharma et al., 2020; Rane et al., 2020a) properties and extracts have very profound applications in treating a variety of disorders for a long time (Man et al., 2008; Firenzuoli et al., 2007), as these products having plant nature are often thought to have lesser toxicity and devoid of adverse effects as compared to synthetic ones (Ayabar et al., 2008; Ayyanar et al., 2008). Medicines originated from herbs have a fascinatingly expanding area nowadays. Diabetes mellitus (DM) is the third leading cause of death in the world, after cancer and cardiovascular disease (CVD) (the two leading causes of death). Flavonoids are a group of naturally occurring polyphenolic chemicals seen in plants. They have diverse uses due to its many biological activities, including their antidiabetic properties. Flavonoids initially originated in green algae about 500 million years ago as a result of the merger of 2 biogenetic passageways, cinnamate, and the older polyketide direction. Plants have gotten increasingly sophisticated as a result of evolution. Bear fruits, leafy vegetables and plant having goods such as red wine and brew contain flavonoids (Geraets et al., 2007). The use of naturally derived medications to treat diabetes has greater favorable actions and do not show any cause negative side effects. So, the plant derived medications prevent cells from damage during diabetes conditions and lower blood glucose levels (Liu et al., 2007). DM has traditionally been known to be cured through the variety of remedy plants, with medical plant withdrawn being reported to enhance diabetic's management. As a result, the hunt for very much effective and guarded antidiabetic drug has now become a priority.

6.2 CLASSIFICATION DIABETES MELLITUS (DM)

Its classifications are of three types which are discussed in subsections.

6.2.1 TYPE I DIABETES

Insulin dependent DM, juvenile is another name for this form of diabetes. When the pancreas, a big gland behind the stomach, fails to generate insulin, diabetes develops. The body's cells are unable to utilize glucose (sugar), which is essential for energy synthesis, if insulin is not present. It has a property to reduce its self-fat in the form of substitute. Insulin injections are necessary regularly to manage the glucose level in the blood of the diabetic person.

6.2.2 TYPE II DIABETES

Non-insulin-dependent DM or adult-onset diabetes are terms used to describe this form of diabetes. Insulin resistance causes this type of diabetes. So, it is a condition in which cells are not able to utilize insulin correctly. Glucose (sugar) builds up in the bloodstream as a result of this.

6.2.3 TYPE III GESTATIONAL DIABETES

Gestational diabetes is a type of diabetes that manifests itself as an increase blood glucose level at the time of gestation period and resolves as soon as the baby departs from the body of the mother. It starts around mid-pregnancy that confirms the changes in the hormones of the mother. As around many 1,200 plants are used to cure high blood sugar all around the world. The medical community faces a difficult dilemma in treating diabetes without complications or adverse effects (Maruthupandian et al., 2011). Traditional medicines have been used to treat diabetes for a long time. Approximately 30% of traditionally used plants are a significant origin of medicinal compounds in the cure of diabetes and other human diseases (Umashanker et al., 2011). Diabetes affects more than 4,000 million individuals around the world. The number will rise as the disease's incidence rises rapidly as an outcome of an increase in population, senescence, urban build up areas and growing obesity and physical inactivity rates. This disease is now a serious global public health issue; recently accessible diabetes medications are expensive and have side effects. They are recently receiving attention as a source material for the creation of novel antidiabetic medicines. Emerging data from epidemiological, animal, and *in-vitro* research has demonstrated the benefits of several dietary (Keservani et al., 2010a, 2020) flavonoids in the cure and management of type-2 diabetes and its consequences (Figure 6.1).

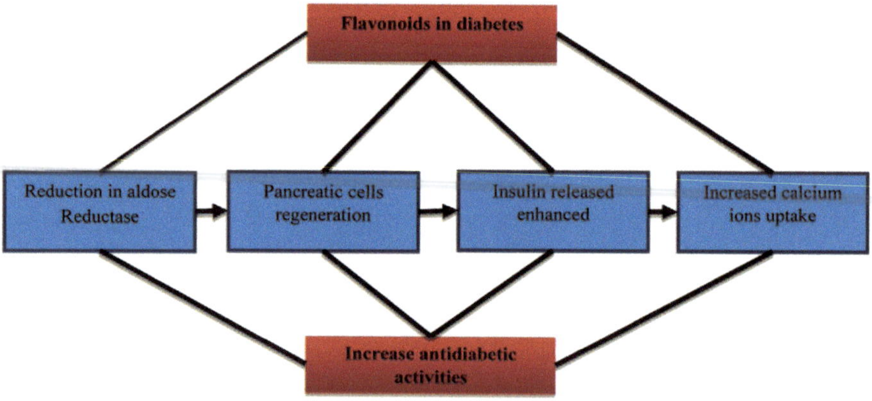

FIGURE 6.1 Different areas of flavonoids in diabetes mellitus.

6.3 DIETARY FLAVONOIDS

Flavonoids (Keservani et al., 2010b; Keservani & Sharma, 2014) are plant secondary metabolites present in a wider range of beery fruits, leafy vegetables and seeds with its derivatives including cocoa, coffee, tea, soya meals and red wine (Panche et al., 2016). They are produced by plants to defend them from invasion of microbes, oxidative injury, and UV damage. It also contributes to the smells, color, and taste of foods. Flavonoids are classified as phenolic chemicals because as contain a basic polyphenolic structure that consists of 2 benzene rings (A and B) joined by an oxygenated heterocyclic ring (C) (Wang et al., 2018). Flavonoids are divided into six subclasses based on the functional groups present on the C-ring (methyl, hydroxyl, glycan, acetyl, or others), the degree of C-ring oxidation, and the connection position of the B-ring, namely flavones, flavanones, flavonols, flavanols, isoflavones, and anthocyanidins. As well as it has distinct markings of hydroxylation and conjugation of phenolic rings distinguish specific molecules from each subclass structurally. Condensed tannins (also known as proanthocyanidins or procyanidins) and hydrolysable tannins are two types of flavonoids that appear as oligomers and polymers. The glycoside-bound form of most flavonoids is found in plants, which adds to its complication and enormous number of unique molecules that are been identified. As well as it has more than 9,000 flavonoids have been identified (Wang et al., 2018), despite the fact that new molecules in this class are continually being discovered. Although flavonoids are consumed in different ways by different cultures, it is believed that the average overall intake in Western countries is around 435 mg per day, and that this may even increase in persons who eat a diet rich in plant-based foods (Peterson et al., 2015; Zamora-Ros et al., 2016). Flavonoids' biological activity, on the other hand, is determined not only by their intake, but also by their bioavailability (Williamson et al., 2010, 2017). Flavonoids' potential bioactivity is determined by their absorption, distribution, metabolism, and excretion in human beings after they are consumed. In a nutshell, flavonoids are metabolized in the small intestine to produce sulfates, glucuronides, and methylated metabolites as they have more water solubility. The conjugated metabolites are then transported into the portal vein and liver, where they undergo phase II metabolism before entering into the bloodstream and then eliminated into urine. Few plasma metabolites can be released into bile and reabsorbed into duodenum, extending their half-life in the systemic circulation. After regular ingestion of flavonoids or flavonoid-rich meals, pure flavonoids and their conjugated forms further detectable in plasma to lesser quantities. Furthermore, because a significant percentage of dietary flavonoids is not absorbed in the small intestine, the colon plays a crucial role in their bioavailability. These unabsorbed flavonoids make their way to the large intestine, where the microbiota breaks them down into a range of tiny phenolic acids and aromatic compounds as they are comfortably absorbed into the colon (Mayorge et al., 2019). These colonic metabolites (mostly phenyl-propionic) are detectable in plasma at higher concentrations than raw flavonoids and conjugated forms (Del Rio et al., 2013; Martin et al., 2016). As a result, parent chemicals and their metabolites, as well as colonic metabolites produced by the gut microbiota, are thought to be the primary contributors to flavonoids' biological activity. Numerous studies showing flavonoids' positive effects on metabolic illnesses such as T2D and obesity have boosted interest in scientific flavonoids in current decades (Gentile et al., 2018; Hussain et al., 2016). The natural chemicals and their metabolites have a variety of biological functions, including anti-oxidant and anti-inflammatory capabilities, which provide them a wide range of health-promoting properties (Hussain et al., 2020). Furthermore, flavonoids show the ability to interact directly with proteins (Singh et al., 2017b) interact directly with proteins such as important cellular receptors or signaling pathway components, affecting a variety of functions in many cells and tissues. As a result, they can lower insulin resistance in tissues that are sensitive with tissues via a variety of methods, including insulin signaling pathway modulation.

6.4 DIETARY FLAVONOIDS' MECHANISM OF ACTION

6.4.1 FLAVONOIDS ARE PARP INHIBITORS

They hinder poly(ADP-ribose)-polymerase *in vivo* and *in vitro*. Nicotinamide adenine dinucleotide (NAD) is found in oxidized (NAD+) and reduced (NADH) forms in all live cells. The major action of NAD+ in cells is to modulate cellular redox state by transporting electrons from one process to another one. With this it has a role in other cellular processes (for example, as a substrate for a post-translational modification enzyme) (Ido, 2007). Hyperglycemia lowers NAD+ levels by increasing glucose flow through the polyol pathway. The mechanism is triggered as intercellular glucose levels are high (Gabbay, 1973). During normal glycemia, just 3% of total glucose passes through the polyol pathway. Phosphorylation of maximum glucose to glucose-6 phosphate and hexokinase will occur. Under hyperglycemic conditions, 10 times more glucose enters the polyol route due to hexokinase saturation (Yabe-Nishimura, 1998). The initial and rate-limiting enzyme in the system, AR, converts glucose to sorbitol utilizing NADPH as a cofactor. Sorbitol is converted to fructose by sorbitol dehydrogenase, which uses NAD+ as a cofactor. The osmotic stress caused by sorbitol accumulation, as well as the redox imbalance caused by NADPH and NAD+ depletion, contribute to damage of cells and organ destruction, eventually leading to cataract genesis, neuropathy, and other diabetes consequences (Reddy et al., 2008; Oyama et al., 2006; Srivastava et al., 2005). NAD+ depletion can also be caused by poly(ADP-ribose) polymerase activity. The nuclear enzyme poly(ADP-ribose)-polymerase has shown that the control of many critical physiological functions such as DNA repair, gene transcription, cell cycle, progression, cell death, chromatin function and genomic stability (Pacher et al., 2008). Single-strand DNA breaks, which can be caused by hyperglycemia, are detected and signaled by poly(ADP-ribose polymerase. Poly(ADP-ribose)-polymerase attaches to Deoxy ribo nucleic acid and synthesizes a poly(ADP-ribose) chain as a signal for DNA repair enzymes when single-strand DNA breaks are detected. NAD+ is required as a substrate for the production of these poly(ADP-ribose) monomers. Then the result of overactivation of poly(ADP-ribose)-polymerase, cellular NAD+ reserves are depleted (Moonen et al., 2005). Several studies (Minchenko et al., 2003; Szab'o et al., 2006; Li et al., 2004) have revealed that PARP activation shows a advancing role in pathogenesis of diabetes sequelae such as neuropathy and retinopathy. As a result, this method demonstrates the significance of dietary flavonoids with antidiabetic properties. Flavonoids protect cells from hyperglycemia in a variety of ways. First, flavonoids can prevent NAD+ levels from dropping by inhibiting PARP-1 overactivation. Flavonoids can also limit the function of AR, preventing a further drop in NAD+ and NADH levels. Flavonoids can also protect against the harmful effects of oxidative stress due to their antioxidant (Singh et al., 2017a) qualities. Flavonoids protect cells against harm caused by high glucose levels by a combination of these mechanisms (Figure 6.2).

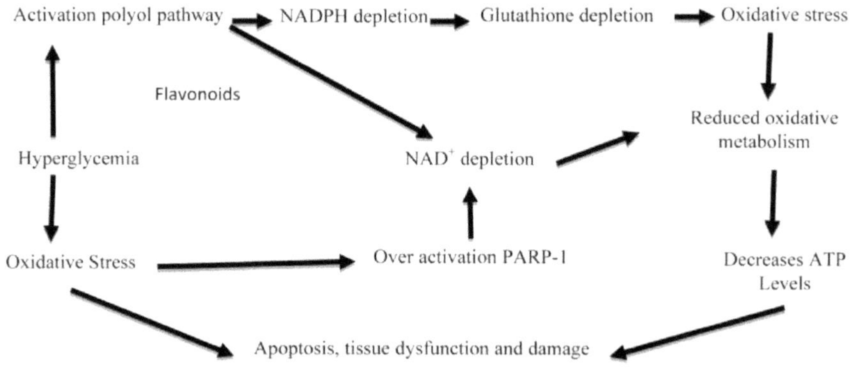

FIGURE 6.2 Mechanism inhibitor PARP-1 by flavonoids.

6.4.2 DIETARY FLAVONOIDS' MECHANISM OF ACTION ON GLUCOSE-TRANSPORTING PROTEINS

The glucose transport through the membrane plasma is major cellular nutrient transporters in eukaryotic cells. A glucose transport protein catalyzes this reaction. Solute carrier 2 (SLC2) genes encode glucose transporter protein, which belongs to the major facilitate of membrane transporters. In human beings, 14 glucose transporter proteins are shown. Then segregated into three groups based on sequence similarity as in Table 6.1.

TABLE 6.1 Classes of Flavonoids on Glucose Transported Proteins

SL. No.	Class	Mechanism of Action Glucose Transported Proteins
1.	First	Glucose transporter protein 1–4 and 14.
2.	Second	Glucose transporter proteins 5, 7, 9, and 11.
3.	Third	Glucose transport proteins 6, 8, 10, 12 and high affinity myo-inositol transporter.

The 14 glucose transport proteins have 500 amino acid residues and are made up of 12 trans-membrane segments, a single N-linked glycosylation site and a rather large central, cytoplasmic linker area. Their N and C termini are present in cytoplasm. They are found in nearly every type of human cell. The glucose rate entrance into the cell is determined by tissue-specific expression of one or more glucose transport proteins.

6.4.2.1 CLASS I – GLUCOSE TRANSPORTER 1 (GLUT1)

It is a major physiological substrate of glucose transporter-1 protein, it also transports galactose, mannose, glucosamine, and reduced ascorbate. The activation of glucose transport-1 mostly caused by cell stressors (Rubin et al., 2003), osmotic stress (Barnes et al., 2002), methylene blue (Louters et al., 2006), and glucose deprivation (Roelofs et al., 2006). The effects of flavonoid-rich medicinal herbs on glucose transporter protein-1 expression levels seem encouraging. Because these compounds dramatically increased glucose absorption by boosting glucose transporter-4 and glucose transporter-1 mRNA expression levels in L6 myotubes, genistein derivatives shown promising results for treating type-2 DM.

6.4.2.2 CLASS I – GLUCOSE TRANSPORTER 2

It is engaged in glucose sensing in pancreatic cells, liver, hypothalamus, and it also activates the glucose-mediated insulin secretion cascade (Burcelin et al., 2000). It may play a role in the development of DM. In diabetic animal models, glucose transporter-2 expression is down regulated in pancreatic cells, but hepatic expression of this glucose transporter is increased. Hesperidin and naringin have been shown to suppress protein expression of glucose transporter-2 in the liver of experimental mice (Jung et al., 2006).

6.4.2.3 CLASS I – GLUCOSE TRANSPORTER 3

Neuron specific glucose transporter is another name for glucose transporter-3. It is involved in the abnormalities in placental function that occur in diabetes pregnancies (Boileau et al., 1995). This protein's primary place of expression is the brain. It is in charge of glucose absorption in the neurons.

6.4.2.4 CLASS I – GLUCOSE TRANSPORTER 4

This is found in tiny sacs which holds fluid in the cytoplasm and can be triggered by insulin and contraction of muscles, causing it to translocate to the plasma membrane, then acts as an exit for glucose uptake. As only 15% of glucose is absorbed by adipose tissue and left out 85% by muscle in fit people, as it plays an important role in glucose sensing (Yang et al., 2005). Several studies have revealed that flavonoids and phenolic compounds play a role in increasing its expression and uptake. In skeletal muscle cells, epigallocatechin gallate (EGCG) shows to improve glucose absorption and stimulate its translocation to the plasma membrane. In adipocytes and skeletal muscle cells, quercetin has antidiabetic effects by upregulating the mRNA expression of Glucose transporter-4 and its translocation to the cell membrane (Jing et al., 2014; Alam et al., 2014). Activation of the insulin receptor causes phosphorylation of the insulin receptor substrate, which causes phosphoinositide 3-kinase to be activated. Phosphoinositide 3-kinase then phosphorylates phosphatidylinositol 4,5-bisphosphate (PIP2) to produce phosphatidylinositol 3,4,5-triphosphate. The phosphoinositide-dependent protein kinase-1 is then activated (PDK1). The activation of protein kinase B by PDK1 leads to the phosphorylation of the Rab GTPase-activating protein AS160, which promotes the translation of GLUT-4 from intracellular storage vesicles to the plasma membrane and improves glucose absorption. Activation of Protein kinase B inhibits glycogen synthase kinase-3, which then phosphorylates and deactivates glycogen synthase. Then it phosphorylates the forehead box protein O1, which inhibits gluconeogenesis in the hepatocyte by deactivating the expression of glucose-6-phosphate and phosphoenol pyruvate carboxy kinase (PEPCK) and suppressing the expression of glucose-6-phosphate and PEPCK (Klover et al., 2004) Adenosine monophosphate-activated protein kinase (AMPK) is a key transporter of cellular metabolism as well as inhibits hepatic glucose synthesis via modulating PEPCK and glucose-6-phosphatase (Mihaylova et al., 2004). AMPK is a serine/threonine kinase that regulates anabolic and catabolic processes as well as maintains the balance of cellular energy. It is activated as a result of cellular energy fluctuations. Unknown components from immunological and metabolic tissues/cells, on the other hand, may work together to regulate this protein kinase (Hardie et al., 1895). When cellular energy needs are high, intracellular ATP is depleted and AMP levels rise. In the case of high cellular energy demand the increase in AMP/ATP is decreased and the AMP level rises. The LKB1: STRAD:MO25 complex is activated as a result of the higher AMP/ATP ratios, which phosphorylates Thr172. The phosphorylation of Thr172 results in the activation of AMPK. The activation of Ca^{++}/calmodulin-dependent protein kinase in response to increased Ca++ levels in the cell cytoplasm is another approach to activate AMPK. By blocking the gluconeogenic enzymes phosphoenol pyruvated carboxy kinase and G-6Pase, as well as translocated GLUT-4 and GLUT-1, active AMPK reduces hepatic glucose synthesis and increases glucose absorption. AMPK as well as stimulates lipid metabolism by lowering malonyl CoA levels in response to acetyl CoA carboxylate (ACC) inhibition and activation of malonyl CoA decarboxylase. The involvement of the CAP/Cb1/TC10 pathway in the translocation of GLUT-4 to the plasma membrane, P13/AKT, has been well characterized. When the insulin receptor is activated, the adaptor protein APS is recruited to the insulin receptor subunit, phosphorylating the Cb1 proto-oncogene. The lipid raft protein flotillin in the plasma membrane interacts with Cb1 and Cb1 associated protein (CAP), causing CrKII to be recruited. CrKII then attaches to C3G, an exchange factor that catalyzes the conversion of GDP to GTP on the lipid-raft protein TC10. The TC10 downstream effectors are known to enhance glucose absorption by translocating GLUT-4 on the plasma membrane. Insulin is released from the pancreas due to increased glucose levels in the bloodstream. Glucose is carried into the cells via GLUT-2. Increased intracellular glucose concentrations cause a rise in ATP generation and an increase in the ATP/ADP ratio. This causes potassium channels on the cell membrane to close. The membrane depolarizes, allowing calcium to flow in a voltage-dependent manner. As increasing Calcium concentration enhances secretion of insulin in the long run (Schnell et al., 2007; Rutter et al., 2001).

6.5 FLAVONOIDS' METABOLITE

The primary enzymes in the bowel, colon, and liver are hydrolyzed and conjugated by flavonoids. Hydrolyzed and conjugated enzymes in the gut convert monomeric flavonoids into O-glucuronides, sulfate ester, and O-methyl ester (Tanveer et al., 2017). The composition of these takes place in 2 stages: first, in the small intestine (first phase), and subsequently into the liver, where phase first ends and phase two begins. The conjugated metabolites are then further processed in the liver to create sulfate and glucuronide derivatives, which are then facilitated and eliminated through bile and urine (Del Rio et al., 2010). Flavonoids that are not absorbed make their way to the colon. After that they are hydrolyzed/fermented by the colonic microbiota (Scalbert et al., 2002). Microbiota hydrolyzes flavonoids glucuronides in the liver into aglycones, which spilt into upcoming into less molecular molecules which are easy in the process of absorption (Spencer et al., 2001).

6.6 FLAVONOIDS WITH ANTIDIABETIC PROPERTIES

6.6.1 FLAVONOLS

Flavonols have an unsaturated carbon ring at carbon 2 and carbon 3 that oxidized at carbon 4 and hydroxylated at carbon 3. Lettuce, berries, kale, etc., are all high in them (Panche et al., 2016).

6.6.1.1 QUERCETIN

The highest abundant flavonoid in human beings is also taken in the form of dietary supplements. It is also known as quercetin dihydrate ($C_{15}H_{10}O_7$). Flowers, apples, tomato seeds, berries, almonds, onions, broccoli, pepper, and shallots are the most common sources (Kelly et al., 2011; Anand et al., 2016). Quercetin serves as a building block for skeleton of other flavonoids such naringenin, rutin, etc. (Hollman et al., 1995). It is involved in a number of biological processes such as glucose homeostasis, insulin sensitization and secretion, glucose utilization in peripheral tissues, and intestinal glucose absorption inhibition (Eid et al., 2017). In the Chinese population, its intake is inversely related to the prevalence of T2DM, implying that it has antidiabetic properties (Yao et al., 2019; Fang et al., 2008). When it is extracted by berries generated an insulin-independent 5′ AMPK pathway in isolated mitochondria, which reduces adenosine diphosphate oxygen consumption via enhancing GLUT 4 translocation and expression.

6.6.1.2 RUTIN

This flavonoid is found in grapes, peaches, limes, lemon, etc. (Kreft et al., 1999). Rutin another name quercetin-3-O-rutinosie, sophorin, and glycosylated quercetin (Huang et al., 2012). The decrease of carbs absorbed from the small intestine, improved uptake of glucose by tissues, inhibit tissue gluconeogenesis, insulin activation from cells and preservation of the islets of Langerhans from degenerative changes are all antidiabetic effects of rutin. As well as inhibits the production of ROS, precursors of advanced glycation end products, sorbitol, and pro-inflammatory cytokines (Ghorbani, 2017). Rutin's hypolipemic and antihyperglycemic effects have been studied in a number of investigations (Stanely et al., 2006). Hepatic enzymes implicated in the gluconeogenic and lipid metabolic processes, such as hexokinase, are activated by rutin (Hao et al., 2012). The flavonoid also reduces protein of urine, blood urea nitrogen, oxidative stress intensity, and fasting blood glucose levels considerably. In diabetic retina, rutin administration revealed anti-apoptotic effects by boosting B-cell lymphoma 2 (Bcl-2) activity and lowering caspase-3 activity (Ola et al., 2015). It comes under the most active flavonoid in lowering FBG, serum lipids and improves tolerance of glucose when compared to further compounds such *Boswellic*, quercetin, and ellagic acid (Jadhav et al., 2012).

6.6.1.3 KAEMPFEROL

Certain fruits and vegetables, such as grapes, apples, onions tea, spinach, etc., have higher content of this flavonoid (Calderon et al., 2011). It is a flavonoid which shows non-toxic property. Kaempferol possesses anti-inflammatory, antioxidant, anti-microbial, and anticancer properties (Chen et al., 2013). Kaempferol extracts by Bauhinia forficate leaves alleviate high blood glucose level and increase glucose absorption as the same way that insulin does (An et al., 2011). Kaempferol has various antidiabetic actions, including enhancing AMP-activated cellular protein expression and activation, decreasing caspase three activities and increase the synthesis of insulin as well as its release from the cells (Zhang et al., 2011). Furthermore it promotes glucose absorption by cells via the protein kinase C and PI3K pathways, as well as the production of additional glucose transporters (Zanatta et al., 2008). So the flavonoid inhibited the genetic expression of PPAR by modulating AMPK activation (Zhang et al., 2011). Furthermore, kaempferol alleviated DM in STZ-induced mice by promoting skeletal muscle glucose metabolism and suppressing hepatic gluconeogenesis (Alkalidy et al., 2018). In another work, kaempferol suppressed RhoA/Rho-kinase mediated pro-inflammatory signaling (i.e., TNF-α, IL-1β, and TGF-1) in NRK-52E and RPTEC cells, reducing diabetic nephropathy (Sharma et al., 2019). Resveratrol is a natural phenol found in grapes, peanuts, etc., that has significant antioxidant and anti-inflammatory properties and may help to prevent diabetes (Hung et al., 2000; Atten et al., 2005).

6.6.1.4 ISORHAMNETIN

Oenanthe javanica, *Hippophae rhamnoides*, and *Ginkgo biloba* all contain an O-methylated bioactive component (Yokozawa et al., 2002). Obesity (Jarouliya & Keservani, 2019) treating properties and reducing the blood sugar level properties are found in this flavonoid. Isorhamnetin's anti-diabetic effect is not confined to lowering blood glucose levels as well as helping to reduce sorbitol accumulation on rat lenses and red blood cells

(Lee et al., 2005). Isorhamnetin glycoside has numerous effects on diabetes, including increasing insulin secretion, the expression of lipid metabolism enzymes, and the expression of endoplasmic reticulum stress indicators, according to an experimental investigation (Rodrigue-Rodriguez et al., 2015; Lee et al., 2008).

6.6.1.5 FISETEN

Some fruits, such as Apples, grapes, strawberries, etc., all have high levels of 3,7,3,'4'-tetrahydroxyflavone (Khan et al., 2013; Arai et al., 2000; Constantin et al., 2010). Fisetin has antidiabetic, anti-inflammatory, and neurotropic properties. When treated with this it lowers the levels of NF-κB p65, Hb A1C, serum nitic oxide and blood glucose in an *in vivo* investigation (Prasath et al., 2013). Fisetin also suppresses the generation of cytokines in monocytes triggered by high glucose levels, which may help to avoid diabetes (Kim et al., 2012). Fisetin's antidiabetic action on hepatic enzymes shows increasing hexokinase activity while decreasing glucose 6 phosphate dehydrogenase and glucose 6-phosphatase activity. It also enhances glucose homeostasis in STZ diabetic mice by inhibiting carbohydrate metabolism enzymes (Prasath et al., 2011). It is shown to slow the progression of diabetic cardiomyopathy in STZ-induced diabetic rats by reducing hyperglycemia or hyperlipidemia-induced oxidative stress, inflammation, and apoptosis (Althunibat et al., 2019). Its therapeutic potential in diabetic neuropathy was demonstrated in preclinical studies by modulating the NF-κB = and Nrf2 signaling pathways (Sandireddy et al., 2019).

6.6.1.6 MORIN

Morin, a flavonoid found largely in historical herbs like Prunus dulcis and Chlorophora, is a natural flavonoid. Glucose tolerance, hyperglycemia, and insulin resistance were significantly improved in animal models after 30 days (Sreedharan et al., 2014). After being treated with morin, diabetic rats had lower levels of lipid peroxides and antioxidants (Sendrayaperumal et al., 2014). Morin effectively reduced inflammatory cytokines such as IL-6 and TNF-α, demonstrating its anti-inflammatory effects (Abuohashish et al., 2014). It restores leptin sensitivity in animal study models, and hepatic insulin reduces liver fat buildup and hyperlipidemia (Wang et al., 2013). Morin has a variety of effects on hepatic enzymes, lowering the activity of G6Pase and Fructose-1,6-diphosphatase while increasing activity of hexokinase and G6PD (Vanitha et al., 2014).

6.6.2 *FLAVANONES*

They are compounds with an oxidized, saturated carbon ring. Citrus fruits contain them as they are recognized by its antioxidant and show property as it protects cells from damage caused by free radicals.

6.6.2.1 HESPERIDIN

The citrus fruits, for example, lemons, limes as well as many more all contain 5,7,3′-trihydroxy-4′-methoxyflavanone, a saturated oxidized organic compound combined with sugar portion of a glycoside (Parhiz et al., 2015). Hesperidin and its glycoside show some pharmacological activities such as vascularity, neuroprotectivity, anti-allergic, anti-carcinogenic properties in addition to diabetes (Visnagri et al., 2014). Its supporting diet serves to regulate the activity of gluconeogenesis and glycolytic hepatic enzymes, as well as improve hyperglycemia in db/db C56BL6 mice (Jung et al., 2004). As seen this flavonoid contains a highly effective hardware in db/db mice, such as raising triglyceride fecal excretion and blocking fat metabolizing enzymes, which improves lipid metabolism (Jung et al., 2006). As increasing glucose transporter type-4 translocation and PPAR, it significantly reduces blood glucose levels (Agrawal et al., 2014). Hesperidin and hesperetin, when administered combined, have differing actions on lipid and glucose destruction and have lipid-reducing properties. As seen in Streptozotocin-induced DM rats, esperidin controls the Klotho/FGF-23 pathway, which has beneficial effects on diabetic damage in kidney and liver (Dokumacioglu et al., 2014).

6.6.2.2 NARINGENIN

Argumes containing fruits, such as oranges, tomatoes, and grapefruits, are high in 5,7,4′-Trihydroxyflavanone, a saturated oxidized glycone (Hasanein et al., 2014). It contains a variety of biological properties, including antioxidant, antidiabetic, anti-inflammatory, anticancer, and antimutagenic properties (Patel et al., 2018). The antidiabetic and anti-obesity properties of naringenin and its glycoside are similar. This flavonoid (25 mg/kg) inhibited α-glucosidase activity in diabetic rats, delaying carbohydrate absorption and lowering postprandial blood glucose levels (Priscilla et al., 2018). Naringin reduces hyperglycemia in type-2 diabetic mice by upregulating LUT4 and regulating the expression of hepatic enzymes involved in gluconeogenesis and glycolysis (Pu et al., 2012; Singh et al., 2018). Naringenin was discovered to limit glucose uptake in an everted rat intestinal sleeve by blocking sodium-glucose co-transporters in the gut (Li et al., 2006). Naringenin's antidiabetic effects in diabetic rats were characterized by antioxidant and stops the process of apoptosis activities, indicating that naringenin has the ability to prevent retinal damage and limit neurodegeneration in diabetic retinopathy (Zhang et al., 2017).

6.6.2.3 ERIODICTYOL

Eriodictyol, found in lemons, has been shown to be effective in preventing obesity and diabetes (Zhang et al., 2012). Eriodictyol has recently been found as a new insulin secretagogue that acts exclusively as a glucose-dependent insulin tropic via the cAMP/PKA route *in-vitro* and *in-vivo* (Hameed et al., 2018). Furthermore, eriodictyol administration can significantly lower oxidative stress in diabetic rats (Miyake et al., 1998). Eriodictyol therapy increased Peroxisome Proliferator-Activated Receptor 2 and the adipocyte-specific fatty acid protein binding expression (Bucolo et al., 2012). As well as its therapy decreased diabetes-related peroxidation of lipids considerably (Shukla et al., 2014). Eriodictyol was recently discovered to protect rat retinal ganglia cells (RGC)-5 from high glucose-induced oxidative stress, pain, swelling, and cell death with activating the Nrf2/HO-1 signaling pathway (Lv et al., 2019).

6.6.3 FLAVONES

When compared to flavonols, flavones have an unsaturated carbon ring at C2-3 and ketone group at carbon four position, but they do not have the hydroxylation at carbon 3. Flowers, leaves, and fruits are all known to manufacture them.

6.6.3.1 APIGENIN

The phytoestrogen aglycoge 5,7,4′-Trihydroxyflavone is predominantly present in almonds, spinach, oranges, celery, etc. (Panda et al., 2007). Apigenin therapy inhibited NF-κB activation, which prevented induced apoptosis (Zang et al., 2006). Apigenin increases AMPK phosphorylation in HepG2 hepatocytes. Apigenin's impact is 200 times stronger than metformin's. The increased translocation of GLUT4 in response to apigenin therapy suggested that it has a hypoglycemic effect (Hossain et al., 2014). Its treatment reduced cellular DNA damage, reactive oxygen species (ROS) formation, protein carboxylation, lipid peroxidation, and reversed cell apoptosis in STZ-induced rats, improving oxidative damage of pancreatic cells (Wang et al., 2017).

6.6.3.1.1 Luteolin

It is a widely known compound due to its anti-inflammatory and anti-diabetic activity (Tuorkey et al., 2016). Fruits and vegetables rich in it include leaves of onion, carrot, pepper, parsley, carrots, apple peel (Miean et al., 2001). In primary mouse adipose cells, it was found to enhance the insulin action and increase the expression of PPAR target genes (Ding et al., 2014). It enhances secretion of insulin in uric acid-damaged pancreatic cells by lowering micro-autologous fat transplantation, a trans activator of the insulin gene via the Nuclear Factor kappa-B signaling pathway (Ding et al., 2014). Lutein (LU) and luteolin-7-O-glucoside shows reduced effect of high blood sugar level that improve blood glucose, Glycated hemoglobin, insulin, and Hemostatic model assessment of insulin resistance levels while inhibiting fat production (Zang et al., 2016). By modifying M1-like macrophage polarization in adipose tissue, luteolin reduces insulin resistance and adipose tissue inflammation of adipose tissue (Baek et al., 2019).

6.6.3.2 TANGERETIN

This flavonoid is found in citrus fruits example tangerine peel, mandarins (Cirmi et al., 2016). In HFD-induced obese mice, tangeretin (200 mg/kg) lowered blood glucose, total cholesterol and weight of body while also regulating adipocytokines such leptin, interleukin-6, and adiponectin (Kim et al., 2012). Tangerine promotes insulin sensitizing factor secretion while lowering insulin resistance factor secretion in 3t3-l1 preadipocytes (Miyata et al., 2011). Furthermore this flavonoid inhibits streptozocin-induced programed cell death in ins-1 cells through inhibiting nuclear factor kappa-B signaling (Liu et al., 2019).

6.6.3.2.1 Chrysin

A variety of natural compounds such as honey, medicinal plants, propolis, pollen by bees as well as *Passifora caerulea* L. and *Tilia tomentosa* have a large amount of Chrysin components (Tran et al., 2012). Due to fast excretion and metabolism, this flavonoid has a lower bioavailability than Apigenin (Mehdi et al., 2018). In STZ-induced rats, chysin administration resulted in an increase in glucose, Malonaldehyde, Triglycerides, Total carbon, and Low density lipoprotein cholesterol, as well as a decrease in High density lipoprotein cholesterol, total protein, superoxide dismutase (SOD), CAT, and GST (Samarghandian et al., 2016). Due to lower levels of pro-inflammatory cytokines in blood, it reduces the chances of nerve damage that occur if you have high blood sugar level in HFD/STZ-induced diabetic mice (Li et al., 2014). In diabetes rats, its therapy lowers the peroxidation of lipids, raises the level of glucose, and lowers blood glucose levels (Sirovina et al., 2013). According to the research, chrysin contains antidiabetic and anti-hypertensive properties (El-Bassosy et al., 2013).

6.6.3.2.2 Wogonin

The flavonoid 5,7-Dihydroxy-8-methoxyflavone came from the roots of *Scutellaria baicalensis* and utilized in the East Asian traditional system of medicines (Huyuh et al., 2017). Anti-inflammatory, neuroprotective, antiviral, antibacterial, and antioxidant effects have been linked to it (Tai et al., 2005). It activates AMP-activated protein kinase and peroxisome proliferator activated receptor that has various favorable impacts on insulin sensitivity, blood glucose, and lipid metabolism (Bak et al., 2014). Its prior treatment reduces vascular inflammatory effects in high glucose-induced vascular inflammation (Ku et al., 2015). Its anti-oxidative and anti-inflammatory properties were discovered in a preclinical investigation to help prevent diabetic cardiomyopathy (Khan et al., 2016).

6.6.3.2.3 Diosmin

Scrophularia nodosa L. was the first plant from which a naturally occurring flavonoid glycoside was isolated in 1925. It can be extracted from a variety of plants or by dehydrogenating the flavanone glycoside hesperidin (Campanero et al., 2010). In type-1 diabetic individuals, diosmin treatment resulted in a drop in Glycated Hemoglobin and enhancement in glutathione peroxidase (GPx) (Manuel et al., 1999).

6.6.3.3 BAICALEIN

Scutellaria baicalensis roots and *Oroxylum indicum* L. fruits contain the 5,6,7-dihydroxyflavone aglycone (Kim et al., 2001). It has a broad extent of anti-inflammatory, anti-neurodegenerative, and anti-cardiovascular estates (Lapchak et al., 2007). Baicalein treatment resulted in significant reductions in Fasting Plasma Glucose, Glycated Hemoglobin, administration of food particles, and weight of body in diabetic rats (Ahad et al., 2014). Tumor Necrosis Factor level levels, advanced glycation end-products, and nuclear factor kappa B activation were all reported to be reduced after treatment with baicalein (El-Bioassy et al., 2014). The mechanism of action of baicalein appears to be upregulation of AMPK signaling pathways, they reduce resistancy to insulin and pain (Pu et al., 2012). Baicalein also protects diabetic cardiomyopathy rats from oxidative stress and inflammation by regulating PI3K/AKT signaling (Ma et al., 2018). The regulation of HMGB1/TLR4/NF-κB signaling by baicalein reduces hepatic inflammation in diabetic db/db mice (Yin et al., 2018).

6.6.4 ISOFLAVONES

They are primarily seen in legumes, soybeans, and microorganisms. Isoflavones are mostly found in genistein and daidzein. As seen, they have been demonstrated to have an antidiabetic impact with boosting pancreatic beta cell insulin output.

6.6.4.1 GENISTEIN

A nature found flavonoid taking place soy isoflavone, 5,7,4′-trihysroxyisoflavone in higher quantity in soybean products and herbs originated in China (Crozier et al., 2009). The antidiabetic benefits of genistein are due to an increase in plants super molecular oil (Mezei et al., 2003). In type-1 diabetes rats, genistein administration improved insulin levels and metabolism of glucose (Yang et al., 2011). Genistein improves hyperglycemia via increasing the cAMP/PKA signaling pathways, according to an *in vivo* research (Babu et al., 2012). The addition of genistein (0.02% in the diet) to the diets of non-obese diabetic rats delayed starting of high blood sugar level and enhanced the balance of insulin and glucagon to maintain blood sugar by preserving cell functioning (Choi et al., 2018). According to a transcriptome study, genistein can affect the regulation of hypothalamus circadian rhythms, and this may be a new target for diabetes and obesity treatment. Furthermore, it contains anti-inflammatory, anti-neuropathic pain, and anti-oxidant properties (Liggins et al., 2000).

6.6.4.2 DAIDZEIN

The phytoestrogen 7,4′-dihydroxyisoflavone is mostly extracted through nuts, vegetables, soyabeans (Ae Park et al., 2006). It improves lipid and glucose metabolism, which has an antidiabetic effect (Cheong et al., 2014). Daidzein offers therapeutic promise for T2DM-related altered glucose, lipid metabolism, and vascular inflammation (Das et al., 2018). Furthermore, daidzein therapy reduces blood glucose, total cholesterol, and AMP-activated protein kinase phosphorylation in the gastrocnemius muscle. In comparison to the control group, pure synthetic form fiercely decreased plasma total cholesterol and blood glucose levels in hamsters (Song et al., 2003).

6.6.5 ANTHOCYANINS

Anthocyanin is a water soluble, unoxidized, unsaturated flavonoid found in abundance in flowers and fruits. When compared to other flavonoids, this flavonoid has a larger dietary intake. Anthocyanins have been shown to have antidiabetic properties in animal models and cell lines in several investigations.

6.6.5.1 CYANIDIN

A flavonoid found in diets originated from plants, harvest grains, fruits, vegetables (Akkarachiyasti et al., 2010), obstructs intestinal-glucosidase and pancreatic-amylase and hence has an antidiabetic effect (Nizamutidinova et al., 2009). Cyanidin restored degenerative alterations in β-cells in STZ-induced diabetic rats by reducing pancreatic apoptosis and enhancing phosphorylation of insulin receptors (Nasri et al., 2011). C3G, a common anthocyanidin in the foodstuffs, enhanced antioxidant status and protected hepatocytes from oxidative stress in the presence of high hyperglycemia induced harm (HG) (Zhu et al., 2012).

6.6.5.2 DELPHINIDIN

Delphinidin, a flavonoid found in ripened fruits, sweet potatoes, red cabbage, tomato, eggplant, etc. (Ko et al., 2015), is abundant in pigmented vegetables and fruits. Anti-inflammatory, antioxidant, and anti-angiogenic properties have been discovered (Lim et al., 2017). Delphinidin was shown to protect endothelial cell function damage associated with high blood glucose level in an *in vivo* investigation (Bertuglia et al., 1995). In high fat-fed mice, delphinidin and cyanidin were found to decrease swelling, pain, and control redox signaling tracts by alleviating insulin resistance (Daveri et al., 2018). The ability of delphinidin to inhibit glucose uptake in mice jejunal tissue and human intestinal cell lines via free fatty acid receptor 1 is responsible for its antidiabetic actions (Hidalgo et al., 2017).

6.6.5.3 PELARGONIDIN

Pelargonidin, a flavonoid seen in berries as it is plentiful (Mazza et al., 2006). Pelargonidin therapy helps to lower the blood sugar level that is increased and a condition in antioxidant level is low (Roy et al., 2008). It shows lower development of thiobarbituric acid reactive substances and the levels of antioxidant defense enzymes in diabetic rats (Mirshekar et al., 2001). Pelargonidin and its aglycoge, pelargonidin-3-galactoside, increased blood glucose release in an *in vitro* research (Jayprakasam et al., 2005).

6.6.6 PROVOCATION UTILIZING FLAVONOIDS

They have shown to be promising options for reducing diabetes etiology with its further effects. It reducing high blood sugar level modulating effects diminish apoptosis and insulin resistance and increasing productivity of insulin and Glucose Transporter 4 translocation.

6.6.6.1 AN APPROXIMATE DEVOURING EXTENT OF FLAVONOIDS

Flavonoids obtained through fruits and vegetables are ingested in small amounts. Furthermore, vegetables and fruits contain a combination of plant metabolites secondary in nature, in addition to flavonoids. As a result, converting the mixture within an elementary cleansed nutritive additive amount is difficult (Ross et al., 2002; Duthie et al., 2003). Efforts have been done around the world to determine an equal human diet food utilization level of flavonoids, however, the estimation methodologies utilized are unproven (Kuhnau et al., 1976). The estimated consumed level of flavonoids by quercetin, myricetin, kaempferol, and luteolin was 26 mg/d, according to a study of 805 males aged 65–84 years in the United States (Sampson et al., 2002). The highest roots of consumption were apples, tea, and onions. In comparison to the United States data (50 g/day), one more study was done in the Netherlands found a two-fold greater level of flavonoids ingested in grown up people (Hertog et al., 1999). In addition, two Dutch investigations found that flavonoids were consumed at a rate of 23 mg per day and 26 mg per day, respectively (Arts et al., 2001; Boker et al., 2002). Daily food habits intake, geography areas of residence, status of economic and social wellbeing, processing of food and preparation method, flavonoids solubility, and ethnicity of the population are all factors that influence flavonoids consumption levels. In Japan, for example, soy-containing foods are widely eaten, and this tells, isoflavone intake is more than that of other flavonoids subclasses. According to one study, orange juice has 81–200 mg/L of soluble flavanones, in comparison with 206–644 mg/L in the cloud, implying this flavonoids concentrations are affected by distilling and depot (Gil Izquierdo et al., 2001). As to the variability of flavonoids' molecular structure and the scarcity of data on their bioavailability, no suggested dosage has been published to yet. Although much progress has been made in studying flavonoids bioavailability, the difficulty of overcoming circumstances such as the rate of passive diffusion of molecules by membrane, solubility ability of all substances, removal of particles and metabolic alteration remains unsolved. The bioavailability of flavanols is very much improved by buckling on absorptive areas, boosting metabolic stability, and increasing intestinal absorption.

6.6.7 FLAVONOIDS' POTENTIAL NEGATIVE SIDE EFFECTS

Flavonoids showed probable genotoxicity and mutagenicity when taken at greater quantities (ranging starting 12.1 nmol to 225.0 nmol) in bacterial and vertebrate animals experiments employing the Ames test (Macgergor et al., 1979). It may also affect amino acid and medication metabolism, as well as the activity of critical metabolizing enzymes (Thilakarathna et al., 2013). By modifying base-pair substitution and frame-shift mutation, quercetin, which is a common flavonol in the human diet, was found to be (Suzuki et al., 1991). As concealment nuclei of liver rats treated with morin and naringenin had a higher level of ROS, such as hydroxyl radicals, which cause DNA damage (Skibola et al., 2000). Flavonoids also operate as a topoisomerase II inhibitor, which makes them cytotoxic. At very less concentrations (10 M), genistein and quercetin are recognized as topoisomerase II inhibitors, accumulating cleavable complexes reported with patients with subsequent leukemia (Dassonneville et al., 1998). Thyroid peroxidation is irreversibly suppressed by genistein, naringenin, kaempferol, and daidzein, which decrease thyroxine production (Doerga et al., 1995). Although no data on the long-term consequences of increasing flavonoid consumption is known, an Asian food including 68 mg flavonol and 20–240 mg isoflavone may enhance the functions of thyroids, reduction in chances of breast cancer mortality, and produce no significant health effects (Egert et al., 2011). The amounts required for many flavonoids to cause mutagenesis and cytotoxic side effects are unlikely to be found in food, although supplementation may result in a higher hazardous level. An example, the suggested dose required of quercetin supplements is between 500 and 1,000 mg per day, which is 20 times higher than what a vegetarian diet might provide (Sahu et al., 1993).

6.6.8 COULD FLAVONOIDS WORK TOGETHER TO CREATE A SYNERGISTIC EFFECT?

While the amount of flavonoids ingested is important for establishing good effects as well as avoiding negative ones, the tables below include certain biologically active water-soluble plant compounds that enhance many progressions which improve diabetes etiology. As found number of DM-associated passageways that are further enhanced by consuming various flavonoids can be used to determine better activity. Four pathways are activated when baicalein is administered: The nuclear factor kappa-B pathway and synthesis of fatty acids are suppressed, while hexokinase activity in the liver area is activated, and death of cell is prevented. Three distinct routes are activated by quercetin: It improves GLUT 4 translocation, lowers lipid peroxidation, and suppresses tyrosine kinase activity. Apoptosis in α-cells can be avoided by giving them cyanidin, kaempferol, or baicalein. The enzymes glucosidase and α-amylase are inhibited by rutin or cyanidin, which limit the absorption of carbohydrates (Singh et al., 2017b) in the areas of GIT Track (small intestine). If swallowing a fusion of different flavonoids that accompany one another by stimulating some more distinct pathways boost their favorable effects on diabetes? The injection of baicalein and quercetin, for example, commences in beneficial effects on diabetes through six primary passageways: Peroximos, AMPK, tyrosine kinase, and NF-κB pathways; glucose transporter; hepatic enzymes; beta cell apoptosis; Peroxisome proliferator-activated receptors; AMP-activated protein kinase; tyrosine kinase; and nuclear factor kappa B pathways. Overactivation of these pathways may be prevented as a result of this proposed combination, while the necessary action to treat diabetes may be taken. These are only suggestions at this moment, and they must be backed up by study. Flavonoids-flavonoids interactions are poorly understood at the moment. Furthermore, other flavonoids had the opposite effect on the same route, resulting in diabetes improvement in both cases. Fisetin, as an example, has an inhibitory outcome on glucose 6 phosphate dehydrogenase, whereas morin has a stimulatory effect, and the literature claims that both benefit diabetes. To understand why this effect occurs, more research is needed—this is due to many regions of macromolecules binding in specific manner, extent at which drug or metabolite enters systemic circulation, tissue exposure or circulation concentrations. The PPAR and NF-κB pathways showed the same pattern but with other flavonoids.

6.6.9 METFORMIN IN COMBINATION OF FLAVONOIDS

It is used to treat high blood sugar level which came from the French lilac that is used to treat diabetes and has a well-established safety profile (Thomas et al., 2017). When metformin and anthocyanins isolated from blueberry bush were evaluated on blood sugar level, blueberry extract (595 mg/g total anthocyanins) resulted in a 33% to 51% drop, compared to a 27% reduction with 1,1-dimethylbiguanide hydrochloride (Grace et al., 2009). According to another study, quercetin activates an insulin-dependent amp-activated protein kinase pathway, which is similar to metformin activation. Further, research investigating in effect of 1,1-dimethylbiguanide hydrochloride and flavone co-treatment on patients that are having illness of breast cancer found a remarkable reduction in cell feasibility and an accelerated the apoptosis process (Zheng et al., 2018).

6.7 FUTURE THOUGHTS

Flavonoids, which are plentiful in fruits and vegetables, have a generally positive impact on diabetes. Consuming fruits and vegetables may help to reduce the level of blood sugar and as well as reduce the chances of newly forming diabetes. In general, combining them with other phytochemicals may boost their blood sugar level lowering effects, although some more amount of future work is required to back up the potential method of lowering blood sugar levels.

6.8 CONCLUSION

In the human era, the use of flavonoids provides an allover complete activity related to biological system such as chances to reduce high blood sugar. The purpose flavonoids in the treatment of diabetes are highly effective with many advantageous activities. The studies proposed have described effective roles of flavonoids in treating diabetes, CVS disorders and other metabolic diseases. The mechanism involved in use of dietary flavonoids is classification of polyphenolic structures as in these flavonoids are divided on the basis of functional group present on it. While seeing the mode of action associated with the dietary flavonoids in high blood sugar level disease could provide DM would provide a sight on the areas of development of drugs and other areas which are involved in therapeutic benefits of the human body. As day by day, the chances of DM is increasing in infants, children, adults, and older age people therefore there is a great need for promoting the assets of nutritional flavonoids in treatment of high blood sugar level and other related disorders. The chapter empowers by showing that the disease can be cured by natural products such as flavonoids and its corresponding nutritional flavonoids in the form of antidiabetics as they will reduce the chances of adverse effects in contact to the synthetic ones. Therefore, the knowledge for more effective and safer flavonoids or reducing blood sugar level compounds has been an active zone in research and development. As they show the different treatment strategies in the production of safer antidiabetic drugs.

KEYWORDS

- **adenosine monophosphate-activated protein kinase**
- **cancer**
- **diabetes mellitus**
- **flavonoids**
- **gestational diabetes**
- **lutein**
- **PARP inhibitors**
- **polyphenolic**
- **retinal ganglia cells**

REFERENCES

Abuohashish, H. M., Al-Rejaie, S. S., Al-Hosaini, K. A., Parmar, M. Y., & Ahmed, M. M., (2013). Alleviating effects of morin against experimentally-induced diabetic osteopenia. *Diabetol. Metab. Syndr.*, *5*, 5. [CrossRef].

Ae Park, S., Choi, M. S., Cho, S. Y., Seo, J. S., Jung, U. J., Kim, M. J., Sung, M. K., Park, Y. B., & Lee, M. K., (2006). Genistein and daidzein modulate hepatic glucose and lipid regulating enzyme activities in C57BL/KsJ-db/db mice. *Life Sci.*, *79*, 1207–1213. [CrossRef] [PubMed].

Agrawal, Y. O., Sharma, P. K., Shrivastava, B., Ojha, S., Upadhya, H. M., Arya, D. S., & Goyal, S. N., (2014). Hesperidin produces cardioprotective activity via PPAR-gamma pathway in ischemic heart disease model in diabetic rats. *PLoS One, 9*, e111212. [CrossRef] [PubMed].

Ahad, A., Mujeeb, M., Ahsan, H., & Siddiqui, W. A., (2014). Prophylactic effect of baicalein against renal dysfunction in type 2 diabetic rats. *Bio-chimie, 106*, 101–110. [CrossRef].

Akkarachiyasit, S., Charoenlertkul, P., Yibchok-Anun, S., & Adisakwattana, S., (2010). Inhibitory activities of cyaniding and its glycosides and synergistic effect with acarbose against intestinal alpha-glucosidase and pancreatic alpha-amylase. *Int. J. Mol. Sci., 11*, 3387–3396. [CrossRef] [PubMed].

Alam, M. M., Meerza, D., & Naseem, I., (2014). Protective effect of quercetin on hyperglycemia, oxidative stress and DNA damage in alloxan induced type 2 diabetic mice. *Life Sciences, 109*, 8–14.

Al-Dosari, D. I., Ahmed, M. M., Al-Rejaie, S. S., Alhomida, A. S., & Ola, M. S., (2017). Flavonoid naringenin attenuates oxidative stress, apoptosis and improves neurotrophic effects in the diabetic rat retina. *Nutrients, 9*, 1161. [CrossRef].

Alkhalidy, H., Moore, W., Wang, Y., Luo, J., McMillan, R. P., Zhen, W., Zhou, K., & Liu, D., (2018). The flavonoid kaempferol ameliorates streptozotocin-induced diabetes by suppressing hepatic glucose production. *Molecules, 23*, 2338. [CrossRef] [PubMed].

Althunibat, O. Y., Al Hroob, A. M., Abukhalil, M. H., Germoush, M. O., Bin-Jumah, M., & Mahmoud, A. M., (2019). Fisetin ameliorates oxidative stress, inflammation and apoptosis in diabetic cardiomyopathy. *Life Sci., 221*, 83–92. [CrossRef].

An, G., Gallegos, J., & Morris, M. E., (2011). The bioflavonoid kaempferol is an Abcg2 substrate and inhibits Abcg2-mediated quercetin effect. *Drug Metab. Dispos., 39*, 426–432. [CrossRef] [PubMed].

Anand, D. A. V., Arulmoli, R., & Parasuraman, S., (2016). Overviews of biological importance of quercetin: A bioactive flavonoid. *Pharmacogn. Rev., 10*, 84–89. [CrossRef] [PubMed].

Arai, Y., Watanabe, S., Kimira, M., Shimoi, K., Mochizuki, R., & Kinae, N., (2000). Dietary intakes of flavonols, flavones and isoflavones by Japanese women and the inverse correlation between quercetin intake and plasma LDL cholesterol concentration. *J. Nutr., 130*, 2243–2250. [CrossRef] [PubMed].

Arts, I. C., Hollman, P. C., Feskens, E. J., Bueno De, M. H. B., & Kromhout, D., (2001). Catechin intake and associated dietary and lifestyle factors in a representative sample of Dutch men and women. *Eur. J. Clin. Nutr., 55*, 76–81. [CrossRef] [PubMed].

Atten, M. J., Godoy-Romero, E., Attar, B. M., Milson, T., Zopel, M., & Holian, O., (2005). Resveratrol regulates cellular PKC alpha and delta to inhibit growth and induce apoptosis in gastric cancer cells. *Investing. New Drugs, 23*, 111–119. [CrossRef].

Ayabar, M. J., Sanchez, R. A. N., Grav, A., & Sanchez, S. S., (2001). Hyperglycemic effect of the water extract of *Smallantus sonchifolius* (yacon) leaves in normal and diabetic rats. *J. Ethnopharmacol., 74*, 125–132.

Ayyanar, M., Sankarasivaraman, K., & Ignacimuthu, S., (2008). Traditional herbal medicines used for the treatment of diabetes among two major tribal groups in South Tamil Nadu, India. *Ethnobotanical Leaflets, 12*, 276–280.

Babu, P. V., Si, H., Fu, Z., Zhen, W., & Liu, D., (2012). Genistein prevents hyperglycemia-induced monocyte adhesion to human aortic endothelial cells through preservation of the cAMP signaling pathway and ameliorates vascular inflammation in obese diabetic mice. *J. Nutr., 142*, 724–730. [CrossRef].

Baek, Y., Lee, M. N., Wu, D., & Pae, M., (2019). Luteolin improves insulin resistance in postmenopausal obese mice by altering macrophage polarization. *Curr. Dev. Nutr., 13*, FS12–FS19. [CrossRef].

Bak, E. J., Kim, J., Choi, Y. H., Kim, J. H., Lee, D. E., Woo, G. H., & Yoo, Y. J., (2014). Wogonin ameliorates hyperglycemia and dyslipidemia via PPARalpha activation in db/db mice. *Clin. Nutr., 33*, 156–163. [CrossRef].

Barnes, K., Ingram, J. C., Porras, O. H., Barros, L. F., Hudson, E. R., Fryer, L. G., et al., (2002). Activation of GLUT1 by metabolic and osmotic stress: Potential involvement of AMP-activated protein kinase (AMPK). *Journal of Cell Science, 115*, 2433–2442.

Bertuglia, S., Malandrino, S., & Colantuoni, A., (1995). Effects of the natural flavonoid delphinidin on diabetic microangiopathy. *Arzneimittelforschung, 45*, 481–485. [PubMed].

Boileau, P., Mrejen, C., Girard, J., & Hauguel-De, M. S., (1995). Expression of GLUT3 placental glucose transporter in diabetic rats. *The Journal of Clinical Investigation, 96*, 309–317.

Boker, K. L., Van Der, S. Y. T., De Kleijn, M. J., Jacques, P. F., Grobbee, D. E., & Peeters, P. H., (2002). Intake of dietary phytoestrogens by Dutch women. *J. Nutr., 132*, 1319. [CrossRef] [PubMed].

Bucolo, C., Leggio, G. M., Drago, F., & Salomone, S., (2012). Eriodictyol prevents early retinal and plasma abnormalities in streptozotocin-induced diabetic rats. *Biochem. Pharmacol., 84*, 88–92. [CrossRef] [PubMed].

Burcelin, R., Dolci, W., & Thorens, B., (2000). Glucose sensing by the hepatoportal sensor is GLUT2-dependent: *In Vivo* analysis in GLUT2-null mice. *Diabetes, 49*, 1643–1648.

Calderon-Montano, J. M., Burgos-Moron, E., Perez-Guerrero, C., & Lopez-Lazaro, M., (2011). A review on the dietary flavonoid kaempferol. *Mini Rev. Med. Chem., 11*, 298–344. [CrossRef].

Campanero, M. A., Escolar, M., Perez, G., Garcia-Quetglas, E., Sadaba, B., & Azanza, J. R., (2010). Simultaneous determination of diosmin and diosmetin in human plasma by ion trap liquid chromatography-atmospheric pressure chemical ionization tandem mass spectrometry: Application to a clinical pharmacokinetic study. *J. Pharm. Biomed. Anal., 51*, 875–881. [CrossRef].

Chen, A. Y., & Chen, Y. C., (2013). A review of the dietary flavonoid, kaempferol on human health and cancer chemoprevention. *Food Chem., 138*, 2099–2107. [CrossRef].

Cheong, S. H., Furuhashi, K., Ito, K., Nagaoka, M., Yonezawa, T., Miura, Y., & Yagasaki, K., (2014). Daidzein promotes glucose uptake through glucose transporter 4 translocation to plasma membrane in L6 myocytes and improves glucose homeostasis in type 2 diabetic model mice. *J. Nutr. Biochem., 25*, 136–143. [CrossRef] [PubMed].

Choi, M. S., Jung, U. J., Yeo, J., Kim, M. J., & Lee, M. K., (2018). Genistein and daidzein prevent diabetes onset by elevating insulin level and altering hepatic gluconeogenic and lipogenic enzyme activities in non-obese diabetic (NOD) mice. *Diabetes Metab. Res. Rev., 24*, 74–81. [CrossRef].

Cirmi, S., Ferlazzo, N., Lombardo, G. E., Maugeri, A., Calapai, G., Gangemi, S., & Navarra, M., (2016). Chemopreventive agents and inhibitors of cancer hallmarks: May citrus over new perspectives? *Nutrients, 8*, 698. [CrossRef].

Constantin, R. P., Constantin, J., Pagadigorria, C. L., Ishii-Iwamoto, E. L., Bracht, A., Ono Mde, K., & Yamamoto, N. S., (2010). The actions of fisetin on glucose metabolism in the rat liver. *Cell Biochem. Funct., 28*, 149–158. [CrossRef].

Crozier, A., Del Rio, D., & Clifford, M. N., (2010). Bioavailability of dietary flavonoids and phenolic compounds. *Mol. Asp. Med., 31*, 446–467. [CrossRef].
Crozier, A., Jaganath, I. B., & Clifford, M. N., (2009). Dietary phenolics: Chemistry, bioavailability and effects on health. Nat. *Prod. Rep., 26*, 1001–1043. [CrossRef].
Das, D., Sarkar, S., Bordoloi, J., Wann, S. B., Kalita, J., & Manna, P., (2018). Daidzein, its effects on impaired glucose and lipid metabolism and vascular inflammation associated with type 2 diabetes. *Biofactors, 44*, 407–417. [PubMed].
Dassonneville, L., & Bailly, C., (1998). Chromosomal translocations and secondary leukemias induced by topoisomerase II inhibitors. *Bull. Cancer, 85*, 254–261. [PubMed].
Daveri, E., Cremonini, E., Mastaloudis, A., Hester, S. N., Wood, S. M., Waterhouse, A. L., Anderson, M., et al., (2018). Cyanidin and delphinidin modulate inflammation and altered redox signaling improving insulin resistance in high fat-fed mice. *Redox Biol., 18*, 16–24. [CrossRef] [PubMed].
Del Rio, D., Calani, L., Scazzina, F., Jechiu, L., Cordero, C., & Brighenti, F., (2010). Bioavailability of catechins from ready-to-drink tea. *Nutrition, 26*, 528–533. [CrossRef].
Del Rio, D., Rodriguez-Mateos, A., Spencer, J. P., Tognolini, M., Borges, G., & Crozier, A., (2013). Dietary (poly)phenolics in human health: Structures, bioavailability, and evidence of protective effects against chronic diseases. *Antioxid. Redox Signal., 18*, 1818–1892. [CrossRef].
Ding, L., Jin, D., & Chen, X., (2010). Luteolin enhances insulin sensitivity via activation of PPARgamma transcriptional activity in adipocytes. *J. Nutr. Biochem., 21*, 941–947. [CrossRef].
Ding, Y., Shi, X., Shuai, X., Xu, Y., Liu, Y., Liang, X., & Su, D., (2014). Luteolin prevents uric acid-induced pancreatic beta-cell dysfunction. *J. Biomed. Res., 28*, 292–298.
Doerge, R., & Divi, D. L., (1995). Porphyrin pi-cation and protein radicals in peroxidase catalysis and inhibition by anti-thyroid chemicals. *Xenobiot. Fate Foreign Compd. Biol. Syst., 25*, 761–767. [CrossRef] [PubMed].
Dokumacioglu, E., Iskender, H., & Musmul, A., (2019). Effect of hesperidin treatment on Klotho/FGF-23 pathway in rats with experimentally-induced diabetes. *Biomed. Pharmacother., 109*, 1206–1210. [CrossRef] [PubMed].
Duthie, G. G., Gardner, P. T., & Kyle, J. A., (2003). Plant polyphenols: Are they the new magic bullet? *Proc. Nutr. Soc., 62*, 599–603. [CrossRef].
Egert, S., & Rimbach, G., (2011). Which sources of flavonoids: Complex diets or dietary supplements? *Adv. Nutr. (Bethesda Md.),* 8–14. [CrossRef] [PubMed].
Eid, H. M., & Haddad, P. S., (2017). The antidiabetic potential of quercetin: Underlying mechanisms. *Curr. Med. Chem., 24*, 355–364.
El-Bassossy, H. M., Abo-Warda, S. M., & Fahmy, A., (2013). Chrysin and luteolin attenuate diabetes-induced impairment in endothelial-dependent relaxation: Effect on lipid profile, AGEs and NO generation. *Phytother. Res., 27*, 1678–1684. [CrossRef].
El-Bassossy, H. M., Hassan, N. A., Mahmoud, M. F., & Fahmy, A., (2014). Baicalein protects against hypertension associated with diabetes: Effect on vascular reactivity and stiffness. *Phytomedicine, 21*, 1742–1745. [CrossRef].
Fang, X. K., Gao, J., & Zhu, D. N., (2008). Kaempferol and quercetin isolated from *Euonymus alatus* improve glucose uptake of 3T3-L1 cells without adipogenesis activity. *Life Sci., 82*, 615–622. [CrossRef].

Firenzuoli, F., & Gori, L., (2007). Herbal medicine today: Clinical and research issues. *Evidence-Based Complementary and Alternative Medicine, 4*(1), 37–40.

Gabbay, K. H., (1973). The sorbitol pathway and the complications of diabetes. *The New England Journal of Medicine, 288*(16), 831–836.

Gentile, D., Fornai, M., Pellegrini, C., Colucci, R., Blandizzi, C., & Antonioli, L., (2018). Dietary flavonoids as a potential intervention to improve redox balance in obesity and related co-morbidities: A review. *Nutr. Res. Rev., 31*, 239–247. [CrossRef].

Geraets, L., Moonen, H. J. J., Brauers, K., Wouters, E. F. M., Bast, A., & Hageman, G. J., (2007). Dietary flavones and flavonoles are inhibitors of poly (ADP-ribose) polymerase-1 in pulmonary epithelial cells. *Journal of Nutrition, 137*(10), 2190–2195.

Ghorbani, A., (2017). Mechanisms of antidiabetic effects of flavonoid rutin. *Biomed. Pharmacother., 96*, 305–312. [CrossRef] [PubMed].

Gil Izquierdo, A., Gil, M., Ferreres, F., & Tomás-Barberán, F., (2001). *In vitro* availability of flavonoids and other phenolics in orange juice. *J. Agric. Food Chem., 49*, 1035–1041. [CrossRef] [PubMed].

Grace, M., Ribnicky, D., Kuhn, P., Poulev, A., Logendra, S., Yousef, G., Raskin, I., & Lila, M., (2009). Hypoglycemic activity of a novel anthocyanin-rich formulation from lowbush blueberry, *Vaccinium angustifolium* Aiton. *Phytomed. Phytomed., 16*, 406–415. [CrossRef] [PubMed].

Hameed, A., Hafizur, R. M., Hussain, N., Raza, S. A., Rehman, M., Ashraf, S., Ul-Haq, Z., et al., (2018). Eriodictyol stimulates insulin secretion through cAMP/PKA signaling pathway in mice islets. *Eur. J. Pharmacol., 5*, 245–255. [CrossRef] [PubMed].

Hao, H. H., Shao, Z. M., Tang, D. Q., Lu, Q., Chen, X., Yin, X. X., & Chen, H., (2012). Preventive effects of rutin on the development of experimental diabetic nephropathy in rats. *Life Sci., 91*, 959–967. [CrossRef] [PubMed].

Hardie, D. G., (2011). AMP-activated protein kinase: An energy sensor that regulates all aspects of cell function. *Genes & Development, 25*, 1895–908.

Hasanein, P., & Fazeli, F., (2014). Role of naringenin in protection against diabetic hyperalgesia and tactile allodynia in male Wistar rats. *J. Physiol. Biochem., 70*, 997–1006. [CrossRef].

Hertog, M. G., Hollman, L., & Katan, M. B., (1999). Dietary antioxidant flavonoids of 28 vegetables and 9 fruits common consumed in the Netherlands. *J. Agric. Food Chem., 40*, 2379–2383. [CrossRef].

Hidalgo, J., Teuber, S., Morera, F. J., Ojeda, C., Flores, C. A., Hidalgo, M. A., Núñez, L., et al., (2017). Delphinidin reduces glucose uptake in mice jejunal tissue and human intestinal cells lines through FFA1/GPR40. *Int. J. Mol. Sci., 18*, 750. [CrossRef] [PubMed].

Hollman, P. C., De Vries, J. H., Van, L. S. D., Mengelers, M. J., & Katan, M. B., (1995). Absorption of dietary quercetin glycosides and quercetin in healthy ileostomy volunteers. *Am. J. Clin. Nutr., 62,* 1276–1282. [CrossRef] [PubMed].

Hossain, C. M., Ghosh, M. K., Satapathy, B. S., Dey, N. S., & Mukherjee, B., (2014). Apigenin causes biochemical modulation, glut4 and cd38 alterations to improve diabetes and to protect damages of some vital organs in experimental diabetes. American journal of pharmacology and toxicology. *Am. J. Pharmacol. Toxicol., 9*, 39–52. [CrossRef].

Huang, W. Y., Zhang, H. C., Liu, W. X., & Li, C. Y., (2012). Survey of antioxidant capacity and phenolic composition of blueberry, blackberry, and strawberry in Nanjing. *J. Zhejiang Univ. Sci. B, 13*, 94–102. [CrossRef] [PubMed].

Hung, L. M., Chen, J. K., Huang, S. S., Lee, R. S., & Su, M. J., (2000). Cardio protective effect of resveratrol, a natural antioxidant derived from grapes. *Cardiovascular. Res., 47*, 549–555. [CrossRef].

Hussain, T., Tan, B., Murtaza, G., Liu, G., Rahu, N., Saleem-Kalhoro, M., Hussain-Kalhoro, D., Adebowale, T. O., Usman-Mazhar, M., Rehman, Z. U., et al., (2020). Flavonoids and type 2 diabetes: Evidence of efficacy in clinical and animal studies and delivery strategies to enhance their therapeutic efficacy. *Pharmacol. Res., 152*, 104629. [CrossRef].

Hussain, T., Tan, B., Yin, Y., Blachier, F., Tossou, M. C., & Rahu, N., (2016). Oxidative stress and inflammation: What polyphenols can do for us? *Oxid. Med. Cell Longev., 16*, 7432797. [CrossRef].

Huynh, D. L., Sharma, N., Kumar, S. A., Singh, S. S., Zhang, J. J., Mongre, R. K., & Kee, J. D., (2017). Anti-tumor activity of wogonin, an extract from *Scutellaria baicalensis*, through regulating different signaling pathways. *Chin. J. Nat. Med., 15*, 15–40. [CrossRef].

Ido, Y., (2007). Pyridine nucleotide redox abnormalities in diabetes. *Antioxidants and Redox Signaling, 9*(7), 931–942.

Jadhav, R., & Puchchakayala, G., (2012). Hypoglycemic and antidiabetic activity of flavonoids: Boswellic acid, ellagic acid, quercetin, rutin on streptozotocin-nicotinamide induced type 2 diabetic rats. *Int. J. Pharm. Pharm. Sci., 4*, 251–256.

Jarouliya, U., & Keservani, R. K., (2019). Pathways leading to child obesity. In: Debasis, B., (ed.), *Global Perspectives on Childhood Obesity: Current Status, Consequences and Prevention* (2nd edn., pp. 137–146) Press, Elsevier, Chapter 12. ISBN: 9780128128404.

Jayaprakasam, B., Vareed, S. K., Olson, L. K., & Nair, M. G., (2005). Insulin secretion by bioactive anthocyanins and anthocyanidins present in fruits. *J. Agric. Food Chem., 53*, 28–31. [CrossRef].

Jing, D., Xian, Z., Lei, Z., Hui-Xi, B., Xu, N., Bin, B., et al., (2014). Quercetin reduces obesity-associated ATM infiltration and inflammation in mice: A mechanism including AMPKα1/SIRT1. *Journal of Lipid Research, 55*, 363–374.

Jung, U. J., Lee, M. K., Jeong, K. S., & Choi, M. S., (2004). The hypoglycemic effects of hesperidin and naringin are partly mediated by hepatic glucose-regulating enzymes in C57BL/KsJ-db/db mice. *J. Nutr., 134*, 2499–2503. [CrossRef] [PubMed].

Jung, U. J., Lee, M. K., Park, Y. B., Kang, M. A., & Choi, M. S., (2006). Effect of citrus flavonoids on lipid metabolism and glucose-regulating enzyme mRNA levels in type-2 diabetic mice. *Int. J. Biochem. Cell Biol., 38*, 1134–1145. [CrossRef] [PubMed].

Kelly, G. S., (2011). Quercetin. Monograph. *Altern. Med. Rev., 16*, 172–194. [PubMed].

Keservani, R. K., & Sharma, A. K., (2014). Flavonoids: Emerging trends and potential health benefits. *Journal of Chinese Pharmaceutical Sciences, 23*(12), 815.

Keservani, R. K., Kesharwani, R. K., Sharma, A. K., Vyas, N., & Chadoker, A., (2010b). Nutritional supplements: An overview. *International Journal of Current Pharmaceutical Review and Research, 1*(1), 59–75.

Keservani, R. K., Kesharwani, R. K., Vyas, N., Jain, S., Raghuvanshi, R., & Sharma, A. K., (2010a). Nutraceutical and functional food as future food: A review. *Der Pharmacia Lettre, 2*(1), 106–116.

Keservani, R. K., Sharma, A. K., & Kesharwani, R. K., (2020). *Nutraceuticals and Dietary Supplements: Applications in Health Improvement and Disease Management*. CRC Press. ISBN: 9781771888738.

Khan, N., Syed, D. N., Ahmad, N., & Mukhtar, H., (2013). Fisetin: A dietary antioxidant for health promotion. *ARS, 19*, 151–162. [CrossRef] [PubMed].

Khan, S., Zhang, D., Zhang, Y., Li, M., & Wang, C., (2016). Wogonin attenuates diabetic cardiomyopathy through its anti-inflammatory and anti-oxidative properties. *Mol. Cell. Endocrinol., 15,* 101–108. [CrossRef].

Kim, H. J., Kim, S. H., & Yun, J. M., (2012). Fisetin inhibits hyperglycemia-induced proinflammatory cytokine production by epigenetic mechanisms. *eCAM*, 639469. [CrossRef].

Kim, M. S., Hur, H. J., Kwon, D. Y., & Hwang, J. T., (2012). Tangeretin stimulates glucose uptake via regulation of AMPK signaling pathways in C2C12 myotubes and improves glucose tolerance in high-fat diet-induced obese mice. *Mol. Cell. Endocrinol., 358*, 127–134. [CrossRef].

Kim, Y. O., Leem, K., Park, J., Lee, P., Ahn, D. K., Lee, B. C., & Kim, H., (2001). Cytoprotective effect of *Scutellaria baicalensis* in CA1 hippocampal neurons of rats after global cerebral ischemia. *J. Ethnopharmacol., 77*, 183–188. [CrossRef].

Klover, P. J., & Mooney, R. A., (2004). Hepatocytes: Critical for glucose homeostasis. *The International Journal of Biochemistry & Cell Biology, 36*(5), 753–758.

Ko, H., Jeong, M. H., Jeon, H., Sung, G. J., So, Y., Kim, I., & Choi, K. C., (2015). Delphinidin sensitizes prostate cancer cells to TRAIL-induced apoptosis, by inducing DR5 and causing caspase-mediated HDAC3 cleavage. *Oncotarget, 6*, 9970–9984. [CrossRef].

Kreft, S., Knapp, M., & Kreft, I., (1999). Extraction of rutin from buckwheat (*Fagopyrum esculentum* Moench) seeds and determination by capillary electrophoresis. *J. Agric. Food Chem., 47*, 4649–4652. [CrossRef].

Ku, S. K., & Bae, J. S., (2015). Baicalin, baicalein and wogonin inhibits high glucose-induced vascular inflammation *in vitro* and *in vivo*. *BMB Rep., 48*, 519–524. [CrossRef].

Kuhnau, J., (1976). The flavonoids. A class of semi-essential food components: Their role in human nutrition. *World Rev. Nutr. Diet., 24*, 117–191. [PubMed].

Lapchak, P. A., Maher, P., Schubert, D., & Zivin, J. A., (2007). Baicalein, an antioxidant 12/15-lipoxygenase inhibitor improves clinical rating scores following multiple infarct embolic strokes. *Neuroscience, 150*, 585–591. [CrossRef].

Lee, J., Jung, E., Lee, J., Kim, S., Huh, S., Kim, Y., & Park, D., (2008). Isorhamnetin represses adipogenesis in 3T3-L1 cells. *Obesity (Silver Spring), 17*, 226–232. [CrossRef] [PubMed].

Lee, Y. S., Lee, S., Lee, H. S., Kim, B. K., Ohuchi, K., & Shin, K. H., (2005). Inhibitory effects of isorhamnetin-3-O-beta-D-glucoside from *Salicornia herbacea* on rat lens aldose reductase and sorbitol accumulation in streptozotocin-induced diabetic rat tissues. *Biol. Pharm. Bull., 28*, 916–918. [CrossRef] [PubMed].

Li, F., Szab´o, C., Pacher, P., et al., (2004). Evaluation of orally active poly (ADP-ribose) polymerase inhibitor in streptozotocin diabetic rat model of early peripheral neuropathy. *Diabetologia, 47*(4), 710–717.

Li, J. M., Che, C. T., Lau, C. B., Leung, P. S., & Cheng, C. H., (2006). Inhibition of intestinal and renal Na+-glucose cotransporter by naringenin. *Int. J. Biochem. Cell Biol., 38*, 985–995. [CrossRef].

Li, R., Zang, A., Zhang, L., Zhang, H., Zhao, L., Qi, Z., & Wang, H., (2014). Chrysin ameliorates diabetes-associated cognitive deficits in Wistar rats. *Neurol. Sci., 35*, 1527–1532. [CrossRef].

Liggins, J., Bluck, L. J., Runswick, S., Atkinson, C., Coward, W. A., & Bingham, S. A., (2000). Daidzein and genistein content of fruits and nuts. *J. Nutr. Biochem., 11*, 326–331. [CrossRef].

Lim, W., & Song, G., (2017). Inhibitory effects of delphinidin on the proliferation of ovarian cancer cells via PI3K/AKT and ERK 1/2 MAPK signal transduction. *Oncol. Lett., 14*, 810–818. [CrossRef].

Liu, I. M., Tzeng, T. F., Liou, S. S., & Lan, T. W., (2007). Improvement of insulin sensitivity in obese Zucker rats by myricetin extracted from *Abelmoschus moschatus*. *Planta Med., 73*, 1054–1060.

Liu, Y., Han, J., Zhou, Z., & Li, D., (2019). Tangeretin inhibits streptozotocin-induced cell apoptosis via regulating NF-κB pathway in INS-1 cells. *J. Cell. Biochem., 120*, 3286–3293. [CrossRef] [PubMed].

Louters, L. L., Dyste, S. G., Frieswyk, D., Tenharmsel, A., Vander, K. T. O., Walters, L., et al., (2006). Methylene blue stimulates 2-deoxyglucose uptake in L929 fibroblast cells. *Life Sciences, 78*, 586–591.

Lv, P., Yu, J., Xu, X., Lu, T., & Xu, F., (2019). Eriodictyol inhibits high glucose-induced oxidative stress and inflammation in retinal ganglial cells. *J. Cell. Biochem., 120*, 5644–5651. [CrossRef] [PubMed].

Ma, L., Li, X. P., Ji, H. S., Liu, Y. F., & Li, E. Z., (2018). Baicalein protects rats with diabetic cardiomyopathy against oxidative stress and inflammation injury via phosphatidylinositol 3-kinase (PI3K)/AKT pathway. *Med. Sci. Monit., 24*, 5368–5375. [CrossRef].

MacGregor, J., & Jurd, L., (1979). Mutagenicity of plant flavonoids: Structural requirements for mutagenic activity in *Salmonella typhimurium*. *Mutat. Res., 54*, 297–309. [CrossRef].

Man, M. Q., & Shi, Y. M., (2008). Chinese herbal medicine (Tuhuai extract) exhibits topical antiproliferative and anti-inflammatory activity in murine disease models. *Experimental Dermatology, 17*(8), 681–687.

Manuel, Y., Keenoy, B., Vertommen, J., & De Leeuw, I., (1999). The effect of flavonoid treatment on the glycation and antioxidant status in type 1 diabetic patients. *Diabetes Nutr. Metab., 12*, 256–263.

Martín, M. A., Goya, L., & Ramos, S., (2016). Antidiabetic actions of cocoa flavanols. *Mol. Nutr. Food Res., 60*, 1756–1769. [CrossRef].

Mayorga-Gross, A. L., & Esquivel, P., (2019). Impact of cocoa products intake on plasma and urine metabolites: A review of targeted and non-targeted studies in humans. *Nutrients, 11*, 116). [CrossRef].

Mazza, G., (2006). Compositional and functional properties of saskatoon berry and blueberry. *Int. J. Fruit Sci., 5*, 101–120. [CrossRef].

Mehdi, S. N., Sana, Z., Md, K., Md, R. M., Sana, Z., Md, K., & Md, R. M., (2018). Chrysin: A promising anticancer agent its current trends and future perspectives. *Eur. J. Exp. Biol., 8*. [CrossRef].

Mezei, O., Banz, W. J., Steger, R. W., Peluso, M. R., Winters, T. A., & Shay, N., (2003). Soy isoflavones exert antidiabetic and hypolipidemic effects through the PPAR pathways in obese Zucker rats and murine RAW264.7 cells. *J. Nutr., 133*, 1238–1243. [CrossRef].

Miean, K. H., & Mohamed, S., (2001). Flavonoid (myricetin, quercetin, kaempferol, luteolin, and apigenin) content of edible tropical plants. *J. Agric. Food Chem., 49*, 3106–3112. [CrossRef].

Mihaylova, M. M., & Shaw, R. J., (2011). The AMPK signaling pathway coordinates cell growth, autophagy and metabolism. *Nat. Cell Biol., 13*, 1016–1023.

Minchenko, A. G., Stevens, M. J., White, L., et al., (2003). Diabetes induced over expression of endothelin-1 and endothelin receptors in the rat renal cortex is mediated via poly (ADP-ribose) polymerase activation. *The FASEB Journal, 17*(11), 1514–1516.

Mirshekar, M., Roghani, M., Khalili, M., & Baluchnejadmojarad, T., (2011). Chronic oral pelargonidin alleviates learning and memory disturbances in streptozotocin diabetic rats. *Iran. J. Pharm. Res., 10*, 569–575.

Miyake, Y., Yamamoto, K., Tsujihara, N., & Osawa, T., (1998). Protective effects of lemon flavonoids on oxidative stress in diabetic rats. *Lipids, 33*, 689–695. [CrossRef] [PubMed].

Miyata, Y., Tanaka, H., Shimada, A., Sato, T., Ito, A., Yamanouchi, T., & Kosano, H., (2011). Regulation of adipocytokine secretion and adipocyte hypertrophy by polymethoxy flavonoids, nobiletin and tangeretin. *Life Sci., 88*, 613–618. [CrossRef] [PubMed].

Moonen, H. J. J., Geraets, L., Vaarhorst, A., Bast, A., Wouters, E. F. M., & Hageman, G. J., (2005). Theophylline prevents NAD+ depletion via PARP-1 inhibition in human pulmonary epithelial cells. *Biochemical and Biophysical Research Communications, 338*(4), 1805–1810.

Nasri, S., Roghani, M., Baluchnejadmojarad, T., Rabani, T., & Balvardi, M., (2011). Vascular mechanisms of cyanidin-3-glucoside response in streptozotocin-diabetic rats. *Pathophysiology, 18*, 273–278. [CrossRef] [PubMed].

Nizamutdinova, I. T., Jin, Y. C., Chung, J. I., Shin, S. C., Lee, S. J., Seo, H. G., & Kim, H. J., (2009). The antidiabetic effect of anthocyanins in streptozotocin-induced diabetic rats through glucose transporter 4 regulation and prevention of insulin resistance and pancreatic apoptosis. *Mol. Nutr. Food Res., 53*, 1419–1429. [CrossRef] [PubMed].

Ola, M. S., Ahmed, M. M., Ahmad, R., Abuohashish, H. M., Al-Rejaie, S. S., & Alhomida, A. S., (2015). Neuroprotective effects of rutin in streptozotocin-induced diabetic rat retina. *J. Mol. Neurosci., 56*, 440–448. [CrossRef] [PubMed].

Oyama, T., Miyasita, Y., Watanabe, H., & Shirai, K., (2006). The role of polyol pathway in high glucose-induced endothelial cell damages. *Diabetes Research and Clinical Practice, 73*(3), 227–234.

Pacher, P., & Szabo, C., (2008). Role of the peroxynitrite-poly (ADPribose) polymerase pathway in human disease. *The American Journal of Pathology, 173*(1), 2–13.

Panche, A. N., Diwan, A. D., & Chandra, S. R., (2016). Flavonoids: An overview. *J. Nutr. Sci., 5*, e47. [CrossRef].

Panche, A. N., Diwan, A. D., & Chandra, S. R., (2016). Flavonoids: An overview. *J. Nutr. Sci., 5*, e47. [CrossRef] [PubMed].

Panda, S., & Kar, A., (2007). Apigenin (40,5,7-trihydroxyflavone) regulates hyperglycaemia, thyroid dysfunction and lipid peroxidation in alloxan-induced diabetic mice. *J. Pharm. Pharmacol., 59*, 1543–1548. [CrossRef] [PubMed].

Parhiz, H., Roohbakhsh, A., Soltani, F., Rezaee, R., & Iranshahi, M., (2015). Antioxidant and anti-inflammatory properties of the citrus flavonoids hesperidin and hesperetin: An updated review of their molecular mechanisms and experimental models. *Phytother. Res., 29*, 323–331. [CrossRef].

Patel, K., Singh, G. K., & Patel, D. K., (2018). A review on pharmacological and analytical aspects of naringenin. *Chin. J. Integr. Med., 24*, 551–560. [CrossRef].

Peterson, J. J., Dwyer, J. T., Jacques, P. F., & McCullough, M. L., (2015). Improving the estimation of flavonoid intake for study of health outcomes. *Nutr. Rev., 73*, 553–576. [CrossRef].

Prasath, G. S., & Subramanian, S. P., (2011). Modulatory effects of fisetin, a bioflavonoid, on hyperglycemia by attenuating the key enzymes of carbohydrate metabolism in hepatic and renal tissues in streptozotocin-induced diabetic rats. *Eur. J. Pharmacol., 668*, 492–496. [CrossRef].

Prasath, G. S., Sundaram, C. S., & Subramanian, S. P., (2013). Fisetin averts oxidative stress in pancreatic tissues of streptozotocin-induced diabetic rats. *Endocrine, 44*, 359–368. [CrossRef] [PubMed].

Priscilla, D. H., Roy, D., Suresh, A., Kumar, V., & Thirumurugan, K., (2014). Naringenin inhibits alpha-glucosidase activity: A promising strategy for the regulation of postprandial hyperglycemia in high fat diet fed streptozotocin induced diabetic rats. *Chem. Biol. Interact., 210*, 77–85. [CrossRef].

Pu, P., Gao, D. M., Mohamed, S., Chen, J., Zhang, J., Zhou, X. Y., & Jiang, H., (2012). Naringin ameliorates metabolic syndrome by activating AMP-activated protein kinase in mice fed a high-fat diet. *Arch. Biochem. Biophys., 518*, 61–70. [CrossRef].

Pu, P., Wang, X. A., Salim, M., Zhu, L. H., Wang, L., Chen, K. J., & Li, H. L., (2012). Baicalein, a natural product, selectively activating AMPKalpha (2) and ameliorates metabolic disorder in diet-induced mice. *Mol. Cell. Endocrinol., 362*, 128–138. [CrossRef].

Rane, B. R., Patil, A. S., Keservani, R. K., & Jain, A. S., (2020a). Novel approaches in herbal formulation. In: Rajesh, K. K., Raj, K. K., & Anil, K. S., (eds.) *Enhancing the Therapeutic Efficacy of Herbal Formulations Through Novel Drug Delivery Systems* (pp. 43–68). IGI Global International Publisher, Pennsylvania, USA, Chapter 2. ISBN13: 9781799844532.

Reddy, G. B., Satyanarayana, A., Balakrishna, N., et al., (2008). Erythrocyte aldose reductase activity and sorbitol levels in diabetic retinopathy. *Molecular Vision, 14*, 593–601.

Rodriguez-Rodriguez, C., Torres, N., Gutierrez-Uribe, J. A., Noriega, L. G., Torre-Villalvazo, I., Leal-Diaz, A. M., & Tovar, A. R., (2015). The effect of isorhamnetin glycosides extracted from opuntia *Ficus-indica* in a mouse model of diet induced obesity. *Food Funct., 6*, 805–815. [CrossRef] [PubMed].

Roelofs, B., Tidball, A., Lindborg, A. E., TenHarmsel, A., Vander, K. T. O., & Louters, L. L., (2006). Acute activation of glucose uptake by glucose deprivation in L929 fibroblast cells. *Biochimie, 88*(12), 1941–1946.

Ross, J. A., & Kasum, C. M., (2002). Dietary flavonoids: Bioavailability, metabolic effects, and safety. *Ann. Rev. Nutr., 22*, 19–34. [CrossRef].

Roy, M., Sen, S., & Chakraborti, A. S., (2008). Action of pelargonidin on hyperglycemia and oxidative damage in diabetic rats: Implication for glycation-induced hemoglobin modification. *Life Sci., 82*, 1102–1110. [CrossRef].

Rubin, D., & Ismail-Beigi, F., (2003). Distribution of Glut1 in detergent-resistant membranes (DRMs) and non-DRM domains: Effect of treatment with azide. *American Journal of Physiology Cell Physiology, 285*, C377–383.

Rutter, G. A., (2001). Nutrient–secretion coupling in the pancreatic islet β-cell: Recent advances. *Molecular Aspects of Medicine, 22*, 247–84.

Sahu, S. C., & Gray, G., (1993). Interactions of flavonoids, trace metals, and oxygen: Nuclear DNA damage and lipid peroxidation induced by myricetin. *Cancer Lett., 70*, 73–79. [CrossRef].

Samarghandian, S., Azimi-Nezhad, M., Samini, F. F., & Arkhondeh, T., (2016). Chrysin treatment improves diabetes and its complications in liver, brain, and pancreas in streptozotocin-induced diabetic rats. *Can. J. Physiol. Pharmacol., 94*, 388–393. [CrossRef] [PubMed].

Sampson, L., Rimm, E., Hollman, P. C., De Vries, J. H., & Katan, M. B., (2002). Flavonol and flavone intakes in US health professionals. *J. Am. Diet. Assoc., 102*, 1414–1420. [CrossRef].

Sandireddy, R., Yerra, V. G., Komirishetti, P., Areti, A., & Kumar, A., (2016). Fisetin imparts neuroprotection in experimental diabetic neuropathy by modulating Nrf2 and NF-B pathways. *Cell. Mol. Neurobiol., 36*, 883–892. [CrossRef] [PubMed].

Scalbert, A., Morand, C., Manach, C., & Remesy, C., (2002). Absorption and metabolism of polyphenols in the gut and impact on health. *Biomed. Pharmacother., 56*, 276–282. [CrossRef].

Schnell, S., Schaefer, M., & Schofl, C., (2007). Free fatty acids increase cytosolic free calcium and stimulate insulin secretion from beta-cells through activation of GPR40. *Molecular and Cellular Endocrinology, 263*, 173–80.

Sendrayaperumal, V., Iyyam, P. S., & Subramanian, S., (2014). Design, synthesis and characterization of zinc-morin, a metal flavonol complex and evaluation of its antidiabetic potential in HFD-STZ induced type 2 diabetes in rats. *Chem. Biol. Interact., 219*, 9–17. [CrossRef] [PubMed].

Sharma, A. K., Keservani, R. K., & Gautam, S. P., (2020). *Herbal Product Development*. Apple Academic Press, CRC Press, Taylor & Francis Group. ISBN: 9781771888776.

Sharma, D., Gondaliya, P., Tiwari, V., & Kalia, K., (2019). Kaempferol attenuates diabetic nephropathy by inhibiting RhoA/Rho-kinase mediated inflammatory signaling. *Biomed. Pharmacother., 109*, 1610–1619. [CrossRef].

Shukla, S., Fu, P., & Gupta, S., (2014). Apigenin induces apoptosis by targeting inhibitor of apoptosis proteins and Ku70-Bax interaction in prostate cancer. *Apoptosis, 19*, 883–894. [CrossRef] [PubMed].

Singh, A. K., Raj, V., Keshari, A. K., Rai, A., Kumar, P., Rawat, A., Maity, B., Kumar, D., Prakash, A., De, A., et al., (2018). Isolated mangiferin and naringenin exert antidiabetic effect via PPAR/GLUT4 dual agonistic action with strong metabolic regulation. *Chem. Biol. Interact., 280*, 33–44. [CrossRef].

Singh, P., Kesharwani, R. K., & Keservani, R. K., (2017a). Antioxidants and vitamins: Roles in cellular function and metabolism. In: *Sustained Energy for Enhanced Human Functions and Activity* (pp. 385–407). Academic Press. ISBN:9780128054130.

Singh, P., Kesharwani, R. K., & Keservani, R. K., (2017b). Protein, carbohydrates, and fats: Energy metabolism. In: *Sustained Energy for Enhanced Human Functions and Activity* (pp. 103–115). Academic Press. ISBN:9780128054130.

Sirovina, D., Orsolic, N., Koncic, M. Z., Kovacevic, G., Benkovic, V., & Gregorovic, G., (2013). Quercetin vs chrysin: Effect on liver histopathology in diabetic mice. *Hum. Exp. Toxicol., 32*, 1058–1066. [CrossRef].

Skibola, C., & Smith, M., (2000). Potential health impacts of excessive flavonoid intake. *Free Radic. Biol. Med., 29*, 375–383. [CrossRef].

Song, T., Lee, S. O., Murphy, P. A., & Hendrich, S., (2003). Soy protein with or without isoflavones, soy germ and soy germ extract, and daidzein lessen plasma cholesterol levels in golden Syrian hamsters. *Exp. Biol. Med. (Maywood), 228*, 1063–1068. [CrossRef] [PubMed].

Spencer, J. P., Schroeter, H., Rechner, A. R., & Rice-Evans, C., (2001). Bioavailability of flavan-3-ols and procyanidins: Gastrointestinal tract influences and their relevance to bioactive forms *in vivo*. *Antioxid. Redox Signal., 3*, 1023–1039. [CrossRef].

Sreedharan, V., Venkatachalam, K. K., & Namasivayam, N., (2009). Effect of morin on tissue lipid peroxidation and antioxidant status in 1, 2-dimethylhydrazine induced experimental colon carcinogenesis. *Investig. New Drugs, 27*, 21–30. [CrossRef] [PubMed].

Srivastava, S. K., Ramana, K. V., & Bhatnagar, A., (2005). Role of aldose reductase and oxidative damage in diabetes and the consequent potential for therapeutic options. *Endocrine Reviews, 26*(3), 380–392.

Stanley, M. P. P., & Kamalakkannan, N., (2006). Rutin improves glucose homeostasis in streptozotocin diabetic tissues by altering glycolytic and gluconeogenic enzymes. *J. Biochem. Mol. Toxicol., 20*, 96–102. [CrossRef] [PubMed].

Suzuki, S., Takada, T., Sugawara, Y., Muto, T., & Kominami, R., (1991). Quercetin induces recombinational mutations in cultured cells as detected by DNA fingerprinting. *Jpn. J. Cancer Res. Gann, 82*, 1061–1064. [CrossRef].

Szab´o, C., Biser, A., Benko, R., et al., (2006). Poly (ADPribose) polymerase inhibitors ameliorate nephropathy of type 2 diabetic Leprdb/db mice. *Diabetes, 55*(11), 3004–3012.

Tai, M. C., Tsang, S. Y., Chang, L. Y., & Xue, H., (2005). Therapeutic potential of wogonin: A naturally occurring flavonoid. *CNS Drug Rev., 11*, 141–150. [CrossRef].

Tanveer, A., Akram, K., Farooq, U., Hayat, Z., & Shafi, A., (2017). Management of diabetic complications through fruit flavonoids as a natural remedy. *Crit. Rev. Food Sci. Nutr., 57*, 1411–1422. [CrossRef].

Thilakarathna, S. H., & Rupasinghe, H. P., (2013). Flavonoid bioavailability and attempts for bioavailability enhancement. *Nutrients, 5*, 3367–3387. [CrossRef].

Thomas, I., & Gregg, B., (2017). Metformin; a review of its history and future: From lilac to longevity: Thomas and Gregg. *Pediatr. Diabetes, 18*, 10–16. [CrossRef] [PubMed].

Tran, V. H., Duke, R. K., Abu-Mellal, A., & Duke, C. C., (2012). Propolis with high flavonoid content collected by honey bees from *Acacia paradoxa*. *Phytochemistry, 81*, 126–132. [CrossRef] [PubMed].

Tuorkey, M. J., (2016). Molecular targets of luteolin in cancer. *Eur. J. Cancer Prev., 25*, 65–76. [CrossRef].

Vanitha, P., Uma, C., Suganya, N., Bhakkiyalakshmi, E., Suriyanarayanan, S., Gunasekaran, P., & Ramkumar, K. M., (2014). Modulatory effects of morin on hyperglycemia by attenuating the hepatic key enzymes of carbohydrate metabolism and beta-cell function in streptozotocin-induced diabetic rats. *Environ. Toxicol. Pharmacol., 37*, 326–335. [CrossRef].

Visnagri, A., Kandhare, A. D., Chakravarty, S., Ghosh, P., & Bodhankar, S. L., (2014). Hesperidin, a flavanoglycone attenuates experimental diabetic neuropathy via modulation of cellular and biochemical marker to improve nerve functions. *Pharm. Biol., 52*, 814–828. [CrossRef].

Wang, N., Yi, W. J., Tan, L., Zhang, J. H., Xu, J., Chen, Y., Qin, M., Yu, S., Guan, J., & Zhang, R., (2017). Apigenin attenuates streptozotocin-induced pancreatic β cell damage by its protective effects on cellular antioxidant defense. *In Vitro Cell. Dev. Biol. Anim., 53*, 554–563. [CrossRef].

Wang, T. Y., Li, Q., & Bi, K. S., (2018). Bioactive flavonoids in medicinal plants: Structure, activity and biological fate. *Asian J. Pharm. Sci., 13*, 12–23. [CrossRef].

Wang, X., Zhang, D. M., Gu, T. T., Ding, X. Q., Fan, C. Y., Zhu, Q., & Kong, L. D., (2013). Morin reduces hepatic inflammation-associated lipid accumulation in high fructose-fed rats via inhibiting sphingosine kinase 1/sphingosine 1-phosphate signaling pathway. *Biochem. Pharmacol., 86*, 1791–1804. [CrossRef].

Wang, Y., Chen, S., & Yu, O., (2011). Metabolic engineering of flavonoids in plants and microorganisms. *Appl. Microbiol. Biotechnol., 91*, 949–956. [CrossRef].

Williamson, G., & Clifford, M. N., (2017). Role of the small intestine, colon and microbiota in determining the metabolic fate of polyphenols. *Biochem. Pharmacol., 139*, 24–39. [CrossRef].

Williamson, G., Kay, C. D., & Crozier, A., (2018). The bioavailability, transport, and bioactivity of dietary flavonoids: A review from a historical perspective. *Compr Rev. Food Sci. Food Saf., 17*, 1054–1112. [CrossRef].

Yabe-Nishimura, C., (1998). Aldose reductase in glucose toxicity: A potential target for the prevention of diabetic complications. *Pharmacological Reviews, 50*(1), 21–33.

Yang, Q., Graham, T. E., Mody, N., Preitner, F., Peroni, O. D., Zabolotny, J. M., et al., (2005). Serum retinol binding rotein 4 contributes to insulin resistance in obesity and type 2 diabetes. *Nature, 436*, 356–362.

Yang, W., Wang, S., Li, L., Liang, Z., & Wang, L., (2011). Genistein reduces hyperglycemia and islet cell loss in a high-dosage manner in rats with alloxan-induced pancreatic damage. *Pancreas, 40*, 396–402. [CrossRef].

Yao, Z., Gu, Y., Zhang, Q., Liu, L., Meng, G., Wu, H., Xia, Y., Bao, X., Shi, H., Sun, S., et al., (2019). Estimated daily quercetin intake and association with the prevalence of type 2 diabetes mellitus in Chinese adults. *Eur. J. Nutr., 58*, 819–830. [CrossRef].

Yin, H., Huang, L., Ouyang, T., & Chen, L., (2018). Baicalein improves liver inflammation in diabetic db/db mice by regulating HMGB1/TLR4/NF-κB signaling pathway. *Int. Immunopharmacol., 55*, 55–62. [CrossRef].

Yokozawa, T., Kim, H. Y., Cho, E. J., Choi, J. S., & Chung, H. Y., (2002). Antioxidant effects of isorhamnetin 3,7-di-O-beta-D-glucopyranoside isolated from mustard leaf (*Brassica juncea*) in rats with streptozotocin-induced diabetes. *J. Agric. Food Chem., 50*, 5490–5495. [CrossRef].

Zamora-Ros, R., Knaze, V., Rothwell, J. A., Hémon, B., Moskal, A., Overvad, K., Tjønneland, A., Kyrø, C., Fagherazzi, G., Boutron-Ruault, M. C., et al., (2016). Dietary polyphenol intake in Europe: The European prospective investigation into cancer and nutrition (EPIC) study. *Eur. J. Nutr., 55*, 1359–1375. [CrossRef].

Zanatta, L., Rosso, A., Folador, P., Figueiredo, M. S., Pizzolatti, M. G., Leite, L. D., & Silva, F. R., (2008). Insulinomimetic effect of kaempferol 3-neohesperidoside on the rat soleus muscle. *J. Nat. Prod., 71*, 532–535. [CrossRef].

Zang, M., Xu, S., Maitland-Toolan, K. A., Zuccollo, A., Hou, X., Jiang, B., & Cohen, R. A., (2006). Polyphenols stimulate AMP-activated protein kinase, lower lipids, and inhibit accelerated atherosclerosis in diabetic LDL receptor-deficient mice. *Diabetes, 5*, 2180–2191. [CrossRef].

Zang, Y., Igarashi, K., & Li, Y., (2016). Antidiabetic effects of luteolin and luteolin-7-O-glucoside on KK-A(y) mice. *Biosci. Biotechnol. Biochem., 80*, 1580–1586. [CrossRef].

Zhang, W. Y., Lee, J. J., Kim, Y., Kim, I. S., Han, J. H., Lee, S. G., & Myung, C. S., (2012). Effect of eriodictyol on glucose uptake and insulin resistance *in vitro. J. Agric. Food Chem., 60*, 7652–7658. [CrossRef].

Zhang, Y., & Liu, D., (2011). Flavonol kaempferol improves chronic hyperglycemia-impaired pancreatic beta-cell viability and insulin secretory function. *Eur. J. Pharmacol., 670*, 325–332. [CrossRef] [PubMed].

Zhang, Z., Ding, Y., Dai, X., Wang, J., & Li, Y., (2011). Epigallocatechin-3-gallate protects pro-inflammatory cytokine induced injuries in insulin-producing cells through the mitochondrial pathway. *Eur. J. Pharmacol., 670*, 311–316. [CrossRef] [PubMed].

Zheng, Z., Zhu, W., Yang, B., Chai, R., Liu, T., Li, F., & Li, G., (2018). The co-treatment of metformin with flavone synergistically induces apoptosis through inhibition of PI3K/AKT pathway in breast cancer cells. *Oncol. Lett., 15*, 5952–5958. [CrossRef].

Zhu, W., Jia, Q., Wang, Y., Zhang, Y., & Xia, M., (2012). The anthocyanin cyanidin-3-O-beta-glucoside, a flavonoid, increases hepatic glutathione synthesis and protects hepatocytes against reactive oxygen species during hyperglycemia: Involvement of a cAMP-PKA-dependent signaling pathway. *Free Radic. Biol. Med., 52*, 314–327. [CrossRef].

CHAPTER 7

ANTIDIABETIC BENEFITS OF DIETARY FLAVONOIDS

RICHA SINGH and DEEPIKA DHAWAN

Department of Dietetics and Applied Nutrition, Amity Medical School, Amity University Haryana, Gurugram, Haryana, India

ABSTRACT

Flavonoids are a group of plant-based chemicals that shield us from several health problems. One of the emerging health problems is diabetes, which affects more than 422 million people worldwide and contributes to 1.5 million mortalities. The covert role of flavonoids was not acknowledged in clinical practice, but in recent years, the beneficial effects and possible mechanisms have been researched and appreciated. An attempt has been made in this regard to identify the potential flavonoids present in plants and to highlight their pathophysiological role in managing diabetes. A literature review was undertaken, and six flavonoid classes were identified, including flavones, flavonols, flavanones, isoflavonoids, anthocyanins, and chalcones. Among these classes, a number of flavonoids have beneficial roles in managing hyperglycemia, promoting insulin secretion and proliferation of β-cells, reducing insulin resistance, managing carbohydrate metabolism and gluconeogenesis, and increasing the transportation of glucose into the cells. Some flavonoids may imitate the role of oral hyperglycemic drugs and help in carbohydrate metabolism. In addition to this, it was found that it may reduce the risk of long-term diabetic complications and delay the progression of associated diseases.

Advances in Flavonoids for Human Health and Prevention of Diseases. Nisha Sharma, Deepika Saini, Rajesh K. Kesharwani, Prakash Chandra Gupta, & Raj K. Keservani (Eds.)

It had a significant role in managing oxidative stress and inflammation. Flavonoids also possess anti-viral and anti-bacterial properties that may enhance the immune system that is compromised in diabetic patients with persistent hyperglycemia. Hence, it may be inferred that flavonoids have a scope of its use as an adjunct medicine for diabetic patients in the future.

7.1 INTRODUCTION

The word flavonoids are derived from the Latin word "*Flavus*" which refers to yellow color (David et al., 2016). Flavonoids are plant based valuable chemicals that are naturally occurring in the foods we eat. Chemically these are polyphenolic compounds, which can be isolated from various parts of the plants like bark, roots, stem, leaves, flower, and fruits. These are non-nitrogenous substances, present universally in the plants. About 1 million metabolites are produced by the plants and out of that, around 6,000 such metabolites belong to the category of flavonoids. There are a number of secondary metabolites which are found in plants like alkaloids, glycosides, terpenoids, tannins, and one of its own kind is flavonoids (Teles et al., 2018). The basic chemical structure of flavonoids constitutes 15 Carbon skeleton, known as 2 phenylbenzo-pyrone derivatives. They occur in three states in plants – aglycones, glycosides, and methylated form. The six classes of flavonoids include – flavones, flavonols, flavonones, isoflavonoids, anthocyanins, and chalcones (Panche et al., 2016).

Flavones include flavonoids like luteolin, tangeritin, apigenein, and luteolin glucosides. These may be derived from foods like red wine, red peppers, tomato skin, buckwheat, fruit peels, chamomile, gingko biloba and herbs like celery, parsley, mint, etc. (Panche et al., 2016). Another class of flavonoids is flavonols. These include myrecetin, fisetin, kaempferol, and quercetin. A wide variety of foods are found to be associated with flavonols, like onions, kale, lettuce, berries, grapes, apples, peaches including beverages such as tea and wine (Neveu et al., 2010). In fact, cocoa, chocolate, and related food products have also been found to be good sources of flavonols.

The third class of flavonoids is flavanones which are found to be present abundantly in citrus fruits–lemon, lime, tangerine, grapefruits, oranges, etc. The subclass includes hesperitin, naringenin, eriodictyol, etc. Bitter taste of citrus fruits is attributed to the presence of these compounds. Tomato and cherries are also associated with flavanones (Iwshina et al.,

2013). Isoflavonoids are widely distributed in the Fabaceae family and are another class of flavonoids. Soya beans and its products are the major source of isoflavonoids. They may also be found in some fruits and nuts. Pinto beans, black beans, lima beans, alfalfa, and red clover are also rich in isoflavonoids. The subclass of isoflavoinoids include daidzein and genistein (Liggins, 2000). Anthocyanins are mainly present in deep pigmented fruits and vegetables like red cabbage, beetroot, cranberries, black currants, blackberries, blueberries, cherries, skin of eggplants, etc. Delphinidin, Cyanidin, Malvidin, Pelargonidin, and Peonidin are the common anthocyanins present in a wide variety of food products (Giusti & Wrostad, 2003). Chalcones is also another category of flavonoids, abundantly present in licorice, Chinese ginger, tomatoes, pears, shallots, potatoes, bean sprouts and some wheat products. Phloridzin, arbutin, phloretin, chalconaringenin are some popular chalcones (Carlo, 1999).

Flavonoids are non-nutrient substances but render a number of biological functions in the body. As mentioned above, flavonoids are widely distributed in nature and present abundantly in a variety of foods. After ingestion of foods rich in flavonoids, they are hydrolyzed in the intestine into its constituent metabolites through enzymatic activity, which takes place during the normal process of digestion. Some of the flavonoids are utilized by the gut microbiota and they further synthesize a variety of metabolites. These flavonoids and its precursors thus produced are absorbed by the bloodstream. These compounds have a variety of beneficial effects on our body. During normal metabolism, harmful nitrogen and oxygen species are produced in our body. Flavonoids being antioxidant in nature neutralizes them and thus, prevents cellular damage. They also have an inhibitory effect on initiation and progression of cancer by scavenging the activity of free radicals and hence have an anti-carcinogenic role. During infections and inflammation, certain harmful substances and allergens are produced. Flavonoids reduce the histamine-mediated response as a result of allergy and hence, possess anti-allergic properties.

Flavonoids have numerous cardio protective benefits with a property of vasorelaxation. The various functions include lowering down the oxidation of LDL (Low Density Lipoprotein) cholesterol and hence further reducing endothelial toxicity, which might be induced by LDL oxidation. Here, copper ions are the free radicals which are chelated as a result of the activity of flavonoids and hence they prevent oxidation of LDL cholesterol, thereby reducing the risk of atherosclerosis (Ríha et al., 2014). Lipid lowering properties of flavonoids has been manifested by significant reduction in lipid hydroperoxides – which are considered to be highly atherogenic (Kim et al., 2010). Quercetin, a subclass of flavonoids, possesses the tendency to reduce vascular superoxide radical by direct scavenging or inhibiting the enzymes involved in the generation of these radicals and even impedes the expression of genes responsible for its production (Zeka et al., 2017).

The anti-oxidative benefits of flavonoids are mainly attributed to its structure:

- It has ortho-dihydroxy structure in the B ring of the flavonoid that allows greater oxidative stability.
- The C2-C3 double bond allows electron delocalization over the flavonoid structure.
- Strong radical absorption is due to the presence of 3-OH and 5-OH groups in its structure.

These features allow radical scavenging activity and improve endothelial dysfunction (Amić et al., 2007).

Flavonoids are also known to control blood sugar levels and hence prove to be useful in the management of type-2 diabetes. One of the oral hypoglycemic agents used in historic times is alkyl biguanides synthalin A and synthalin B. These are derivatives of guanidine derived from Goat's rue (*Galega officinalis*). It is surprising that this led to the further development of metformin–a known drug for treatment of type-2 diabetes (Hussain et al., 2013). But these oral hypoglycemic drugs (OHDs) have been known to have possible side effects ranging from gastrointestinal symptoms and lactic acidosis.

Flavonoids are known to modulate carbohydrate digestion by enhancing insulin signaling and secretion. Studies show that it may improve glucose uptake by the liver and muscles and allow adipose deposition. It also promotes β cell proliferation and regeneration with an adjunct effect of reduced apoptosis. Hence, it supports management of glucose metabolism in the liver and stabilizes blood sugar levels.

7.2 DIETARY FLAVONOIDS–A BOON FOR DIABETIC PATIENTS

Diabetes is a complicated metabolic disorder which is adding to the global burden of disease progressively. The characteristic feature of diabetes is elevated blood glucose levels. The glucose remains elevated in the blood due to low insulin production, high glucose synthesis or reduced insulin receptor sensitivity. As per the data published by WHO (World Health Organization, 2019), about 463 million people are affected with diabetes. This figure is projected to increase to 578 million in the next 10 years and further to 700 million in the next 25 years (Saeedi et al., 2019). It has been seen that the apoptosis process of β cells of pancreas initiates as early as 9–12 years before the onset of the actual condition of hyperglycemia. Study related evidence suggests regenerating and proliferating effects of flavonoids on the β cells of the pancreas.

Another fact reveals that in diabetes, the insulin production in response to raised blood glucose levels is hampered. Fortunately, a diet rich in flavonoids improve insulin secretion and have positive effects on pancreatic sensitivity. Flavonoids regulate lipid oxidation, which may decrease the supply of free fatty acids, known to cause liver insulin resistance. Inflammatory activity of deleterious adipokines (known cause for insulin resistance) that are elevated in type-2 diabetes can also be managed with dietary flavonoids (Keservani et al., 2020; Rane et al., 2020a–c), owing to their anti-inflammatory properties (Hudish et al., 2019; Thomassian, 2009).

Polyphenols, a group of flavonoids may alter carbohydrate metabolism by reducing α amylase and sucrase activity. Another class of flavonoids known as catechins present in tea, coffee, cacao, etc., and isoflavones derived from soya products alter transport of glucose in the body. The activity of sodium 1-GLUT (glucose cotransporter 1) is reduced, which is responsible for the transport of glucose to the liver. Flavonoids derived from banana flower activates insulin receptor tyrosine kinase and hence helps in the regulation of insulin resistance. Quercetin found in a variety of foods like onions, berries, apples, broccoli, oranges, tangerine, etc., is also found to induce regeneration of pancreatic cells (Saad & Bushra, 2013). Apart from this, there are a number of flavonoids which have anti-diabetic efficiency, and these have been discussed in subsections.

7.2.1 REUTIN

Reutin reduces catalyzing activity of enzymes involved in carbohydrate metabolism. It suppresses glucose absorption from the small intestine by hindering the activity of enzymes α glucosidases and α amylase, involved in the digestion of carbohydrates. Hence, it forestalls the rise in post-prandial blood glucose levels. The flavonoid reutin imitates the hormone insulin and triggers glucose uptake by muscles. This series of events is initiated by activating transporter GLUT-4. In diabetic patients, one of the major reasons for hyperglycemia is triggered by a surge in gluconeogenesis–synthesis of glucose from sources other than carbohydrates. This process is supported by a number of enzymes like glucose-6-phosphatase. Reutin reduces the activity of this enzyme in liver and kidneys up to some extent. Thereby, downsizing the process of gluconeogenesis. Reutin also plays a potent role in decreasing β cell degeneration. It also activates AMP (Adenosine Monophosphate)-activated protein kinase and insulin receptor substrate 2, which further enhances the efficiency of carbohydrate metabolism by mediating insulin signaling process. Hence, foods rich in reutin should be included in the diet of diabetic patients (Ghorbani, 2017).

In a study conducted by Kamalakkannan & Prince (2006), on diabetes induced rats found that rutein supplementation for 45 days decreased fasting blood glucose and HbA1C levels. Further, improvements include an increase in insulin, C-peptide, hemoglobin, and protein (Keservani et al., 2015) levels. Many studies show the protective effect of reutin against a number of diabetic complications like neuropathy, nephropathy, retinopathy, liver damage and progression of cardiovascular complications. A wide variety of foods which are rich sources of reutin are citrus fruit and fruit rinds, buckwheat, asparagus, green tea, elderflower tea, etc.

7.2.2 QUERCETIN

Study based evidence reveals that this fraction of flavonoid promotes insulin secretion via pathways involving calcium ion signaling. The functionality and recovery of β cells of the pancreas are enhanced due to increased activity of vascular endothelial growth factor (VEGF) and its receptor in the presence of Quercetin. The pancreatic cell damage is profound in diabetes and that might be aggravated due to hypercholesterolemia. However, Quercetin helps in reduction of β cell dysfunction (Huan et al., 2021). Oxidative stress due to diabetes can be managed by intake of quercetin and its rich foods. Quercetin is an antioxidant known to scavenge free oxygen radicals and arrest lipid peroxidation. Along with ascorbic acid, the antioxidant potential of quercetin is amplified, which has positive effects on neurovasculature structure in the skin as the integrity of the neurovascular system is compromised in diabetic patients.

Diabetic nephropathy is a common complication of diabetes that involves sclerosis, fibrosis, and atrophy of the glomerular membranes of kidneys. Research shows that quercetin safeguards and postpones such pathological events in the body, thereby decreasing the risk for kidney damage. This flavonoid also reduces the risk of neurodegenerative diseases and consequent loss of cognition and memory in geriatric population suffering from long-term complications of diabetes.

Patients suffering from diabetes are susceptible to gastric ulcers and related complications. However, quercetin may have ameliorative effect by protecting the gastrointestinal mucosa caused by side effects of OHDs. Foods that are rich sources of quercetin include berries, cherries, coffee, green tea, cocoa, chocolate, capers, tomato, red leaf lettuce, onions, etc.

7.2.3 GENISTEIN

Genistein is a flavonoid which belongs to the class of isoflavones, commonly found in soya and soya-based products. Association of genistein and its positive effects are evident through much research. It is a phytoestrogenic compound which has potent health benefits in post-menopausal women. Studies report its role in controlling blood glucose levels in post-menopausal women; it may also help in the regeneration of pancreatic β cells by triggering c-AMP signaling (Fu et al., 2009).

A study involving 12-week supplementation of soya in post-menopausal women, helped to lower hyperinsulinemia, insulin resistance, HbA1C levels and total and LDL cholesterol (Jayagopal et al., 2002). Although there may be a confounding factor, i.e., Daidzein present in Soya products which may affect the bioavailability of genistein. Another study based on animal model supplemented 54 mg/day of genistein to post-menopausal women lead to a reduction in fasting blood glucose and showed better glucose tolerance and sensitivity for insulin. Thus, it may be inferred that isolated genistein may show better effects on type-2 diabetes patients in comparison to genistein administered in the form of soya products (Villa et al., 2009).

Genistein has also been found beneficial for the management of diabetic complications like diabetic nephropathy. A study on diabetes induced rats showed improvements in renal oxidative stress, blood urea nitrogen and creatinine levels. Anti-inflammatory role of genistein has also been propounded through reduction in certain inflammatory markers which shows decrement of inflammation. These markers include nuclear factor kappa B, phosphorylated inhibitory kappa B α, C-reactive protein, monocyte chemotactic protein-1 (MCP-1), cyclooxygenase-2 and tumor necrosis factor α (TNF-α). On the other hand, improvements were also observed in biomarkers indicating oxidative stress, that includes nuclear related factor E2, heme oxygenase-1, glutathione peroxidase (GPx), superoxide dismutase (SOD) isoforms. The improvements were constant and not dose dependent. Genistein supplementation also reduces the risk of fibrosis by hindering its markers that includes protein kinase C, protein kinase C-beta II, and transforming growth factor β-I (TGF-β-I) (Kim & Lim, 2013).

7.2.4 ISORHAMNETIN

Isorhamnetin is a fraction of flavonoid which is found in berries, apples, cherries, etc. It is also present in many medicinal plants like sea buckthorn and water dropwort. It is related to quercetin as it is a metabolite of this flavonoid subclass. The potent effect of isorhamnetin is well documented through splendid research data, which are especially based on animal experiments. It has been found to possess properties that may improve cardiovascular and cerebrovascular health, prevention of obesity, immune-modulatory effect, etc. Immunity is compromised in many metabolic disorders including diabetes, isorhamnetin has anti-viral, anti-bacterial properties and is a potent agent in inflammation and oxidative stress. Diabetic complications may have deleterious consequences on kidney health – which can be managed by isorhamnetin rich foods. In addition to being renal friendly, it also protects against hyperuricemia. The liver is a vital organ with diverse metabolic functions in the body. This flavonoid also has a positive effect on hepatocytes.

Isorhamnetin has a crucial role in maintaining the homeostasis, especially in blood glucose control. This process is facilitated by activation of tyrosine kinase/transcription factor signaling pathway which promotes GLUT4 transportation, thereby stabilizing plasma glucose levels. An enzyme alpha-glucosidase, which is involved in the breakdown of starch and disaccharide molecules present in food. The activity of this enzyme is known to be inhibited by Isorhamnetin. Endothelial dysfunction and vascular damage are common conditions in diabetes which might arise due to the formation of highly reactive carbonyl compounds like methylglyoxal. However, isorhamnetin has been found to have positive effects on protection against vascular and endothelial flaws. Hypertension concurrently occurs with diabetes, being an associated risk factor. Consumption of isorhamnetin helps to stabilize blood pressure (BP) and manage hypertension in individuals with hyperglycemia (Li et al., 2016; Gong et al., 2020).

7.2.5 APIGENIN

Apigenin is a flavonoid which is abundantly present in many fruits and vegetables. Foods like parsley, celery, chamomile, spinach, artichokes, oregano, oranges, tea, onion, wheat sprouts, etc. This sub class of flavonoid has a significant role in glucose metabolism in the body by regulating insulin production and glycogen synthesis. It may help to increase levels of GLUT-4 and helps transport of glucose to the skeletal muscles. It has been reported in an animal-based study, that apigenin does not allow maturation of microRNA and thus enhances glucose resistance. Apigenin-6-C-β-fucopyranoside (derivative of apigenin) allows steady glucose uptake through the insulin receptor signaling pathway. By virtue of its nature, apigenin has antioxidant properties which protect against long term diabetic complications. Endothelial damage caused as a result of hyperglycemia can be corrected with apigenin as it supplies nitric oxide (NO) and heals endothelial cells. Enzymatic activity of aldose reductase (AR) can also be controlled by apigenin, which in turn helps in removing Sorbitol from cell membranes through diffusion. It has been manifested that sorbitol accumulation in diabetic patients can further aggravate complications like cataract, neuropathy, nephropathy, retinopathy, etc.

Apigenin reduces the effect of free radicals in the body. Thus, preventing pancreatic injury induced by free radicals. Also, leads to a reduction in oxidative stress and risk of fibrosis. Accumulation of reactive oxygen species (ROS) in the body is a common phenomenon in diabetic patients. While apigenin triggers enzymatic activity of antioxidant enzymes–GPx and SOD (Barky et al., 2020). It may be concluded that apigenin is a potential nutraceutical (Keservani et al., 2010a, b) with diverse health benefits to diabetic patients (Salehi et al., 2019).

7.2.6 BAICALEIN

Baicalein belongs to the class of flavonoids (Keservani & Sharma, 2014) called Flavones. A number of medicinal plants, especially belonging to Asian regions of the world are considered to have numerous phenolic compounds, including baicalein. Chinese traditional medicine *Huangqin* is a rich source of baicalein and is used extensively for its several therapeutic benefits. Another medicinal plant which is also a source of baicalein is *Scutellariae baicalensis Georgi.* Baicalein has the ability to detoxify free radicals by chelating redox-active metal ions. Also improves the activity of antioxidant enzymes in the cortex. It also suppresses genes responsible for inflammation and hence an anti-inflammatory agent. It also binds chemokines like interleukin-8 (that is linked to increased risk of diabetes). Cancer is a common risk factor for diabetes, and it is seen through animal studies that supplementation with baicalein may prevent cancer and its initiation process.

The integrity of the mitochondrial inner membrane is also maintained with baicalein, especially in diabetic patients where its health is compromised and affected. Studies also show that it helps to manage serum cholesterol and free fatty acid levels. Baicalein improves the concentration of insulin in the blood and to conserve β cells of the pancreas. The blood glucose levels are regulated through baicalein using the AMP signaling pathway (Fu, 2012). Although literature about the relationship between baicalein and diabetes is limited but a few studies based on animal research report some beneficial effects of diabetes. In a study on diabetes induced rats, 150 mg/kg/d baicalein was administered after five days of streptozotocin injection. After 24 weeks, regeneration of β cells was observed and it also proved to be beneficial in reducing inflammation. Another beneficial effect was prevention of vascular abnormalities and retinopathy (Yang et al., 2009).

7.2.7 FISETIN

Fisetin is a flavonol compound which has a potent bioactive role. This flavonoid can be isolated from a variety of food sources such as strawberry, grapes, cherries, onions, persimmon, cucumber, etc. The concentration of fisetin may fall between 2 and 160 μg/g of the fruit/vegetable (Khan et al., 2013). Anti-diabetic effect of fisetin can be manifested in a way that it improves glycolysis and inhibits the process of gluconeogenesis. It enhances the uptake of glucose by hepatocytes and promotes the activity of glycolytic enzymes – hexokinase, pyruvate kinase and lactate dehydrogenase in hepatic and renal tissues. In diabetic patients, the body tends to use non-carbohydrate sources as a means of energy, as carbohydrate metabolism is no longer sufficient to meet the energy needs of the body. The enzymes involved in gluconeogenesis include glucose-6-phosphatase, glucose-6-phosphate dehydrogenase and fructose-1,6 bisphophatase. With the supplementation of fisetin, the activity of these enzymes is inhibited.

The production of harmful cytokines is profound in chronic hyperglycemia in monocytes and several other factors are activated that cause inflammation and may cause metabolic disorders. However, fisetin possesses anti-inflammatory effects by altering the expression of cytokines and other related factors. Administration of fisetin has also been linked to reducing the vascular permeability of albumin in the cells, which is seen as an ill effect of persistent hyperglycemia. Diabetic patients are naturally predisposed to cardiovascular diseases (CVDs) (Keservani et al., 2016a) and are susceptible to the development of thrombosis and atherosclerosis. Using fisetin supplementation can be helpful in subsiding the consequent progression to heart diseases (Sari & Soysal, 2020). Thus, it may be concluded that fisetin being anti-diabetic, also protects diabetic patients from inflammation and related complications.

7.2.8 DAIDZEIN

Daidzein is a known isoflavone with immense nutritional potential. It is abundantly found in soyabeans and legumes and chemically known as 7-hydroxy-3-(4-hydroxyphenyl)-4H-chromen-4-one. It is of multi-faceted potential as it is known to be anti-cancerous, anti-fibrosis, and anti-diabetic properties. Its structure and functionality in the human body is similar to that of estrogen. Daidzein is a flavonoid which undergoes further alteration by the intestinal microflora and is converted to dihydrodaidzein, that is further metabolized to equol. This metabolite has a strong estrogenic and antioxidant activity. However, as per studies conducted on population, this process is dependent upon intestinal microflora and genetic makeup of an individual (Frankenfeld et al., 2009). As, previously discussed, oxidative stress manifest in condition of hyperglycemia and this may further aggravate cell damage and inflammation. Park et al. (2016) reported that daidzein supplementation reduced oxidative stress damage due to hyperglycemia.

It is of great therapeutic benefits especially to diabetic post-menopausal women, susceptible to reduced bone mass index and other post-menopausal symptoms. It may also reduce the risk of reproductive cancer, like breast and ovarian cancer (Rawat et al., 2019). Hypertension is another malady which concomitantly occurs with diabetes and can also be managed with daidzein. An animal study found that daidzein dose of 100 mg/kg significantly lowered systolic and diastolic BP. It is also beneficial for individuals with compromised cardiovascular health as it delays cardiac tissue necrosis and fibrosis (Laddha & Kulkarni, 2021). Daidzein works in conjugation with genistein and assists a number of pathways like AMP activate protein kinase and calcineurin. These pathways further enhance the insulin secretion by the pancreatic cells (Fu & Liu, 2009).

Excellent sources of daidzein include green soyabean, tempeh, and other soya products. Other fair sources are pistachios, mung beans, asparagus, white grapefruit, alfalfa sprouts, beans, passion fruit, garlic, walnuts, etc. (Popa & Rusu., 2017; Keservani et al., 2016b).

7.2.9 TANGERITIN

Tangeritin is a polymethoxylated flavone isolated from peels of citrus fruits with numerous pharmacological functions in the body. Chemically it is known as 5,6,7,8,4′-pentamethoxy flavone and like other flavonoids also benefit diabetic patients (Arafa et al., 2021). This was revealed in an animal-based model in which high fat diet was given to obese-diabetic rats and 200 mg/kg of tangeritin was administered. It helped to manage hyperglycemia and altered weight gain (Kim et al., 2012). Studies also report that tangeritin promotes insulin secretion and long-term management of hyperglycemia by regulating HbA1c levels. It also interferes in the carbohydrate metabolism by modulating important enzymes such as hexokinase, pyruvate kinase, lactate dehydrogenase, glucose-6-phosphatase, fructose-1,6-bisphosphatase, glucose-6-phosphate dehydrogenase involved in glycolysis.

Studies on tangeritin is scarce but literature reports its comparative health benefits with oral hypoglycemic drug like glibenclamide for diabetic patients (Sundaram et al., 2014). Cell culture-based studies report that tangeritin reduces the expression of harmful adipokines such as interleukin-6, TGF-β-I, VEGF in hyperglycemia conditions in human eye-cell cultures and reduce the incidence of retinopathy in diabetic population. These adipokines are known to trammel insulin signaling and cause insulin resistance (Qin & Jiang, 2020).

Tangeritin also exhibits anti-inflammatory and anti-oxidative properties that may enhance the immunity of individuals with diabetes as known to be compromised in hyperglycemia. Consistent rise in glucose levels may cause hypoxia and further increase the risk of nephropathy. Tangeritin may have a protective effect against diabetic-hypoxia and stress related fibrosis. Diabetic patients with advance in age are at high risk of metabolic syndrome due to chronic oxidative stress and organ damage. However, consumption of tangeritin may protect vital organs against cellular damage caused by high blood glucose and related inflammatory processes (Nery et al., 2021).

7.2.10 CYANIDIN

Cyanidin, belongs to the group of anthocyanins, which are potent flavonoids present in dark colored fruits and vegetables. It is a water-soluble pigment present abundantly in grapes, black berries, blue berries, red cabbage, cherries, cranberry, acai berries, raspberry, and even black rice (Matsukawa et al., 2016). In plants, this particular flavonoid is present in bound form as cyanidin 3-O-β-glucoside. Cyanidin has been shown to have positive effects on insulin sensitivity. In a study conducted by Guo et al. (2007), it was found that the TNF-α induced insulin resistance was reduced significantly in a dose dependent manner. Another study based on animal experimentation involved administration of high fat diet which led to induction of hyperglycemia, hyperinsulinemia, and hyperleptinemia. Here, the inclusion of dietary cyanidin curbed the concentration of liver total lipids and triacylglycerols by inhibiting the enzymes involved in the process (Takanori et al., 2003).

Another entity of cyanidin that is cyanidin-3-O-β glucoside has been found to be most potent anthocyanin and has been found to possess health promoting properties. Associative progression and occurrence of diabetes and CVDs are very well documented. Evidence related to its protective role can be seen in the case of a study involving hypercholesterolemic mice in which administration of cyanidin-3-O-β glucoside led to reduction in atherogenic activity and increased endothelial function. Since endothelial dysfunction is one of the common complications in diabetes, therefore, the role of cyanidin is well documented to ameliorate the same (Liu et al., 2014). Activating phosphorylation of insulin receptors, activity, and transport of GLUT 4 are also supported by anthocyanins. This in turn, enhances glucose uptake and utilization (Różańska & Regulska-Ilow, 2018).

7.2.11 KAEMPFEROL

Kaempferolisyellowcolorflavonoid,chemicallyknownas3,5,7-trihydroxy-2-(4-hydroxyphenyl)-4H-1-benzopyran-4-one. It is abundant in plant kingdom found in green leafy vegetables like spinach, kale, dill, chives, and other vegetables like broccoli, cabbage, leeks, tomatoes, beans, etc. (Dabeek & Marra, 2019). It is a potent antioxidant scavenger especially super oxide and hydroxide radicals at very low concentrations in the cells makes it powerful antioxidant for diabetic patients. It also empowers antioxidant enzymes activity such as SOD, catalase (CAT), and heme oxygenase-1 that protect the body from oxidative stress. Diabetic patients are at high risk of atherosclerosis due to increased levels of circulating LDL cholesterol. Kaempferol protects LDL cholesterol from oxidation. It also has a higher efficiency of penetration into cells and outperforms other flavonoids to provide beneficial metabolic activity.

Kaempferol upregulates glucose uptake by cells via protein kinase C and PI3K pathways and allows better glucose tolerance on supplementation. It also promotes glycogenesis in muscles and liver keeping the body energized. It also protects β-cell mass and reduce the risk of diabetes by delaying its onset. Literature reveals that Bauhinia forficate leaves rich sources of kaempferol supplementation, helps in stabilizing blood sugar levels and promotes insulin sensitivity by cells as it imitates insulin in the body. It triggers expression of adenosine monophosphate, reduces cell death by curbing caspase 3 activities. It suppresses gluconeogenesis activity in hepatic cells, generally high in insulin resistance diabetes. Animal experimentation studies report that it reduces diabetic complications especially diabetic nephropathy (Calderón-Montaño et al., 2011; Al-Ishaq et al., 2019).

7.2.12 MORIN

The chemical name of Morin is 3,5,7,2′,4′-pentahydroxyflavone. It is a bioflavonoid present in many herbs and fruits and especially *Moracae* family. Its food sources include almonds, onion, apples, guava, osage orange and cereal grains. It is also found in beverages like tea, coffee, red wine. The protective biological activity of morin shows it to be cytoprotective antioxidant, anti-inflammatory, and anti-mutagenic agent (Francis et al., 1989). Morin has cellular protective activity, it safeguards various cells of the human body against oxidative damage – such as myocytes, hepatocytes, erythrocytes, endothelial cells, etc. (Numair et al., 2014). The anti-inflammatory activity of this flavonoid is attributed to inhibition of protein glycosylation which further reduces inflammation indicators in the central nervous system.

Sendraperumal, reported improvement in glucose tolerance, hyperglycemia, and insulin in rats which were orally administered morin for a period of 30 days (Sendraperumal et al., 2014). Here, a complex mixture of zinc-morin was given in a dose of 5 mg/kg body weight/day. Also, there was a reduction in TNF alpha level, which shows its anti-inflammatory property as well.

Psidium guajava (Guava) possesses anti-diabetic activities, this has been manifested from the methanolic extract of the fruit, which contains a good amount of morin (Basha & Vinoji, 2012). Since, gluconeogenesis is enhanced in diabetic patients, morin has been shown to arrest this activity and enhanced activation of insulin receptors. Normalization of blood glucose levels can be attributed to the ability of morin to imitate insulin in the body (Choudhury et al., 2017). Thus, research-based evidence suggests the probable utilization of Morin as a potent substitute for synthetic OHDs.

7.2.13 LUTEOLIN

Luteolin is an emerging anti-cancerous compound known as 3,4,5,7-tetrahydroxy flavone and is heat stable antioxidant. It is abundant in various parts of plants like chrysanthemum flowers, peels of apple, leaves of onion, cabbage, carrots, bell peppers, broccoli, parsley, etc. Like any other flavonoid it too has numerous biological health benefits and is involved cell signaling mechanism (Imran et al., 2019). It exhibits its role due to its structure comprising hydroxyl group and double bond between second and third carbon atom. Post-digestion it is further metabolized into free form luteolin and other metabolites such as glucuronides or sulfate that possess anti-inflammatory role in the body. Akt2 kinase is involved in regulating glycogenesis, gluconeogenesis, and transportation of glucose in insulin signaling process. Luteolin improves insulin cell sensitivity by promoting Akt2 kinase activity in diabetes.

Apart from this luteolin stimulates the activity of enzyme heme-oxygenase-1 that prevents oxidative stress in kidneys and protect against nephropathy. It helps maintain optimum blood glucose levels and weight gain in obese patients. Luteolin is also known to be permeable to the brain tissues and improve the health of nerves and prevent diabetic neuropathy. It also alleviates health of the retina that is hampered in hyperglycemia and reduce the risk of diabetic retinopathy in chronic diabetes. Research indicates that regular consumption of luteolin stabilizes inflammatory markers and protects renal, hepatic, cardiac, and lung tissues from damage caused by elevated levels of glucose and side-effects of medications that pose serious threat to diabetic patients (Sangeetha, 2019).

7.2.14 CHRYSIN

Chrysin belongs to the flavonoid class–flavone. It is a hydroxylated flavonoid. It is also known as 5,7-dihydroxyflavone and possesses a plethora of biological activities ranging from prevention of oxidative stress, neurodegeneration, inflammation, and progression of cancer. Food products like honey, propolis, passionflower, etc. However, since the solubility of chrysin is less, which directly affects its bioavailability. Chrysin has been shown to enhance the detoxification capabilities of the liver and prevents hepatic necrosis. Its role in protection of the liver from the effects of long-term alcoholism and also non-alcoholic fatty liver disease cannot be neglected.

The antioxidant effect of chrysin leads to reduction in malodialdehyde and increases the levels of SOD, CAT, and glutathione. Complications of diabetes could be aggravated due to increase oxidative stress, which can be managed with administration of Chrysin (Jung, 2016). It may be used in the form of nanocapsules for patients suffering from diabetes and dyslipidemia. There is ongoing research regarding the potential of chrysin as a carrier of glucose into the cells that may benefit diabetic patients in the future (Goracy et al., 2021). Certain animal-based studies show chrysin to exhibit effects similar to metformin and reduce glucose and triglyceride levels. The production of inflammatory cytokines are a characteristic feature of diabetes, here chrysin interferes with its production (Espinosa et al., 2017).

7.2.15 DELPHINIDIN

The chemical name for delphinidin is 3,3,'4,'5,5,'7-hexahydroxyflavylium. It is a flavonoid compound belonging to the subclass–anthocyanidin and is abundantly present in berries and red wine (Dorman et al., 2021). Delphinidin possesses antioxidant and anti-inflammatory properties which is well documented through a number of research conducted in this regard. A study conducted by Daveri et al. (2018) showed mitigating effects of high fat diet induced on mice. Attenuation of insulin resistance, dyslipidemia, steatosis, and liver inflammation was seen with supplementation of cyanidin and delphinidin.

Another human based study involved the provision of small doses of delphinidin to the subjects for a period of one week. The results showed better glucose control as delphinidin reduced the functionality of glucose transporter and enhanced insulin sensitivity (Alvarado et al., 2016). GLP-1 (Glucagon like peptide-1) is a gut hormone secreted by the intestinal endocrine cells. They are known to promote insulin secretion in the post-prandial state and induces multiplication of the β cells of the pancreas. Study based evidence suggests delphinidin as a potent stimulator of GLP-1 (Tani et al., 2017). As previously mentioned, the major food sources of delphinidin are blue and red color pigmented plants that include berries, grapes, purple sweet potato, red cabbage, etc.

7.2.16 WOGONIN

Wogonin is a flavone with a chemical name 5,7-dihydroxy-8-methoxy-flavone. This flavonoid falls under the category of flavones and has been found to possess anti-inflammatory properties. Research have shown that wogonin may prevent vascular inflammation which is seen in hyperglycemia induced rats. Atherosclerosis is a common health problem seen with progression in type-2 diabetes. It has been seen that this flavonoid is a dominant compound present in *Sculletaria bacelensis Georgi*, which is a dried root commonly used for its therapeutic use in Chinese herbal medicine. The therapeutic properties include anti-viral, anti-inflammatory, antioxidant, and anti-carcinogenic properties (Rad et al., 2021). It is also known to reduce the risk of cardiomyopathy in Diabetic patients. It also prevents cell death and reduce the risk of CVDs (Khan & Kamal, 2019).

7.3 CONCLUSION

Plant based foods have a potential role in health and disease. The food that we consume contains many nutrients such as proteins, carbohydrates, fats, and minerals that are utilized in cell metabolism. During this process of cell metabolism, a number of toxic compounds and free radicals are produced. We as humans do not produce any specific compound to stabilize/neutralize the effect of these toxic constituents. Only plant-based foods contain nature-gifted compounds known as phytonutrients. These phytonutrients have antioxidant ability that prevents deleterious effects of toxins like cellular apoptosis, endothelial dysfunction, plaque formation, insulin resistance, proliferation of cancerous cells, all of them involved in the development of some or other non-communicable diseases. Evidence-based studies show the potent role of phytochemicals like flavonoids in mitigating a number of lifestyle and non-communicable diseases like diabetes, CVDs, atherosclerosis, etc.

The revered role of flavonoids in alleviating diabetes has been demonstrated in several animals, cell-culture, and human based experiments. Flavonoids have an important role in glucose metabolism as it regulates glycogenesis and promotes the cellular uptake of glucose. It activates protein kinase C and phosphoinositide kinase 3 pathway that helps in the transportation of glucose and prevents hyperglycemia. It is known to enhance the activity of insulin receptors and insulin cell sensitivity by stimulating the activity of enzyme AKt2 Kinase. Many flavonoid subclasses suppress the process of gluconeogenesis and promotes glucose metabolism. Other flavonoids also help in the regeneration and proliferation of β cells of the pancreas and prevents its premature death. Reduced insulin secretion is also one of the major clinical symptoms in diabetes, and flavonoids facilitates its secretion by activating the glucose transporter GLUT-1,4 and also GLP-1 (a gut interim).

Flavonoids not only have a role in management of diabetes but also involved in other physiological processes that are affected in diabetes. Some of them includes management of inflammation by reducing the production inflammatory cytokines, increase production of antioxidant enzymes like SOD, CAT, and glutathione – thus reducing/prevent oxidative stress and stabilizing free radicals. This effect is seen in all organs of the body, while their functioning is reduced in Diabetes. It protects from hepatic necrosis, neurodegeneration, nephropathy, retinopathy, cardiomyopathy, and cancer. Renal nephropathy is common in diabetics patients, due to persistent spike in glucose, kidneys are over-burdened. Flavonoids are known to enhance the functioning of hemeoxygenase 1 which protects against renal damage and nephropathy. Many experimental studies have shown similar ameliorative effects.

These phytonutrients also have been tested to play an important role in managing dyslipidemia. It has lipid lowering effects on the body and may prevent the development of atherosclerosis by reducing the level of lipid hydroperoxides–which are potentially atherogenic. Anti-hypercholestrolemic properties of flavonoids are owing to its ability to lower LDL Cholesterol. Flavonoids imitate the drug metformin and other OHDs in the body and participate in glucose metabolism. OHDs have long-term implications and may induce health problems in later life. The potential use of nanocapsules containing extracts of flavonoids is being tested as a potential therapy for the management of diabetes. Hence, focused research should be done to understand the therapeutic role of flavonoids and promote their use in clinical practice.

KEYWORDS

- **diabetes**
- **flavonoids**
- **gluconeogenesis**
- **hyperglycemia**
- **insulin**
- **oral hypoglycemic drugs**
- **reactive oxygen species**
- **transforming growth factor β-I**
- **tumor necrosis factor α**

REFERENCES

Al-Ishaq, R. K., Abotaleb, M., Kubatka, P., Kajo, K., & Büsselberg, D., (2019). Flavonoids and their anti-diabetic effects: Cellular mechanisms and effects to improve blood sugar levels. *Biomolecules, 9*(9), 430.

Al-Numair, K. S., Chandramohan, G., Alsaif, M. A., Veeramani, C., & El Newehy, A. S., (2014). Morin, a flavonoid, on lipid peroxidation and antioxidant status in experimental myocardial ischemic rats. *African Journal of Traditional, Complementary, and Alternative Medicines, 11*(3), 14–20.

Amić, D., Davidović-Amić, D., Beslo, D., Rastija, V., Lucić, B., & Trinajstić, N., (2007). SAR and QSAR of the antioxidant activity of flavonoids. *Curr. Med. Chem., 14*, 827–845.

Barky, A. R. E., Ezz, A. A., & Mohammed, T. M., (2020). The potential role of apigenin in diabetes mellitus. *International Journal of Clinical Case Reports and Reviews, 3*(1).

Basha, S. K., & Vinoji, S. K., (2012). *In vitro* antidiabetic activity of *Psidium guajava* leaves extracts. *Asian Pacific Journal of Tropical Disease*.

Calderón-Montaño, J. M., Burgos-Morón, E., Pérez-Guerrero, C., & López-Lázaro, M., (2011). A review on the dietary flavonoid kaempferol. *Mini Reviews in Medicinal Chemistry, 11*(4), 298–344.

Choudhury, A., Chakraborty, I., Banerjee, T. S., Vana, D. R., & Adapa, D., (2017). Efficacy of morin as a potential therapeutic phytocomponent: Insights into the mechanism of action. *Int. J. Med. Res. Health Sci., 6*, 175–194.

Dabeek, W. M., & Marra, M. V., (2019). Dietary quercetin and kaempferol: Bioavailability and potential cardiovascular-related bioactivity in humans. *Nutrients, 11*(10), 2288.

Daniela, S. P., & Marius E. R., (2017). *Isoflavones: Vegetable Sources, Biological Activity, and Analytical Methods for Their Assessment, Superfood and Functional Food–The Development of Superfoods and Their Roles as Medicine*. IntechOpen. Available from: https://www.intechopen.com/chapters/53352 (accessed on 13 February 2023).

David, A. V. A., Arulmoli, R., & Parasuraman, S. (2016). Overviews of biological importance of quercetin: A bioactive flavonoid. *Pharmacognosy Reviews, 10*(20), 84–89.

Di Carlo, G., Mascolo, N., Izzo, A. A., & Capasso, F., (1999). Flavonoids: Old and new aspects of a class of natural therapeutic drugs. *Life Sci., 65*(4), 337–53.

Dormán, G., Flachner, B., Hajdú, I., & András, C., (2021). *Nutraceuticals* (2nd edn., pp. 315–343). Academic Press.

El-Shaimaa, A. A., Noura, T. S., & Manal, A. B., (2021). Therapeutic implications of a polymethoxylated flavone, tangeretin, in the management of cancer via modulation of different molecular pathways. *Advances in Pharmacological and Pharmaceutical Sciences*.

Elena, D., Eleonora, C., Angela, M., Shelly, N. H., Steven, M. W., Andrew, L. W., Mauri, A., et al., (2018). Cyanidin and delphinidin modulate inflammation and altered redox signaling improving insulin resistance in high fat-fed mice. *Redox Biology, 18*, 16–24.

Francis, A. R., Shetty, T. K., & Bhattacharya, R. K., (1989). Modulating effect of plant flavonoids on the mutagenicity of N-methyl-N'-nitro-N-nitrosoguanidine. *Carcinogenesis, 10*(10), 1953–5.

Frankenfeld, C. L., Atkinson, C., Thomas, W. K., Gonzalez, A., Jokela, T., Wähälä, K., Schwartz, S. M., Li, S. S., & Lampe, J. W., (2009). High concordance of daidzein-metabolizing phenotypes in individuals measured 1 to 3 years apart. *Br. J. Nutr., 94*, 873–876.

Fu, Y., (2012). *Baicalein, a Novel Anti-Diabetic Compound.* (Master's Dissertation), Virginia Polytechnic Institute and State University. Virginia Tech Works. https://vtechworks.lib.vt.edu/handle/10919/76854 (accessed on 13 February 2023).

Fu, Z., & Liu, D., (2009). Long-term exposure to genistein improves insulin secretory function of pancreatic beta-cells. *Eur. J. Pharmacol., 616*, 321–327.

Fu, Z., Zhang, W., Zhen, W., Lum, H., Nadler, J., Bassaganya-Riera, J., Jia, Z., et al., (2010). Genistein induces pancreatic beta-cell proliferation through activation of multiple signaling pathways and prevents insulin-deficient diabetes in mice. *Endocrinology, 151*(7), 3026–3037.

Ghorbani, A., (2017). Mechanisms of antidiabetic effects of flavonoid rutin. *Biomedecine & Pharmacotherapie., 96*, 305–312.

Giusti, M., & Wrolstad, R., (2003). Acylated anthocyanins from edible sources and their applications in food systems. *Biochem Eng J., 14*, 217–225.

Gong, G., Guan, Y. Y., Zhang, Z. L., Rahman, K., Wang, S. J., Zhou, S., Luan, X., & Zhang, H., (2020). Isorhamnetin: A review of pharmacological effects. *Biomedicine & pharmacotherapy = Biomedecine & pharmacotherapie, 128*, 110301.

Guo, H., Ling, W., Wang, Q., Liu, C., Hu, Y., & Xia, M., (2007). Cyanidin 3-glucoside protects 3T3-L1 adipocytes against H(2)O(2)- or TNF-alpha-induced insulin resistance by inhibiting c-Jun NH(2)-terminal kinase activation. *Biochem. Pharmacol.*.

György, D., Beáta, F., István, H., & Csaba, A. (2021). *Target Identification and Polypharmacology of Nutraceuticals*. (pp. 315–343). Academic Press.

Huan, Y., Hengyang, P., Xinyue, W., Xinmei, X., Tingting, K., Jing, Z., Leilei, D., & Gang, F., (2021). The therapeutic effects and mechanisms of quercetin on metabolic diseases: Pharmacological data and clinical evidence. *Oxidative Medicine and Cellular Longevity*.

Hudish, L. I., Reusch, J. E., & Sussel, L., (2019). β Cell dysfunction during progression of metabolic syndrome to type 2 diabetes. *The Journal of Clinical Investigation, 129*(10), 4001–4008.

Hussain, S., & Marouf, B., (2013). Flavonoids as alternatives in treatment of type 2 diabetes mellitus. *Academia Journal of Medicinal Plants, 1*, 31–36.

Imran, M., Rauf, A., Abu-Izneid, T., Nadeem, M., Shariati, M. A., Khan, I. A., Imran, A., et al., (2019). Luteolin, a flavonoid, as an anticancer agent: A review. *Biomedicine & Pharmacotherapy = Biomedecine & Pharmacotherapie, 112*, 108612.

Iwashina, T., (2013). Flavonoid properties of five families newly incorporated into the order *Caryophyllales. Bull Natl Mus Nat Sci., 39*, 25–51.

Javad, S. R., Jesus, H. B., Luis, A. S., Shabnum, S., Seyed, A. A., Farzad, K., Muhammad, I. A., et al., (2021). The therapeutic potential of wogonin observed in preclinical studies. *Evidence-Based Complementary and Alternative Medicine*.

Jayagopal, V., Albertazzi, P., Kilpatrick, E. S., Howarth, E. M., Jennings, P. E., Hepburn, D. A., & Atkin, S. L., (2002). *Diabetes Care, 25*, 1709–1714.

Jorge, L. A., Andrés, L., Olivera-Nappa, A., Ana-María, S., Hernán, R., Carolina, L., & Pilar, V., (2016). Delphinidin-rich maqui berry extract (Delphinol®) lowers fasting and postprandial glycemia and insulinemia in prediabetic individuals during oral glucose tolerance tests. *BioMed Research International*.

Jung, J., (2016). Emerging utilization of chrysin using nanoscale modification. *Journal of Nanomaterials*.

Kamalakkannan, N., & Prince, P. S., (2006). Anti-hyperglycaemic and antioxidant effect of rutin, a polyphenolic flavonoid, in streptozotocin-induced diabetic Wistar rats. *Basic & Clinical Pharmacology & Toxicology, 98*(1), 97–103.

Keservani, R. K., & Sharma, A. K., (2014). Flavonoids: Emerging trends and potential health benefits. *Journal of Chinese Pharmaceutical Sciences, 23*(12), 815.

Keservani, R. K., Kesharwani, R. K., Sharma, A. K., Vyas, N., & Chadoker, A., (2010b). Nutritional supplements: An overview. *International Journal of Current Pharmaceutical Review and Research, 1*(1), 59–75.

Keservani, R. K., Kesharwani, R. K., Vyas, N., Jain, S., Raghuvanshi, R., & Sharma, A. K., (2010a). Nutraceutical and functional food as future food: A review. *Der Pharmacia Lettre, 2*(1), 106–116.

Keservani, R. K., Sharma, A. K., & Jarouliya, U., (2015). Protein and peptide in drug targeting and its therapeutic approach. *ARS Pharmaceutica (Internet), 56*(3), 165–177.

Keservani, R. K., Sharma, A. K., & Kesharwani, R. K., (2016a). Nutraceutical and functional foods for cardiovascular health. In: *Food Process Engineering* (pp. 257–278, 291–312). Apple Academic Press, CRC Press. ISBN: 9781771884020.

Keservani, R. K., Sharma, A. K., & Kesharwani, R. K., (2016b). Medicinal effect of nutraceutical fruits for the cognition and brain health. *Scientifica*, 10. Article ID 3109254. doi: 10.1155/2016/3109254.

Keservani, R. K., Sharma, A. K., & Kesharwani, R. K., (2020). *Nutraceuticals and Dietary Supplements: Applications in Health Improvement and Disease Management*. CRC Press. ISBN: 9781771888738.

Khan, N., Syed, D. N., Ahmad, N., & Mukhtar, H., (2013). Fisetin: A dietary antioxidant for health promotion. *Antioxidants & Redox Signaling, 19*(2), 151–162.

Khan, S., & Kamal, M. A., (2019). Can wogonin be used in controlling diabetic cardiomyopathy? *Current Pharmaceutical Design, 25*(19), 2171–2177.

Kim, J. W., Lim, S. C., Lee, M. Y., Lee, J. W., Oh, W. K., Kim, S. K., & Kang, K. W., (2010). Inhibition of neointimal formation by trans-resveratrol: Role of phosphatidyl inositol 3-kinase-dependent Nrf2 activation in heme oxygenase-1 induction. *Mol. Nutr. Food Res., 54*, 1497–1505.

Kim, M. S., Hur, H. J., Kwon, D. Y., & Hwang, J. T., (2012). Tangeretin stimulates glucose uptake via regulation of AMPK signaling pathways in C2C12 myotubes and improves glucose tolerance in high-fat diet-induced obese mice. *Molecular and Cellular Endocrinology, 358*(1), *127–134.*

Laddha, A. P., & Kulkarni, Y. A., (2021). Daidzein mitigates myocardial injury in streptozotocin-induced diabetes in rats. *Life Sciences*, 284.

Li, W., Chen, Z., Yan, M., He, P., Chen, Z., & Dai, H., (2016). The protective role of isorhamnetin on human brain microvascular endothelial cells from cytotoxicity induced by methylglyoxal and oxygen-glucose deprivation. *Journal of Neurochemistry, 136*(3), 651–659.

Liggins, J., Bluck, L. J., Runswick, S., Atkinson, C., Coward, W. A., & Bingham, S. A., (2000). Daidzein and genistein contents of vegetables. *Br. J. Nutr., 84*(5), 717–725.

Li-Ping, Y., Hui-Li, S., Le-Meng, W., Xiu-Juan, G., Hong-Liang, D., Mark, O. M. T., Lin, Z., & Shun-Min, L., (2009). Baicalein reduces inflammatory process in a rodent model of diabetic retinopathy. *Invest. Ophthalmol. Vis. Sci., 50*(5), 2319–2327.

Liu, Y., Li, D., Zhang, Y., Sun, R., & Xia, M., (2014). Anthocyanin increases adiponectin secretion and protects against diabetes-related endothelial dysfunction. *American Journal of Physiology. Endocrinology and Metabolism, 306*(8), E975–E988.
Min, J. K., & Yunsook, L., (2013). Protective effect of short-term genistein supplementation on the early stage in diabetes-induced renal damage. *Mediators of Inflammation.*
Neveu, V., Perez-Jimenez, J., Vos, F., Crespy, V., Du, C. L., Mennen, L., Knox, C., Eisner, R., & Cruz, J., (2010). Phenol-explorer: An online comprehensive database on polyphenol contents in foods. *Database (Oxford).*
Panche, A. N., Diwan, A. D., & Chandra, S. R., (2016). Flavonoids: An overview. *Journal of Nutritional Science, 5*, e47.
Park, M. H., Ju, J. W., Kim, M., & Han, J. S., (2016). The protective effect of daidzein on high glucose-induced oxidative stress in human umbilical vein endothelial cells. *Z. Naturforsch C J. Biosci., 71*(1, 2), 21–28.
Qin, D., & Jiang, Y. R., (2020). Tangeretin inhibition of high-glucose-induced IL-1β, IL-6, TGF-β1, and VEGF expression in human RPE cells. *Journal of Diabetes Research*, 9490642.
Ramírez-Espinosa, J. J., Saldaña-Ríos, J., García-Jiménez, S., Villalobos-Molina, R., Ávila-Villarreal, G., Rodríguez-Ocampo, A. N., Bernal-Fernández, G., & Estrada-Soto, S., (2017). Chrysin induces antidiabetic, anti-dyslipidemic and anti-inflammatory effects in athymic nude diabetic mice. *Molecules (Basel, Switzerland), 23*(1), 67.
Rane, B. R., Bharath, M. S., Patil, R. R., Keservani, R. K., & Jain, A. S., (2020b). Novel approaches in nutraceuticals. In: Rajesh, K. K., Raj, K. K., & Anil, K. S., (eds.), *Enhancing the Therapeutic Efficacy of Herbal Formulations Through Novel Drug Delivery Systems* (pp. 241–266). IGI Global International Publisher, Pennsylvania, USA, Chapter 11. ISBN13: 9781799844532.
Rane, B. R., Patil, A. S., Keservani, R. K., & Jain, A. S., (2020a). Novel approaches in herbal formulation. In: Rajesh, K. K., Raj, K. K., & Anil, K. S., (eds.) *Enhancing the Therapeutic Efficacy of Herbal Formulations Through Novel Drug Delivery Systems* (pp. 43–68). IGI Global International Publisher, Pennsylvania, USA, Chapter 2. ISBN13: 9781799844532.
Rane, B. R., Tadavi, S. A., & Keservani, R. K., (2020c). Naturopathy. In: Anil, K. S., Raj, K. K., & Surya, P. G., (eds.), *Herbal Product Development* (pp. 321–347). Apple Academic Press, CRC Press, Taylor & Francis Group, Chapter 12. ISBN: 9781771888776.
Rawat, S., Pathak, S., Gupta, G., Singh, S. K., Singh, H., Mishra, A., & Gilhotra, R., (2019). Recent updates on daidzein against oxidative stress and cancer. *EXCLI Journal, 18*, 950–954.
Ríha, M., Karlícková, J., Filipský, T., Jahodár, L., Hrdina, R., & Mladenka, P., (2014). *In vitro* copper-chelating properties of flavonoids. *Free Radical Biology & Medicine, 75*(1), S46.
Różańska, D., & Regulska-Ilow, B., (2018). The significance of anthocyanins in the prevention and treatment of type 2 diabetes. *Advances in Clinical and Experimental Medicine: Official Organ Wroclaw Medical University, 27*(1), 135–142.
Saeedi, P., Petersohn, I., Salpea, P., Malanda, B., Karuranga, S., Unwin, N., Colagiuri, S., Guariguata, L., Motala, A. A., Ogurtsova, K., Shaw, J. E., Bright, D., Williams, R., & IDF Diabetes Atlas Committee, (2019). Global and regional diabetes prevalence estimates for 2019 and projections for 2030 and 2045: Results from the International Diabetes Federation Diabetes Atlas. *Diabetes Research and Clinical Practice* (9th edn., Vol. 157, p. 107843).

Salehi, B., Venditti, A., Sharifi-Rad, M., Kręgiel, D., Sharifi-Rad, J., Durazzo, A., Lucarini, M., et al., (2019). The therapeutic potential of apigenin. *International Journal of Molecular Sciences, 20*(6), 1305.

Sangeetha, R., (2019). Luteolin in the management of type 2 diabetes mellitus. *Curr. Res. Nutr. Food Sci., 7*(2).

Sari, E. N., & Soysal, Y., (2020). Molecular and therapeutic effects of fisetin flavonoid in diseases. *J. Basic Clin. Health Sci., 4*, 190–196.

Sendrayaperumal, V., Iyyam, P. S., & Subramanian, S., (2014). Design, synthesis and characterization of zinc-morin, a metal flavonol complex and evaluation of its antidiabetic potential in HFD-STZ induced type 2 diabetes in rats. *Chem. Biol. Interact., 219*, 9–17.

Sreedharan, V., Venkatachalam, K. K., & Namasivayam, N., (2009). Effect of morin on tissue lipid peroxidation and antioxidant status in 1, 2-dimethylhydrazine induced experimental colon carcinogenesis. *Investig. New Drugs, 27*, 21–30.

Stompor-Gorący, M., Bajek-Bil, A., & Machaczka, M., (2021). Chrysin: Perspectives on contemporary status and future possibilities as pro-health agent. *Nutrients, 13*(6), 2038.

Sundaram, R., Shanthi, P., & Sachdanandam, P., (2014). Effect of tangeretin, a polymethoxylated flavone on glucose metabolism in streptozotocin-induced diabetic rats. *Phytomedicine: International Journal of Phytotherapy and Phytopharmacology, 21*(6), 793–799.

Takanori, T., Fumihiko, H., Kaji, U., Hiromitsu, A., & Thoshihiko, O., (2003). Dietary cyanidin 3-O-beta-D-glucoside rich purple corn color prevents obesity and ameliorates hyperglycemia in mice". *J. Nutr., 133*, 2125–2130.

Tani, T., Nishikawa, S., Kato, M., & Tsuda, T., (2017). Delphinidin 3-rutinoside-rich blackcurrant extract ameliorates glucose tolerance by increasing the release of glucagon-like peptide-1 secretion. *Food Science & Nutrition, 5*(4), 929–933.

Teles, Y., Souza, M., & Souza, M., (2018). Sulphated flavonoids: Biosynthesis, structures, and biological activities. *Molecules (Basel, Switzerland), 23*(2), 480.

Thomassian, B., (2009). *Cardiac Nursing Textbook* (6th edn.). Wolters Kluwer and Lippincott Williams and Wilkins.

Toshiya, M., Myra, O. V., & Hiroko, I., (2016). The type 2 diabetes-preventive effect of cyanidin-3-glucoside on adipocytes. *Journal of Developments in Sustainable Agriculture, 11*(1), 31–35.

Villa, P., Costantini, B., Suriano, R., Perri, C., Macri, F., Ricciardi, L., Panunzi, S., & Lanzone, A., (2009). *The Journal of Clinical Endocrinology and Metabolism, 94*, 552–558.

Zeka, K., Ruparelia, K., Arroo, R., Budriesi, R., & Micucci, M., (2017). Flavonoids and their metabolites: Prevention in cardiovascular diseases and diabetes. *Diseases (Basel, Switzerland), 5*(3), 19.

CHAPTER 8

APPLICATION OF FLAVONOIDS IN PREGNANCY-INDUCED DISORDERS

TARKESHWAR[1] and PRADIP KUMAR SAINI[2]

[1]*Department of Genetics and Plant Breeding, Acharya Narendra Deva University of Agriculture and Technology, Kumarganj, Ayodhya, Uttar Pradesh, India*

[2]*Department of Crop Physiology, Acharya Narendra Deva University of Agriculture and Technology, Kumarganj, Ayodhya, Uttar Pradesh, India*

ABSTRACT

There are numerous biological resources in nature that are vital to health care. There has been an increase in interest in research on flavonoids from plant sources due to their versatile health effects demonstrated in many epidemiological studies. Flavonoids' structure and function linkages must be evaluated because they are closely related to human dietary elements and health. Flavonoids' bioavailability, metabolism, and biological activity are influenced by their nuclear structure, the total number of hydroxyl groups, and functional group substitution. The principal dietary sources of flavonoids for humans include fruits and vegetables, as well as tea and wine. This chapter focuses on defining the importance of flavonoids, which have a wide range of pharmacological effects, including anti-cancer, antioxidant, anti-hypertensive, and anti-cholinesterase properties. These flavonoids have shown promise in the treatment of pregnancy-related

Advances in Flavonoids for Human Health and Prevention of Diseases. Nisha Sharma, Deepika Saini, Rajesh K. Kesharwani, Prakash Chandra Gupta, & Raj K. Keservani (Eds.)

illnesses. We also attempted to discuss a few research that demonstrates the efficacy of flavonoids in the treatment of pandemic-19. They've also been shown to have more potential in treating neurodegenerative illnesses. Nanomedicine will also play a larger role in the near future, as it aids flavonoids in demonstrating effective targeted drug delivery.

8.1 INTRODUCTION

Flavonoids, popularly known as dietary bio-actives to prevent human diseases, as well as their potential as pharmaceutical leads, has skyrocketed in recent years. Flavonoids are a subclass of plant polyphenols that have been proven in a variety of studies, ranging from cell-based assays to epidemiological and human intervention studies, to have several health-promoting physiological advantages. Beyond their role as biologically active plant-food components, fresh insights into the potential utilization of flavonoid derivatives as effective cancer treatments are provided. Flavonoids have also shown promise in the treatment of obesity and inflammation-related disorders, as well as the prevention of infectious infections like COVID-19. Flavonoids can be found in a variety of foods and natural health products. Flavonoids have a 15-carbon skeleton that is made up of C_6-C_3-C_6, with variable structural configurations for distinct subclasses. Flavanols or catechins (e.g., epigallocatechin (EGC) 3-gallate from green tea), flavones (e.g., apigenin from celery), flavonols (e.g., quercetin glycosides from apples, berries, and onions), flavanones (e.g., naringenin from citrus), anthocyanins (e.g., cyanidin-3-O-glucose (e.g., genistein from soya beans). Regular consumption of dietary flavonoids in enough proportions has been demonstrated to reduce the risk of oxidative stress and chronic inflammation-mediated pathogenesis of human diseases such as cardiovascular disease (CVD), cancer, and neurological disorders (Williams et al., 2018).

Multiple mechanisms of action have been demonstrated for dietary flavonoids physiological benefits, including redox homeostasis, epigenetic regulation, activation of survival genes and signaling pathways, regulation of mitochondrial function and bioenergetics, and modulation of inflammation response. The impact of flavonoids microbial metabolites on optimal health and their role in gut microbiota has begun to emerge. Flavonoid molecules' structural variety is responsible for their diverse physiological modulations. Some flavonoids, however, are poorly absorbed, and their bioavailability could be improved by structural alterations and nanotechnology uses such as encapsulation (Rupasinghe, 2020).

8.2 CLASSIFICATIONS OF FLAVONOIDS

The carbon of the C ring on which the B ring is bonded, as well as the degree of unsaturation and oxidation of the C ring, can be used to separate flavonoids into various subgroups (Figure 8.1). Isoflavones are flavonoids in which the B ring is connected in position 3 of the C ring. Neoflavonoids are those in which the B ring is joined in position 4; those in which the B ring is linked in position 2 can be further split into numerous subgroups based on the structural properties of the C ring. These subgroups are flavones, flavonols, flavanones, flavanonols, flavanols or catechins, anthocyanins, and chalcones (Figure 8.1).

8.2.1 FLAVONES

Flavones are one of the most important flavonoid subclasses. Flavones are found as glycosides in leaves, flowers, and fruits. Flavones can be found in celery, parsley, red peppers, chamomile, mint, and ginkgo biloba. This group of flavonoids includes luteolin, apigenin, and tangeritin. The polymethoxylated flavones tageretin, nobiletin, and sinensetin (Manach et al., 2004) are abundant in citrus fruit peels. They have a double bond between positions 2 and 3 of the C ring, as well as a ketone in position 4. The hydroxyl group in position 5 of the A ring is found in the majority of flavones from vegetables and fruits, while hydroxylation in other positions, most commonly in position 7 of the A ring or 3′ and 4′ of the B ring, varies depending on the taxonomic classification of the particular vegetable or fruit.

8.2.2 FLAVONOLS

Flavonoids with a ketone group are known as flavonols. Proanthocyanins are made up of these basic units. Flavonols can be found in abundance in a wide range of fruits and vegetables. The most studied flavonols are kaempferol, quercetin, myricetin, and fisetin. A wide range of vegetables and fruits like tomatoes, onions, lettuce, kale, grapes, apples, and berries are a great source of flavonols. Besides these fruits and vegetables, some other beverages like tea and red wine are also a good source of flavonols. Flavonol consumption has been linked to a variety of health advantages, including antioxidant potential and a lower risk of vascular disease. Flavonols have a hydroxyl group in position 3 of the C ring, which can be glycosylated, unlike flavones. Flavonols, like flavones, have a wide range of methylation and hydroxylation patterns, and they are the most prevalent and largest subclass of flavonoids in fruits and vegetables, owing to the many glycosylation patterns. Many plant meals, for example, include quercetin (Iwashina, 2013).

8.2.3 FLAVANONES

Flavanones are another important class of compounds that can be found in citrus fruits, including oranges, lemons, and grapes. This group of flavonoids includes hesperitin, naringenin, and eriodictyol. Flavanones are associated with a number of health benefits because of their free radical scavenging properties. Citrus fruit juice and peel contain these chemicals, which give them a bitter taste. Citrus flavonoids have pharmacological actions that include antioxidant, anti-inflammatory, blood lipid-lowering, and cholesterol-lowering properties. The C ring is saturated in flavanones, also known as dihydroflavones; thus, unlike flavones, the double bond between positions 2 and 3 is saturated, and this is the only structural variation between the two subgroups of flavonoids. In the last one and half decades, a number of flavanones has been significantly increased (Iwashina, 2013).

8.2.4 ISOFLAVONOIDS

Flavonoids are divided into isoflavonoids, which constitute a big and separate subgroup. Isoflavonoids are found mostly in soybeans and other leguminous plants and have a limited distribution in the plant kingdom. Microbes have also been found to contain certain isoflavonoids (Matthies et al., 2008). During plant-microbe interactions, they are also discovered to play a crucial function as precursors for the formation of phytoalexins

(Aoki et al., 2013). Isoflavonoids have a lot of potential in terms of disease prevention. Because of their estrogenic action in certain animal models, isoflavones such as genistein and daidzein are widely classified as phytoestrogens. Szkudelska & Nogowski examined the role of genistein in producing hormonal and metabolic alterations, which can affect a range of disease pathways (Szkudelska & Nogowski, 2007).

8.2.5 NEOFLAVONOIDS

Polyphenolic substances known as neoflavonoids are a type of polyphenol. Neoflavonoids contain a 4-phenylchromen backbone with no hydroxyl group substitution at position 2, whereas flavonoids have a 2-phenyl-chromen-4-one backbone. Calophyllolide, derived from *Calophyllum inophyllum* seeds, was the first neoflavone discovered from natural sources in 1951. It is also found in the bark and wood of *Mesua thwaitesii* (Linuma et al., 1987), a Sri Lankan indigenous plant.

8.2.6 FLAVANOLS, FLAVAN-3-OLS, OR CATECHINS

Flavanonols are the 3-hydroxy derivatives of flavanones, commonly known as dihydroflavonols or catechins. They are a multi-substituted and highly diverse subclass. Because the hydroxyl group is always attached to position 3 of the C ring, flavanols are also known as flavan-3-ols. There is no double bond between positions 2 and 3, unlike many flavonoids. Flavanols are prominently found in apples, bananas, blueberries, peaches, pears, etc. (Panche et al., 2016).

8.2.7 ANTHOCYANINS

Anthocyanins are pigments that give plants, flowers, and fruits their colors. The most studied anthocyanins are cyanidin, delphinidin, malvidin, pelargonidin, and peonidin. They're mostly found in the outer cell layers of cranberries, black currants, red grapes, merlot grapes, raspberries, strawberries, blueberries, bilberries, and blackberries, among other fruits. These compounds' stability, combined with their health benefits, allows them to be used in a number of applications in the food industry (Giusti & Wrolstad, 2003). The pH of the surrounding as well as methylation or acylation of the hydroxyl groups on the A and B rings also affects the color of anthocyanin (Iwashina, 2013).

8.2.8 CHALCONES

Chalcones are a subclass of flavonoids that are characterized by the absence of 'ring C' of the basic flavonoid skeleton structure (Figure 8.1). Hence, they can also be referred to as open-chain flavonoids. Phloridzin, phloretin, arbutin, chalconaringenin, etc., are the major examples of chalcones. Chalcones are present in substantial quantity in certain wheat products, tomatoes, pears, strawberries, etc. Chalcones and chalcones derivatives are getting more attention due to its numerous nutritional and biological benefits. The simplest and safest strategy to prevent diseases and adjust activities may be to consume flavonoids through food sources (Panche et al., 2016).

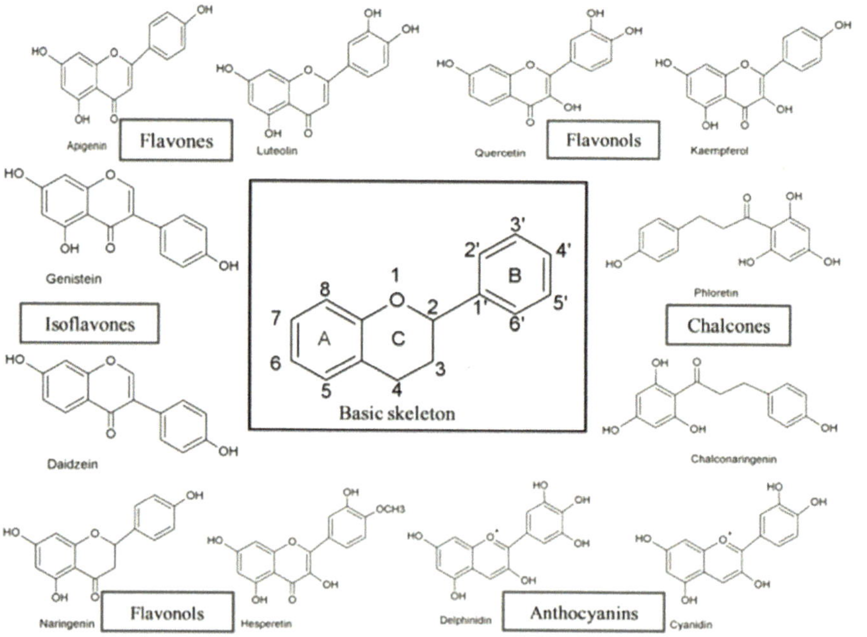

FIGURE 8.1 Basic skeleton structure of flavonoids and their classes

Source: Reprinted from Panche et al., 2016. (http://creativecommons.org/licenses/by/4.0/)

8.3 PLANT FLAVONOIDS IN PROMOTING HUMAN HEALTH, CURING, AND PREVENTING DISEASES

8.3.1 ANTIOXIDANT EFFECTS

Reactive oxygen species (ROS) and reactive nitrogen species (RNS) are formed as by-products of the cellular redox reaction that produces adenosine triphosphate (ATP) to provide energy for the cells using oxygen. ROS and RNS are useful chemicals for cellular processes and immunological responses when they are in balance, but when they are out of balance, they create oxidative stress, which can lead to chronic and degenerative diseases (Valko et al., 2006; Pham-Huy et al., 2008). Natural antioxidant compounds have been used in medical and pharmaceutical products as substitutes for artificial antioxidant chemicals, which have been thought to be one of the principal causes of carcinogenesis (Carocho et al., 2014). Plant secondary metabolites, such as phenolic compounds and flavonoids, which are produced by plants to defend themselves or stimulate development under unfavorable conditions, have long been described as a potential source of natural antioxidant chemicals in medicinal plants. Furthermore, the antioxidant activity of flavonoids, such as radical scavenging activity and/or metal ion chelation capacity, influenced functional group organization, configuration, substitution, and the quantity of hydroxyl groups (Heim et al., 2002). The major phytochemical compounds with antioxidant capabilities found in plants are phenolics and flavonoids (Andreu et al., 2018; Wang et al., 2016; Okpuzor et al., 2009).

8.3.2 ANTIBACTERIAL EFFECT

Interestingly, there are several flavonoids and phenolics that have antibacterial properties; these substances may be found in both non-flowering and flowering medicinal plants. The fern *Asplenium nidus* L. included gliricidin 7-O-hexoside and quercetin-7-O-rutinoside, which were found to be effective against three infections, including *Proteus mirabilis* Hauser, *Proteus vulgaris* Hauser and *Pseudomonas aeruginosa* (Schroeter) Migula (Jarial et al., 2018). Furthermore, a wide range of plant families, including many medicinal plant species used in traditional medicine and food use, produce flavonoid and phenolic chemicals. Nutmeg or *Myristica fragrans* Houtt., is an obvious example; this plant has long been utilized as a flavoring ingredient in Indonesia and other Southeast Asian countries (Nagja et al., 2016). However, a nutmeg seed ethanolic extract containing 30,40,7-trihydroxyflavone was found to be effective against MDR gram-negative bacteria such as *Providencia stuartii* Ewing and *Escherichia coli* (Migula) Castellani and Chalmers (Dzotam et al., 2018). *Pseudarthria hookeri* Wight and Arn, for example, has been utilized in Africa as a traditional herbal medicine for the treatment of pneumonia, abdominal aches, cough, and diarrhea (Tchamgoue et al., 2016). Dzoyem and his colleagues discovered that flavonoids from this medicinal species had the strongest antibacterial activity against both gram-positive and gram-negative bacteria in antibacterial research of this medicinal species. *E. coli*, *Klebsiella pneumonia* (Schroeter) Trevisan, *Pseudomonas aeruginosa* (Schroeter) Migula, *Enterococcus faecalis* (Andrew and Horder) Schleifer and Kilpper-Blazand, and *Staphylococcus aureus* Rosenbach; pseudarflavone A and 6-prenylpinocembrin have them (Dzoyem et al., 2018).

8.3.3 ANTI-CANCER EFFECT

Cancer is undeniably one of the leading causes of mortality worldwide; an imbalance and high level of free radicals such as ROS and RNS can also act as mutagenic or carcinogenic agents, resulting in the development of cancer. Chemotherapy is used in the treatment of cancer all over the world, although it has a number of downsides. Unwanted side effects, for example, can arise during chemotherapeutic treatment. As a result, it's intriguing to look for alternative cancer treatments that don't have any side effects and aren't too pricey. Flavopiridol is a flavonoid-derived medicine obtained from *Dysoxylum binectariferum* (Roxb.) Hook.f. ex Bedd. *Dysoxylum gotadhora* (Buch.-Ham.) Mabb. is the current scientific name. Anticancer medicines derived from phytochemical compounds have been used to treat lymphomas and leukemia (Cragg & Newman, 2005). Furthermore, nutritional supplements have a significant role in the prevention and treatment of certain cancers. Flavonoids, in particular, have long been used as chemo preventive agents in cancer treatment (Ahmed et al., 2016; Brusselmans et al., 2003).

8.3.4 CARDIOPROTECTIVE EFFECTS

Since many decades ago, several studies have looked at the cardioprotective benefits of various types of phenolics and/or flavonoids found in medicinal plants (Kumar & Pandey, 2013). The cardioprotective role of various phenolic compounds against cardiotoxicity of doxorubicin was demonstrated in a comprehensive review by Razavi-Azarkhiavi and his team (Razavi-Azarkhiavi et al., 2016). Doxorubicin is a widely used anticancer medicine for lymphomas, leukemia, and breast cancers in clinical application that has vulnerable side effects such as pericarditis, arrhythmias, myocarditis, and acute heart failure. Many phenolic and flavonoid compounds have been studied and reported their cardioprotective properties via various mechanisms such as inhibition of ROS generation, mitochondrial dysfunction, apoptosis, NF-kB, p53, and DNA damage *in vitro*, *in vivo*, and clinical studies. Many flavonoid and phenolic compounds, such as kaempferol, rutin, luteolin, and resveratrol, were reported to be effective against doxorubicin-induced cardiotoxicity but had no effect on the anticancer activity of the drug (Han et al., 2012). Isorhamnetin was the chemical that drew the most attention, because it had a cardioprotective impact against doxorubicin's cardiotoxicity and increased the drug's anticancer activity (Sun et al., 2013). The phenolic composition of methanolic extracts of the aerial portions of two medicinal plants in Poland, *Centaurea borysthenica* Gruner and *Centaurea daghestanica* (Lipsky) Wagenitz, has recently been studied. The cardioprotective effect of both *C. borysthenica* and *C. daghestanica* extracts on rat cardiomyocytes treated with the anticancer medication doxorubicin was demonstrated by the results of oxidative stress, cell viability, and mitochondrial membrane potential assays. According to this research, the methanolic extracts of *C. daghestanica* had no effect on the efficacy of doxorubicin in this experiment (Korga et al., 2017).

8.3.5 IMMUNE SYSTEM PROMOTING AND ANTI-INFLAMMATORY EFFECTS

In the human immune system, medicinal, and pharmacological substances, nutrients, contaminants, and other environmental elements all play a role.

Flavonoids and other phenolics have been shown to have significant impacts on immune system function and inflammatory processes (Middleton & Kandaswami, 1992). Flavonoids with anti-inflammatory properties have been identified as quercetin, apigenin, hesperidin, and luteolin (Kumar & Pandey, 2013). Rupasinghe's team used *in vitro* human monocytic cell line THP-1 derived macrophages stimulated by lipopolysaccharide (LPS) to investigate the anti-inflammatory properties of Canadian medicinal plant extracts, *Lonicera caerulea* L. or Haskap berry in various cultivars, with a focus on pro-inflammatory cytokines. Haskap berry cultivar Borealis had the highest phenolic, flavonoid, and anthocyanin content (p 0.05) and anti-inflammatory benefits comparable to diclofenac, a COX inhibitory medication (Rupasinghe et al., 2015). Furthermore, scientists are currently interested in the synergistic effects of bioactive chemicals and probiotic bacteria on immunological and health-promoting qualities. The effect of *Lactobacillus paracasei* culture filtrates and *Cynara scolymus* L. or artichoke phenolic extract from edible part of its fresh buds on cytokine production by dendritic cells, for example, was investigated by Sisto's group. The investigation revealed that a culture filtrate formed after a probiotic *L. paracasei* strain grew in media enriched with artichoke phenolic extract had a remarkable anti-inflammatory activity (Sisto et al., 2016).

8.3.6 SKIN PROTECTIVE EFFECT FROM UV RADIATION

UV radiation can cause skin damage if it is exposed to too much of it. It causes a large amount of ROS to be produced, which leads to skin damage (Ichihashi et al., 2003). However, there are a number of skin-protection measures that can be used. One of the most intriguing possibilities that demonstrates good benefits on UV-irradiated skin is phytochemical compounds, particularly phenolics and flavonoids (Svobodova et al., 2003). Flavonoids have photoprotective qualities due to their ability to chelate iron, which can damage lipid and protein on cell membranes, and alter many signaling pathways, such as inhibiting xanthine oxidase, which is a generator of ROS and leads to oxidative stress (Ferrali et al., 1997). Several phenolic compounds have been identified as promising antioxidant agents for the treatment of a variety of skin ailments, including UV-induced diseases (Działo et al., 2016). Apigenin is a significant flavone that protects the skin from UV light and is found in a variety of edible medicinal plants and plant-derived beverages, such as red wine, beer, and chamomile tea (McKay & Blumberg, 2006). Quercetin is a flavonol found in onion skin, apple peel, and the leaves of *Hypericum perforatum* L. (Wach et al., 2007). In hairless mice, topical treatment of quercetin successfully prevented UVB-induced skin damage (Casagrande et al., 2006). Furthermore, in an *in vivo* study using UVB irradiated-skin mice model, *Ginkgo biloba* L. extract (EGb 761), which contains a lot of quercetin derivatives, was able to reduce sunburn symptoms UVB-induced skin; the results indicated that oral intake of EGb 761 may act as a protective and therapeutic agent (Ozkur et al., 2002). Silybin, a main active component of milk thistle (*Silybum marianum* (L.) Gaernt.) fruits, is found in silymarin, a standardized extract of flavonolignans from the milk thistle (*Silybum marianum* (L.) Gaernt.) fruits (Bijak, 2017). In an *in vitro* investigation, topical silymarin therapy accelerated the repair of UVB-induced DNA damage, preventing apoptosis in UVB-exposed human epidermal keratinocytes and fibroblasts (Katiyar et al., 2011).

8.4 FLAVONOIDS' THERAPEUTIC ACTIONS DURING PREGNANCY

Flavonols can be found in foods and beverages such as onions, leeks, broccoli, cherry tomatoes, red wine, and tea. Rutin is a flavonol glycoside made up of quercetin and rutinous disaccharide (quercetin-3-rhamnosyl glucoside). 12th Rutin was assumed to have anticarcinogenic and vasoprotective properties. Flavanols or flavan-3-ols, which include epicatechin and catechin monomers, are extensively consumed in teas, red grapes, red wines, apples, and cocoa-based goods (Mondal & Rahaman, 2020).

The flavanols, like flavonols, have been found to exhibit vasoprotective qualities such as anti-inflammatory and vasodilatory actions. 13th. In terms of the putative health advantages of flavanols and flavonols, the results of epidemiological and intervention research have been mostly favorable. While there is a lot of interest in the potential health advantages of flavonoid consumption, there hasn't been much research done on the possible side effects of taking significant doses of these phytochemicals (Powe et al., 2011; Tomimatsu et al., 2017).

Many flavonoids are accessible as dietary supplements, and it's possible to get quantities of flavonoids that far exceed what most people get from their meals. Because various biological activities, including gene expression and cell signaling, have been found to be affected by these chemicals, they can have pleiotropic effects. There is very little research on the physiological effects of excessive and chronic ingestion, especially when it comes to the toxicity of these nutrients during pregnancy and development. There is uncertainty in the current literature about whether all flavonoids consumed during pregnancy are beneficial or harmful to developmental processes (Słowinski et al., 2002).

Some of these conclusions are based on research into altered fetal development (heart, genitourinary, and limb abnormalities), decreased absorption and other nutrient metabolism, and chronic long-term impacts such as a higher chance of disease onset at a young age (e.g., acute childhood leukemia). While the studies above supplemented diets with flavonoid-rich foods or flavonoid preparations, only a few studies have looked at the effects of chronic flavonoid intake during pregnancy. Excessive intake of three flavonoids (epicatechin, catechin, and rutin) during pregnancy and lactation might result in impaired fetal and neonatal development in a mouse model, according to a hypothesis presented in a study (Baksu et al., 2005).

8.4.1 DIABETES MELLITUS (DM)

Regardless of confounding circumstances, epidemiological research conducted in the United States and Europe demonstrate that dietary flavonoids have a preventive effect against the development of diabetes mellitus (DM). According to European cohort statistics, people who consume more flavonoids on a regular basis had a lower risk of diabetes. Cohort findings in the United States backed with this theory. Liu et colleagues found that taking 500 mg of flavonoids per day was related with a 5% reduction in the incidence of diabetes in a dose-response analysis of prospective studies (Amaral et al., 2015). During pregnancy, the intestinal flora of women changes dramatically, which can interfere with flavonoid absorption and metabolism. While research *in vitro* and in animal models suggest that flavonoids may protect against gestational diabetes mellitus (GDM), no epidemiological studies have been discovered to evaluate the link between flavonoid intake during pregnancy and excess weight and GDM. The consumption of excess body weight flavonoids and GDM were found to be inversely related in these studies.

8.4.2 PREGNANCY INDUCED HYPERTENSION

The rising number of CVD patients, particularly the frequent occurrence of hypertension during pregnancy, highlights the need for new preventative and treatment measures. According to the National High Blood Pressure Education Program Working Group on High Blood Pressure in Childbirth, hypertensive conditions during childbirth are divided into four categories: chronic hypertension, preeclampsia, preeclampsia superimposed on chronic hypertension, and gestational hypertension (Possomato-Vieira & Khalil, 2016). This definition is more specific than the older but more widely used pregnancy-induced hypertension (PIH) definition. PIH is a state of high oxidative stress due to a loss in antioxidant capacity. As an imbalance between free radical production and antioxidant defense, oxidative stress is a recognized critical role in the pathophysiology of many diseases, including obstetric problems. One of the impacts of increased oxidative stress is the impairment of placental blood flow, intrauterine fetal hypoxia, and disruption of oxygen transfer. PIH is a multisystem illness that includes changes in oxidant-antioxidant balance, as well as inflammation and endothelial dysfunction. Furthermore, women are more prone to undergo higher oxidative stress during pregnancy, which can result in negative neonatal outcomes (Calabro et al., 2018).

The pathogenesis of PIH is thought to be linked to a systemic inflammatory response, although there are currently no inflammatory markers that can be used to predict the condition. It is thought that women have an increased inflammatory response during pregnancy as a result of uneven modulation of innate and adaptive immune responses. The etiology of PIH is also thought to be linked to deregulation of endogenous defense pathways. Inflammation is a dynamic process regulated by a variety of mediators that regulate important cellular activities in order to restore tissue homeostasis. Anti-inflammation properties Nonsteroidal anti-inflammatory medicines are thought to have a constrictive impact on the fetal duct (NSAIDs). A growing number of cases of fetal ductal constriction have been reported that are not linked to the use of NSAIDs during pregnancy. We previously showed that maternal intake of polyphenol-rich beverages during pregnancy was linked to fetal ductal constriction, and that discontinuing these substances resulted in echocardiographic improvement in the majority of instances. Herbal teas, grape, and orange derivatives, dark chocolate, fruit, and a number of other foods and beverages are high in flavonoids and have been consumed freely throughout pregnancy (Morales-Cano et al., 2014; Galindo et al., 2012).

We hypothesized that polyphenols or flavonoids commonly consumed in food and beverages may be clinically important in terms of fetal hemodynamic compromise and the potential for neonatal pulmonary hypertension, and during late pregnancy, they inhibit maternal circulating prostaglandin have been known till now. In a study, ductal flow activity and proper ventricular capacity were evaluated in third-trimester fetuses exposed and not exposed to polyphenol-rich foods (PRFs) and beverages via maternal ingestion (Mondal & Rahaman, 2020).

8.4.3 ANTI-INFLAMMATORY EFFECTS

A useful effect on the fetal duct is recognized to have nonsteroidal anti-inflammatory drugs (NSAIDs). A number of cases have been recorded of fetal ductal constriction that were not related to the use of NSAIDs during pregnancy. It has been observed that fetal ductal constriction was related to maternal ingestion of polyphenol containing drink during pregnancy. The avoidance of such substances in most cases resulted in echocardiographic improvement. A huge variety of foods and drinks, like herbal teas, grape, and orange products, chocolate flavored substances, fruit, and many others that are rich in flavonoids are eaten normally during the gestation period (Morales-Cano et al., 2014).

8.4.4 PLANT PHENOLICS AND EXTRACTS IN PREECLAMPSIA ANIMAL MODELS

Preeclampsia (ICD-10 code 014) is a multifactorial pregnancy condition that is potentially fatal to both the mother and the fetus (Sibai et al., 2005). Furthermore, there is currently a shortage of therapy options for this complex condition. Elucidating the etiology of preeclampsia could lead to the development of a new medication or drug combination. Many research studies (Ozarowski et al., 2018) have examined this, emphasizing the central significance of the inflammatory process. Only Wang et al. (2018) recently published a study on baicalin (50, 100, and 150 mg/kg/day, i.p. for 20 days) in experimental preeclampsia with kidney and liver damage produced by the nitric oxide (NO) synthase inhibitor NG-nitro-L-arginine-methyl-ester (LNAME). Baicalin lowered blood pressure (BP) and apoptosis in kidney and liver cells in a dose-dependent manner.

Furthermore, in the liver, baicalin increased the expression of anti-apoptotic proteins such as XIAP and Bcl-2 while decreasing the production of the apoptotic protein caspase-9.

Gong et al. (2016) investigated the effects of curcumin (0.36 mg/kg) given to pregnant rats with LPS-induced preeclampsia. This polyphenol was found to lower BP and urine protein concentrations in this investigation. Curcumin can help with LPS-related issues such as trophoblast invasion problems and spiral artery remodeling. Curcumin also reduced the expression of TLR4 and NF-κB proteins in the placenta, as well as IL-6 and MCP-1 proteins in rat serum, which were both increased by LPS. The effect of curcumin found in this study can explain its mode of action as an anti-inflammatory drug since the toll-like receptor 4 signaling pathway plays a role in the pathophysiology of inflammation during preeclampsia.

In another study, Zhou et al. (2017) found that curcumin at a dose of 100 g/kg/d i.g. decreased expression of inflammatory factors like TNF-, IL-1, and IL-6, as well as levels of chemokines like MCP-1 and MIP-1, and increased the level of phosphorylated Akt by upregulation in the placenta in pregnant mice with LPS-induced preeclampsia. Curcumin was also beneficial in preventing macrophage invasion in this organ. Punicalagin was studied in pregnant rats with L-NAME-induced hypertension, and dosages of 25, 50, or 100 mg/kg were given orally to the animals on days 14–21 of pregnancy (Ozarowski et al., 2021). The greatest dose of punicalagin had a higher effect in lowering systolic and diastolic BP, as well as mean arterial pressure, according to a comparative investigation. Enhancing the anti-oxidant capacity and boosting the amounts of NO in the placenta had a similar dose-dependent impact. Punicalagin caused increased expression and levels of mRNA VEGF, as well as downregulation of expression of vascular endothelial growth factor (VEGF) receptor-1/fms-like tyrosine kinase-1, according to a molecular analysis (Wang et al., 2018). This suggests that punicalagin has a positive effect on angiogenic balance during experimental preeclampsia. The following plant is *Moringa oleifera* Lam., the most widely cultivated species, which grows wild in India (Tshabalala et al., 2019).

Flavonoids (e.g., quercetin, kaempferol, isorhamnetin) (Makita et al., 2016), phenolic acids (e.g., chlorogenic acids) (Makita et al., 2017), glucosinolates (e.g., glucomoringin) (Ramabulana et al., 2016) and tannins (e.g., glucomoringin) (Teixeira et al., 2014) are abundant in all parts of this plant. These molecules have a variety of pharmacological capabilities, including anti-oxidant, anti-inflammatory, anti-cancer, and anti-diabetic effects (Tshabalala et al., 2019). Batmomolin et al. (2020) recently published a study on the efficacy of an ethanolic extract from *M. oleifera* leaves in a rat model of preeclampsia produced by L-NAME (50 mg/kg b.w.) given on days 5–18 of pregnancy. The results were compared to acetylsalicylic acid (1.35 mg/200 g b.w.) when the plant extract was given to rats at various doses (50, 100, and 200 mg/kg b.w.). According to Batmomolin et al. (2020) two doses of *M. oleifera* extract (50 and 100 mg/kg) were more effective than low-dose acetylsalicylic acid in lowering IL-17 levels in rats' serum.

8.4.5 PHENOLICS AND FLAVONOIDS FROM PLANTS FOR MENOPAUSAL AND POST-MENOPAUSAL WOMEN

In pharmacy and medicine, there are advantages and disadvantages to employing synthetic chemical compounds and phytochemical substances. Synthetic chemicals or medicines are simple and quick to generate on a big scale during the drug development process, and they can be modified to fit a variety of patient needs. In contrast, many non-natural (or manufactured chemical compounds) have a number of negative side effects, especially when used over long periods of time (Lahlou et al., 2013). Because of their hazardous side effects, certain synthetic drugs were not approved for clinical usage. Synthetic estrogen, which was once widely used in menopause women for hormone replacement therapy, is an apparent example. Long before a lot of studies identified its unfavorable side effects, such as an increase in the risk of breast, uterine, and ovarian malignancies (Burton & Wells, 2002) this synthetic chemical compound could work well to alleviate menopause symptoms. Estrogen is a sex hormone that is primarily responsible for women's reproductive activities and menstrual cycles. Estrogen is decreased in postmenopausal women due to the ovary's failure to respond to the pituitary. When estrogen levels drop, it causes a slew of postmenopausal symptoms, including CVD. Women who are postmenopausal and have metabolic syndrome (MetS) have a higher risk of CVD (De Gregorio et al., 2017).

Flavonoid supplementation can help improve cardiovascular function in postmenopausal women with metabolic syndrome, according to Squadrito and his colleagues' investigation (Squadrito et al., 2013). Genistein is a well-known flavonoid phytoestrogen that has been shown to improve endothelial function in postmenopausal women with MetS (Irace et al., 2013). The effects of genistein supplementation on cardiac function in postmenopausal women with MetS were explored by Gregorio and his research team (De Gregorio et al., 2017). The study included 120 postmenopausal women patients with type-2 DM who were free of past cardiovascular illness. By employing a computer-generated double-blind randomization, the patients were evenly divided into two groups: those who received genistein supplementation and those who received a placebo. The findings suggested that genistein can help postmenopausal women with MetS improve their cardiac function (De Gregorio et al., 2017; Tungmunnithum et al., 2018).

8.5 USES OF HERBAL EXTRACTS DURING PREGNANCY

The usage of herbal items is one of the ways utilized in complementary and alternative medicine. Several international studies have recently been conducted to determine the prevalence of herbal medication use during pregnancy. Due to a lack of awareness of phytochemical qualities or effective use of herbal medical remedies, as well as a lack of communication with healthcare prices, an increasing number of women are putting their own and their unborn children's health at risk (Carlstrom et al., 2007).

There has been a continuous rise in interest in the use of herbal products, however the current guidelines are frequently confusing or contradictory. Across different geographic, socioeconomic, and cultural settings, the prevalence of herbal medicine use in pregnancy ranges from 7% to 55%, and it differs by ethnic group. Herbal products are frequently preferred over prescription medications because individuals believe herbs are better for the fetus than contemporary drugs. Herbal medical product manufacture is extremely difficult since it involves non-standardized procedures such as plant production, procuring vegetable raw material from all over the world, extract processing, and product creation in accordance with local standards of good manufacturing practice (Duarte et al., 2001; Illamola et al., 2020).

8.6 FUTURE PERSPECTIVES OF FLAVONOID RESEARCH

In conclusion, flavonoids are present in low amounts in many dietary sources, most of them have low water solubility, some of them are unstable under certain conditions, have low intestinal absorption and bioavailability, and may require high doses to show efficacy in human studies, despite their broad and multi-potent pharmacological properties. To illustrate the targeted efficacy and route of action, most *in vitro* studies have used metabolically implausible high flavonoid doses; however, these findings must be verified using standardized *in vivo* experimental paradigms. Prospective human trials are still needed to demonstrate the therapeutic efficacy of standardized flavonoid compounds extracted from plant sources.

To produce cost-effective flavonoid-based natural health products, scale-up, consumer-, and environment-friendly green technologies are required. Flavonoid supplementation should be done with caution in cancer patients because it may interfere with radiotherapy and different chemotherapies. To overcome bioavailability constraints, targeted administration, and improve the therapeutic efficacy of specific flavonoids, multidisciplinary research collaborations are required. Most main flavonoids' phase 2 metabolism and pharmacokinetics have been documented (Rupasinghe, 2020). However, the interaction of unabsorbed flavonoids with the colon microbiota and the metabolites that occur, as well as their involvement in disease prevention and treatment, remain unknown. As a result, new fields in life science emerged viz., flavonoid nanotechnology, flavonoid-microbiome pharmacology, and flavonoid-inspired therapies (Kumar & Pandey, 2013).

8.7 CONCLUSIONS

The use of phytochemicals, particularly flavonoids, in the prevention and treatment of illnesses is widely established. Flavonoids are found naturally in fruits and vegetables. The physical, chemical, and physiological properties of various flavonoids found in nature vary. Flavonoids' structure-function link epitomizes major biological functions. Many flavonoids have been shown to have antibacterial, hepatoprotective, anti-inflammatory, anticancer, and antiviral properties. In developing areas, these substances are more often utilized. Biochemical assays must be used to validate the therapeutic usage of novel drugs. Flavonoids may now be produced in enormous quantities because of genetic changes. More accomplishments will lead to newer insights and, without a doubt, a new era of flavonoid-based pharmacological agents for the treatment of various infectious and degenerative disorders.

KEYWORDS

- **adenosine triphosphate**
- **diabetes mellitus**
- **flavonoids**
- **gestational diabetes mellitus**
- **phenolics**
- **polyphenol-rich foods**
- **preeclampsia**
- **pregnancy disorders**
- **pregnancy-induced hypertension**
- **reactive nitrogen species**

REFERENCES

Agnes, B., Ahsan, A., Arsana, W. I. W., & Sanarto, S., (2020). Ethanolic extract of *Moringa oleifera* leaves improve inflammation, angiogenesis, and blood pressure in rat model of preeclampsia. J. *Appl. Pharm. Sci., 10*, 52–57.

Amaral, L. M., Cunningham, M. W., Cornelius, D. C., et al., (2015). Preeclampsia: Long-term consequences for vascular health. *Vascular Health and Risk Management., 11*, 403–415.

Andreu, L., Nuncio-Jáuregui, N., Carbonell-Barrachina, Á. A., Legua, P., & Hernández, F., (2018). Antioxidant properties and chemical characterization of Spanish *Opuntia ficus-indica* mill. cladodes and fruits. *J. Sci. Food Agric., 98*, 1566–1573.

Aoki, T., Akashi, T., & Ayabe, S., (2000). Flavonoids of leguminous plants: Structure, biological activity, and biosynthesis. *J. Plant Res., 113*, 475–488.

Baksu, B., Davas, I., Baksu, A., et al., (2005). Plasma nitric oxide, endothelin-1 and urinary nitric oxide and cyclic guanosine monophosphate levels in hypertensive pregnant women. *International Journal of Gynecology and Obstetrics, 90*(2), 112–117.

Bijak, M., (2017). Silybin, a major bioactive component of milk thistle (*Silybum marianum* L. Gaernt.)—Chemistry, bioavailability, and metabolism. *Molecules, 22*, 1942.

Brusselmans, K., De Schrijver, E., Heyns, W., Verhoeven, G., & Swinnen, J. V., (2003). Epigallocatechin-3-gallate is a potent natural inhibitor of fatty acid synthase in intact cells and selectively induces apoptosis in prostate cancer cells. *Int. J. Cancer., 106*, 856–862.

Burton, J. L., & Wells, M., (2002). The effect of phytoestrogens on the female genital tract. *J. Clin Pathol., 55*, 401–407.

Calabro, V., Litterio, M. C., Fraga, C. G., et al., (2018). Effects of quercetin on heart nitric oxide metabolism in l–NAME treated rats. *Archives of Biochemistry and Biophysics., 647*, 47–53.

Carlstrom, J., Symons, J. D., Tzu, C. W., et al., (2007). A quercetin supplemented diet does not prevent cardiovascular complications in spontaneously hypertensive rats. *Journal of Nutrition, 137*(3), 628–633.

Carocho, M., Barreiro, M. F., Morales, P., & Ferreira, I. C. F. R., (2014). Adding molecules to food, pros and cons: A review on synthetic and natural food additives. *Compr. Rev. Food Sci. Food Saf., 13*, 377–399.

Casagrande, R., Georgetti, S. R., Verri, W. A., Dorta, D. J., Dos Santos, A. C., & Fonseca, M. J. V., (2006). Protective effect of topical formulations containing quercetin against UVB-induced oxidative stress in hairless mice. *J. Photochem. Photobiol. B Biol., 84*, 21–27.

Cos, P., Ying, L., Calomme, M., Hu, J. P., Cimanga, K., Van, P. B., Pieters, L., Vlietinck, A. J., & Vanden, B. D., (1998). Structure-activity relationship and classification of flavonoids as inhibitors of xanthine oxidase and superoxide scavengers. *J. Nat. Prod., 61*, 71–76.

Cragg, G. M., & Newman, D. J., (2005). Plants as a source of anti-cancer agents. *J. Ethnopharmacol., 100*, 72–79.

De Gregorio, C., Marini, H., Alibrandi, A., Di Benedetto, A., Bitto, A., Adamo, E. B., Altavilla, D., Irace, C., Di Vieste, G., Pancaldo, D., et al., (2017). Genistein supplementation and cardiac function in postmenopausal women with metabolic syndrome: Results from a pilot strain-echo study. *Nutrients, 9*, 584.

Duarte, J., Perez-Palencia, R., Vargas, F., et al., (2001). Antihypertensive effects of the favonoid quercetin in spontaneously hypertensive rats. *British Journal of Pharmacology, 133*(1), 117–124.

Działo, M., Mierziak, J., Korzun, U., Preisner, M., Szopa, J., & Kulma, A., (2016). The potential of plant phenolics in prevention and therapy of skin disorders. *Int. J. Mol. Sci., 17*, 160.

Dzotam, J. K., Simo, I. K., Bitchagno, G., Celik, I., Sandjo, L. P., Tane, P., & Kuete, V., (2018). *In vitro* antibacterial and antibiotic modifying activity of crude extract, fractions and 30, 40, 7-trihydroxyflavone from *Myristica fragrans* houtt against MDR gram-negative enteric bacteria. *BMC Complement. Altern. Med., 18*, 15.

Dzoyem, J. P., Tchamgoue, J., Tchouankeu, J. C., Kouam, S. F., Choudhary, M. I., & Bakowsky, U., (2018). Antibacterial activity and cytotoxicity of flavonoids compounds isolated from *Pseudarthria hookeri* wight & Arn. (*Fabaceae*). *S. Afr. J. Bot., 114*, 100–103.

Ferrali, M., Signorini, C., Caciotti, B., Sugherini, L., Ciccoli, L., Giachetti, D., & Comporti, M., (1997). Protection against oxidative damage of erythrocyte membrane by the flavonoid quercetin and its relation to iron chelating activity. *FEBS Lett., 416*, 123–129.

Galindo, P., Gonzalez-Manzano, S., Zarzuelo, M. J., et al., (2012). Different cardiovascular protective effects of quercetin administered orally or intraperitoneally in spontaneously hypertensive rats. *Food & Function, 3*(6), 643–650.

Giusti, M., & Wrolstad, R., (2003). Acylated anthocyanins from edible sources and their applications in food systems. *Biochem. Eng. J., 14*, 217–225.

Gong, P., Liu, M., Hong, G., Li, Y., Xue, P., Zheng, M., Wu, M., Shen, L., Yang, M., Diao, Z., et al., (2016). Curcumin improves LPS-induced preeclampsia-like phenotype in rat by inhibiting the TLR4 signaling pathway. *Placenta, 41*, 45–52.

Han, X., Gao, S., Cheng, Y., Sun, Y., Liu, W., Tang, L., & Ren, D., (2012). Protective effect of naringenin-7-O-glucoside against oxidative stress induced by doxorubicin in H9c2 cardiomyocytes. *Biosci. Trends., 6*, 19–25.

Heim, K. E., Tagliaferro, A. R., & Bobilya, D. J., (2002). Flavonoid antioxidants: Chemistry, metabolism and structure-activity relationships. *J. Nutr. Biochem., 13*, 572–584.

Ichihashi, M., Ueda, M., Budiyanto, A., Bito, T., Oka, M., Fukunaga, M., Tsuru, K., & Horikawa, T., (2003). UV-induced skin damage. *Toxicology, 189*, 21–39.

Illamola, S. M., Amaeze, O. U., Krepkova, L. V., Birnbaum, A. K., Karanam, A., Job, K. M., Bortnikova, V. V., et al., (2020). Use of herbal medicine by pregnant women: What physicians need to know. *Front. Pharmacol., 10*, 1483. doi: 10.3389/fphar.2019.01483.

Irace, C., Marini, H., Bitto, A., Altavilla, D., Polito, F., Adamo, E. B., Arcoraci, V., Minutoli, L., Di Benedetto, A., Di Vieste, G., et al., (2013). Genistein and endothelial function in postmenopausal women with metabolic syndrome. *Eur. J. Clin. Investig., 43*, 1025–1031.

Iwashina, T., (2013). Flavonoid properties of five families newly incorporated into the order caryophyllales (review). *Bull. Natl. Mus. Nat. Sci., 39*, 25–51.

Jarial, R., Thakur, S., Sakinah, M., Zularisam, A. W., Sharad, A., Kanwar, S. S., & Singh, L., (2018). Potent anticancer, antioxidant and antibacterial activities of isolated flavonoids from *Asplenium nidus. J. King Saud Univ. Sci., 30,* 185–192.

Katiyar, S. K., Mantena, S. K., & Meeran, S. M., (2011). Silymarin protects epidermal keratinocytes from ultraviolet radiation-induced apoptosis and DNA damage by nucleotide excision repair mechanism. *PLoS One, 6,* e21410.

Korga, A., Józefczyk, A., Zgórka, G., Homa, M., Ostrowska, M., Burdan, F., & Dudka, J., (2017). Evaluation of the phytochemical composition and protective activities of methanolic extracts of *Centaurea borysthenica* and *Centaurea daghestanica* (Lipsky) Wagenitz on cardiomyocytes treated with doxorubicin. *Food Nutr. Res., 61*, 1344077.

Kumar, S., & Pandey, A. K., (2013). Chemistry and biological activities of flavonoids: An overview. *The Scientific World Journal, 2013*, 16. Article ID 162750. http://dx.doi.org/10.1155/2013/162750.

Lahlou, M., (2013). The success of natural products in drug discovery. *Pharmacol. Pharm., 4*, 17–31.

Linuma, M., Tanaka, T., Hamada, K., et al., (1987). Revised structure of neoflavone in *Coutarea hexandra. Phytochemistry., 26*, 3096–3097.

Makita, C., Chimuka, L., Cukrowska, E., Steenkamp, P., Kandawa-Schutz, M., Ndhlala, A., & Madala, N., (2017). UPLC-qTOF-MS profiling of pharmacologically important chlorogenic acids and associated glycosides in *Moringa ovalifolia* leaf extracts. *S. Afr. J. Bot., 108*, 193–199.

Makita, C., Chimuka, L., Steenkamp, P., Cukrowska, E., & Madala, E., (2016). Comparative analyses of flavonoid content in *Moringa oleifera* and *Moringa ovalifolia* with the aid of UHPLC-qTOF-MS fingerprinting. *S. Afr. J. Bot., 105*, 116–122.

Manach, C., Scalbert, A., Morand, C., et al., (2004). Polyphenols: Food sources and bioavailability. *Am. J. Clin. Nutr., 79*, 727–747.

Matthies, A., Clavel, T., Gütschow, M., et al., (2008). Conversion of daidzein and genistein by an anaerobic bacterium newly isolated from the mouse intestine. *Appl. Envrion. Microbiol., 74*, 4847–4852.

McKay, D. L., & Blumberg, J. B., (2006). A review of the bioactivity and potential health benefits of chamomile tea (*Matricaria recutita* L.). *Phyther. Res., 20*, 519–530.

Middleton, E., & Kandaswami, C., (1992). Effects of flavonoids on immune and inflammatory cell functions. *Biochem. Pharmacol., 43*, 1167–1179.

Mondal, S., & Rahaman, S. T., (2020). Flavonoids: A vital resource in healthcare and medicine. *Pharm. Pharmacol. Int. J., 8*(2), 91–104.

Morales-Cano, D., Menendez, C., Moreno, E., et al., (2014). The flavonoid quercetin reverses pulmonary hypertension in rats. *PLoS One, 9*(12), e114492.

Nagja, T., Vimal, K., & Sanjeev, A., (2016). Myristica fragrans: A comprehensive review. *Int. J. Pharm. Pharm. Sci., 8*, 27–30.

Okpuzor, J., Ogbunugafor, H., Kareem, G. K., & Igwo-Ezikpe, M. N., (2009). *In vitro* investigation of antioxidant phenolic compounds in extracts of *Senna alata. Res. J. Phytochem., 3*, 68–76.

Ozarowski, M., Karpinski, T. M., Szulc, M., Wielgus, K., Kujawski, R., Wolski, H., & Seremak-Mrozikiewicz, A., (2021). Plant phenolics and extracts in animal models of preeclampsia and clinical trials—Review of perspectives for novel therapies. *Pharmaceuticals, 14*, 269. https://doi.org/ 10.3390/ph14030269.

Ozarowski, M., Mikołajczak, P. A., Kujawski, R., Wielgus, K., Klejewski, A., Wolski, H., & Seremak-Mrozikiewicz, A., (2018). Pharmacological effect of quercetin in hypertension and its potential application in pregnancy-induced hypertension: Review of *in vitro*, *in vivo*, and clinical studies. *Evid. Based Complement. Altern. Med.*, 7421489.

Ozkur, M. K., Bozkurt, M. S., Balabanli, B., Aricioglu, A., Ilter, N., Gurer, M. A., & Inaloz, H. S., (2002). The effect of EGb 761 on lipid peroxide levels and superoxide dismutase activity in sunburn. *Photodermatol. Photoimmunol. Photomed., 18*, 117–120.

Panche, A. N., Diwan, A. D., & Chandra, S. R., (2016). Flavonoids: An overview. *Journal of Nutritional Science, 5*, e47, 1–15. doi: 10.1017/jns.2016.41.

Pham-Huy, L. A., He, H., & Pham-Huy, C., (2008). Free radicals, antioxidants in disease and health. *Int. J. Biomed. Sci., 4*, 89–96.

Possomato-Vieira, J. S., & Khalil, R. A., (2016). Mechanisms of endothelial dysfunction in hypertensive pregnancy and preeclampsia. *Advances in Pharmacology., 77*, 361–431.

Powe, C. E., Levine, R. J., & Karumanchi, S. A., (2011). Preeclampsia, a disease of the maternal endothelium: The role of antiangiogenic factors and implications for later cardiovascular disease. *Circulation, 123*(24), 2856–2869.

Ramabulana, T., Mavunda, R., Steenkamp, P., Piater, L., Dubery, I., & Madala, N., (2016). Perturbation of pharmacologically relevant polyphenolic compounds in *Moringa oleifera* against photo-oxidative damages imposed by gamma radiation. *J. Photochem. Photobiol. B: Biol., 156*, 79–86.

Razavi-Azarkhiavi, K., Iranshahy, M., Sahebkar, A., Shirani, K., & Karimi, G., (2016). The protective role of phenolic compounds against doxorubicin-induced cardiotoxicity: A comprehensive review. *Nutr. Cancer, 68*, 892–917.

Rupasinghe, H. P. V., (2020). Special issue "flavonoids and their disease prevention and treatment potential": Recent advances and future perspectives. *Molecules, 25*, 4746; doi: 10.3390/molecules25204746.

Rupasinghe, H. P. V., Boehm, M. M. A., Sekhon-Loodu, S., Parmar, I., Bors, B., & Jamieson, A. R., (2015). Anti-inflammatory activity of haskap cultivars is polyphenols-dependent. *Biomolecules, 5*, 1079–1098.

Sibai, B., Dekker, G., & Kupferminc, M., (2005). Pre-eclampsia. *Lancet, 365*, 785–799.

Sisto, A., Luongo, D., Treppiccione, L., De Bellis, P., Di Venere, D., Lavermicocca, P., & Rossi, M., (2016). Effect of *Lactobacillus paracasei* culture filtrates and artichoke polyphenols on cytokine production by dendritic cells. *Nutrients, 8*, 635.

Słowinski, T., Neumayer, H., Stolze, T., et al., (2002). Endothelin system in normal and hypertensive pregnancy. *Clinical Science, 103*(s48), 446S–449S.
Squadrito, F., Marini, H., Bitto, A., Altavilla, D., Polito, F., Adamo, E. B., D'Anna, R., Arcoraci, V., Burnett, B. P., Minutoli, L., et al., (2013). Genistein in the metabolic syndrome: Results of a randomized clinical trial. *J. Clin. Endocrinol. Metab., 98,* 3366–3374.
Sun, J., Sun, G., Meng, X., Wang, H., Luo, Y., Qin, M., Ma, B., Wang, M., Cai, D., Guo, P., et al., (2013). Isorhamnetin protects against doxorubicin-induced cardiotoxicity *in vivo* and *in vitro*. *PLoS One, 8*, e64526.
Svobodová, A., Psotová, J., & Walterová, D., (2003). Natural phenolics in the prevention of UV-induced skin damage. A review. *Biomed. Pap., 147*, 137–145.
Szkudelska, K., & Nogowski, L., (2007). Genistein – a dietary compound inducing hormonal and metabolic changes. *J. Steroid Biochem. Mol. Biol., 105*, 37–45.
Tchamgoue, J., Hafizur, R. M., Tchouankeu, J. C., Kouam, S. F., Adhikari, A., Hameed, A., Green, I. R., & Choudhary, M. I., (2016). Flavonoids and other constituents with insulin secretion activity from *Pseudarthria hookeri. Phytochem. Lett., 17*, 181–186.
Teixeira, E. M. B., Carvalho, M. R. B., Neves, V. A., Silva, M. A., & Arantes-Pereira, L., (2014). Chemical characteristics and fractionation of proteins from *Moringa oleifera* Lam. leaves. *Food Chem., 147,* 51–54.
Tomimatsu, T., Mimura, K., Endo, M., et al., (2017). Pathophysiology of preeclampsia: An angiogenic imbalance and long-lasting systemic vascular dysfunction. *Hypertension Research, 40*(4), 305–310.
Tshabalala, T., Ncube, B., Madala, N. E., Nyakudya, T. T., Moyo, H. P., Sibanda, M., & Ndhlala, A. R., (2019). Scribbling the cat: A case of the "miracle" plant. *Moringa oleifera. Plants, 8*, 510.
Tungmunnithum, D., Thongboonyou, A., Pholboon, A., & Yangsabai, A., (2018). Flavonoids and other phenolic compounds from medicinal plants for pharmaceutical and medical aspects: An overview. *Medicines, 5*, 93. doi: 10.3390/medicines5030093.
Valko, M., Rhodes, C. J., Moncol, J., Izakovic, M., & Mazur, M., (2006). Free radicals, metals and antioxidants in oxidative stress-induced cancer. *Chem. Biol. Interact., 160*, 1–40.
Wach, A., Pyrzynska, K., & Biesaga, M., (2007). Quercetin content in some food and herbal samples. *Food Chem., 100*, 699–704.
Wang, J., Cao, X., Ferchaud, V., Qi, Y., Jiang, H., Tang, F., Yue, Y., & Chin, K. L., (2016). Variations in chemical fingerprints and major flavonoid contents from the leaves of thirty-one accessions of *Hibiscus sabdariffa* L. *Biomed. Chromatogr., 30*, 880–887.
Wang, Y., Jia, Y., Yang, X., Liang, B., Gao, H., & Yang, T., (2018). A potential role of baicalin to inhibit apoptosis and protect against acute liver and kidney injury in rat preeclampsia model. *Biomed. Pharmacother., 108*, 1546–1552.
Williamson, G., Kay, C. D., & Crozier, A., (2018). The bioavailability, transport, and bioactivity of dietary flavonoids: A review from a historical perspective. *Compr. Rev. Food Sci. Food Saf., 17*, 1054–1112.
Zhou, J., Miao, H., Li, X., Hu, Y., Sun, H., & Hou, Y., (2017). Curcumin inhibits placental inflammation to ameliorate LPS-induced adverse pregnancy outcomes in mice via upregulation of phosphorylated Akt. *Inflamm. Res., 66,* 177–185.

CHAPTER 9

FLAVONOIDS AND THEIR ANTI-CANCER PROPERTIES

SHIVAM SINGH[1], DEEPIKA SAINI[2], DEEPA DEVI[3], and KAMAL KANT SHARMA[4]

[1]*Scientist B, Government District Hospital Cum Trauma Center, Waidhan, Singrauli, Madhya Pradesh, India*

[2]*Assistant Professor, Chaman Lal Mahavidyalaya, Landhaura, Haridwar, Uttarakhand, India*

[3]*Scientist-B, Parasitology Division, ICMR–National Institute of Malaria Research, Delhi, India*

[4]*Research Scholar, Department of Botany and Microbiology, Gurukula Kangri (Deemed to be University), Haridwar, Uttarakhand, India*

ABSTRACT

Flavonoids are plant-based polyphenolic secondary metabolites that comprise important nutraceutical values required to regulate various metabolic processes. Traditional plant-based drugs consist of medicinally important flavonoids as their major ingredient that are crucial for various therapeutic regimes. Some of these have shown their capability to regulate cell proliferation and division and in key pathways, viz. modulating ROS (reactive oxygen species) scavenging enzyme activities, cooperating in inhibition of the cell cycle, inducing autophagy, apoptosis, and reducing proliferation and intrusiveness in cancer cells. These flavonoids, in a couple of cases, show unmistakable action for ROS-they go about as cancer

Advances in Flavonoids for Human Health and Prevention of Diseases. Nisha Sharma, Deepika Saini, Rajesh K. Kesharwani, Prakash Chandra Gupta, & Raj K. Keservani (Eds.)

prevention agents under regular conditions and are intensely favorable to oxidants in disease cells setting off the apoptotic pathways and downregulating supportive of provocative flagging pathways. Therefore, a deviating multifaceted mechanism of actions, probably interacting with one another, can be hypothesized for such a nutritional molecule. This review emphasizes the anti-cancer activity, its biochemistry, and most probable mechanisms of action of some therapeutically important flavonoids, which can be further used in targeting the development of novel nutraceutical active anti-cancer remedies.

9.1 INTRODUCTION

Flavonoids are polyphenolic buildings combined by plants as bioactive auxiliary metabolites that are answerable for different pharmacological activities (Nabavi et al., 2018). These flavonoids act as potent antioxidants (Vrhovsek et al., 2004) protecting plants from various unfavorable conditions (Nabavi et al., 2018), subsequently, they have acquired consideration from analysts for their conceivable useful impacts in numerous intense and ongoing human disorder (Rodriguez-Garcia et al., 2019). As announced earlier, the fitting admission of flavonoids was related with a decreased danger of degenerative illnesses like cardiovascular infection and disease (Béliveau & Gingras, 2007; Patel, 2008; Xiao et al., 2011). Experimental statistics have revealed that flavonoids inhibited the advancement of latent microtumors and this outcome could be stimulated by the modulation of the ROS neutralizing enzyme systems and/or by directly triggering apoptosis leading to death of cancerous cells (Conney, 2003; Béliveau & Gingras, 2007). For example, phenethyl isothiocyanate, curcumin, resveratrol, and naringenin isolated from cruciferous vegetables, turmeric, grapes, and oranges, respectively have all been revealed to comprise strong proapoptotic activity against various tumor cells (Patel, 2008). Flavonoid rich fruits, as berries, plums, cherries, and apples (De Pascual-Teresa et al., 2000; Vrhovsek et al., 2004) and vegetables, as beans (Mejri et al., 2018), olives (Romani et al., 1999), along with various spinach and allium species (Pandjaitan et al., 2005; Fattorusso et al., 2002), at the point when fused in diet in adequate sum can limit activation of cancer-causing agents, openness to injurious substances, and can likewise augment the admission of specific advantageous supplements like isothiocyanates, unsaturated fats, selenium, alongside a couple of terpenes and terpenoids (Franco et al., 2004; Johnson, 2004). In addition, expanded utilization of isoflavones has likewise been connected with decreased danger of estrogen-related malignant growths and cardiovascular illnesses (Birt et al., 2001) (*see* Table 9.1).

Information from four companion studies and six case-control studies, which have analyzed the relationship of flavonoid consumption with malignant growth hazard uncovered those flavonoids, particularly quercetin, may lessen the danger of cellular breakdown in the lungs in two investigations however a non-huge, expanded danger in a third report (Totta et al., 2005). High versus low quercetin and kaempferol admissions were related with 40 and 50% decrease in hazard, individually, for stomach disease. These discoveries are all around upheld by a few epidemiological investigations (Kasiri et al., 2020). Additionally, different *in vitro* and *in vivo* studies have now uncovered that these flavonoids could apply numerous beneficiary pharmacological properties as calming, immunomodulatory, and solid anti-cancer exercises (Birt et al., 2001).

9.2 FLAVONOID METABOLISM

In order to understand the effectiveness of flavonoids as an anti-cancer agent, it is important to keep an eye on its metabolism (Liskova et al., 2020). Critical information were accessible with regards to the digestion of dietary flavonoids (Keservani et al., 2020; Keservani & Sharma, 2014) in creatures however less significantly in people (Scheline, 1991; Liskova et al., 2020). In summary, the flavonoid metabolism starts from intestine, where its major proportion is left unabsorbed (in proximal intestine) which later reach to colon (Birt et al., 2001; Liskova et al., 2020). The colon microbiota capacities as a metabolic reactor by hydrolyzing glycosylated flavonoids (Keservani et al., 2010a, b) like flavones, isoflavones, flavonols, and anthocyanins into their individual aglycones, besides its heterocyclic oxygen containing ring is additionally hydroxylated coming about phenyl carboxylic acids that could be then retained (Cassidy & Minihane, 2017). Therefore, flavonoids go through its principal digestion in the gastrointestinal epithelium while resultant metabolites are moved into the liver and afterward to kidney, from which ensuing metabolic advances bring about the more-polar mixtures that intervene natural impacts in the objective tissue (Cassidy & Minihane, 2017). Mechanism of such metabolism has not been entirely revealed,

TABLE 9.1 Flavonoid Types with Their Major Dietary Source and Most Studied Anti-Cancer Properties

Flavonoids Groups	Compound	Dietary Source	Anti-cancer Properties
Anthocyanins	Cyanidin, Delphinidin, Malvidin, Pelargonidin, Peonidin	Blueberry, black grapes, cheery, rhubarb, plum, strawberry, red cabbage purple grapes, raspberry seeds	Colorectal cancer.
Flavonol	Quercetagetin, Quercetin, Kaempferol, Myricetin, Isorhamnetin	Olive, tomato broccoli, green, and black tea, blueberry, black grapes, red apple	Ovarian cancer, bladder carcinoma, breast, prostate, gastric, skin, lung, pancreatic, and blood cancers.
Flavanols	Catechin, Epicatechin Epigallocatechin, Silibinin, Silymarin, Taxifolin, Pinobanksin, Proanthocyanidins	Apricot, apple, bean, peach, cider, green tea, berries, grapes, cocoa, chocolates	Oral, rectal, and prostate cancer.
Flavones	Apegenin, Chrysin, Luteolin, Tangeretin, Rutin, Tetramethoxyflavone, Sinensetin, Quercetagetin	Capsicum, parsley, celery, hot peppers, thyme	Lung carcinoma, colorectal cancer, breast cancer, liver cancer, melanoma, osteosarcoma glioblastoma, prostate, and pancreatic cancers.
Flavanones	Naringenin, Eriodictyol, Hesperitin, Dihydroquercetin, Dihydrobinetin	Citrus fruits, orange, grapes, lemon	Melanoma, lung carcinoma and colorectal cancer.
Isoflavones	Daidezein, Genistein, Glycitein	Soya foods, soya cheese, soya flour, tofu, legumes	Breast cancer, prostate cancer, colon, kidney, and thyroid cancers.

but it is currently believed that its membrane transport requires both ATP-dependent and ATP-independent membrane transporters (Passamonti et al., 2009). Additionally, the intestinal microbiome plays an essential role in the metabolism of flavonoids as it is extremely involved in the production of metabolites (Murota et al., 2018; Kawabata et al., 2019). Generally, about 20–200 mg of flavonoids are consumed by human per day, but the intake of tea and other similar beverages regularly, may increase this level up to 1,000 mg/day with no significant toxicity (Mullie et al., 2008). In human, the maximum concentration of plasma flavonoid level typically demarcated in 1–2 h after the intake of flavonoid-rich diet (Lotito et al., 2006). This proliferation in flavonoid plasma level depends on the type of flavonoid ingested; for example, flavonols have a half-life 6–10 times higher as compared to hydroxyflavan catechins (Hollman & Katan, 1997). Moreover, due to its high abundancy, dietary flavonoid as quercetin glucosides, catechin, flavanones, isoflavones, and gallic acid plasma level ranges from ~0.5–7.6 μM, as compared to least available flavonoids as anthocyanins and pro-anthocyanidins (Lotito et al., 2006; Thilakarathna & Rupasinghe, 2013). Few reports showed that the flavonoids in the form of high-potency supplements if ingested in an extraordinarily large amount, can produce potential health risks (Mullie et al., 2008).

9.3 FLAVONOIDS ACTING AS ANTI-CANCER AGENT

Cancer is a heterogeneous group of disease described by unsynchronized proliferation of cells resulted due to unregulated cell cycle leading to the excessive growth of abnormal cells that can metastasize to different body parts (Kroemer & Pouyssegur, 2008; Neagu et al., 2019). The major internal source of cancer are genetic mutations, oxidative stress, hypoxia, and impaired apoptotic activity, whereas the external causes are associated with increased and frequent stress exposure, pollution, radiation, along with tobacco chewing and smoking (Blackadar, 2016). Altered cell cycles, impaired metabolism, mutations, prolonged inflammation, inadequate immune response, angiogenesis leading to induction of metastasis, are the major features of the cancer cells (Neagu et al., 2019). There are various evidences which revealed that cancer can be induced by various degrees of metabolic alterations and dysfunctions in mitochondria (Kroemer & Pouyssegur, 2008; Bock & Tait, 2019). The mitochondria opt a crucial role in cellular energy, metabolic regulation, cell death signaling and ROS generation mitochondria (Bock & Tait, 2019). The key metabolic indication of the cancer/tumor cells is increased aerobic glycolysis (Lebelo et al., 2019), unregulated pH (Chiche et al., 2010), reduced lipid metabolism (Zaidi et al., 2013), restricted enzyme activities (Lu & Wang, 2018) fallowed by increased ROS production (Weinberg et al., 2019). As a result, this sequence of events creates an acidic extracellular environment which favors inflammation (Lee & Kim, 2016) and creates a hyperpolarized mitochondrial environment that is currently associated with the invasiveness and malignancy of cancerous cells (Neagu et al., 2019). Moreover, the glutamine mediated lipid biosynthesis regulates the pathways related to tumorigenesis and metastasis (Vegliante et al., 2018). Along with this, a decrease in membrane cardiolipin level also triggers impaired enzyme activity which supports cancer initiation in normal cells (Seyfried et al., 2014; Zhong et al., 2017).

Several modern studies revealed that flavonoids induce anti-cancer activity, generally by promoting apoptosis, reducing reactive oxygen species (ROS), and regulating mitochondrial functions and cell cycle which in other hand inhibit the proliferation and invasiveness of cancer cell, as discussed later in the chapter (Kopustinskiene et al., 2020; Dziedzic et al., 2021).

9.3.1 EFFECT OF FLAVONOIDS IN APOPTOSIS

Apoptosis is generally known as programmed cell death which is initiated by pro-apoptotic proteins (as caspases and Bcl-2 family proteins) and a sequence of signal transduction pathways (Abotaleb et al., 2018). Apoptosis comprises two key signaling cascades as extrinsic and intrinsic. The extrinsic cascade expedites with caspase 8 as its principal signaling protein, which is a member of tumor necrosis factor (TNF) superfamily, while the intrinsic cascade acting with mitochondrial pathway, initiates with the activation of caspases 9, 7, and 3, which are associated with Bcl-2 family proteins (McArthur & Kile, 2018). The overexpression of oncogenic genes (e.g., *H-RAS*), results in cellular proliferation, p53 suppression, and stimulation of anti-apoptotic proteins (e.g., Bcl-2 family proteins) in cancer cells, whereas pro-apoptotic proteins and caspases are downregulated, which on another hand mediates suppression of apoptosis in cancer cells (Jan & Chaudhry, 2018) (*see* Figure 9.1).

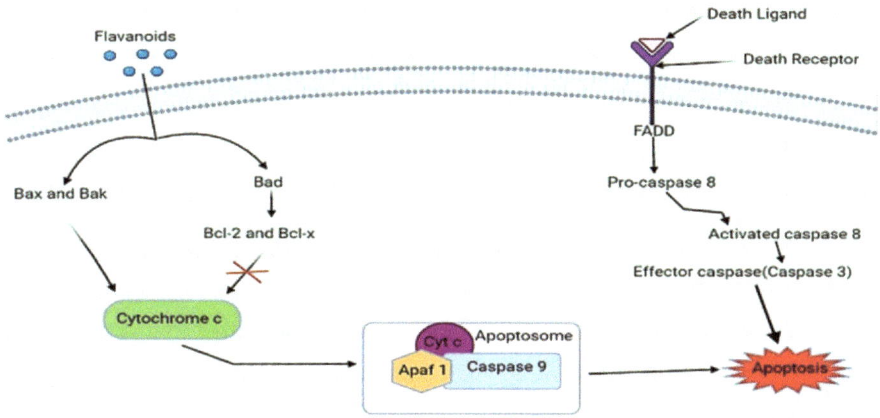

FIGURE 9.1 Two fundamental, elective pathways of apoptosis: extrinsic (death receptor-mediated cascade) and intrinsic (mitochondria-dependent cascade).

Flavonoids can initiate apoptotic signaling cascade activating the apoptotic pathways (Abotaleb et al., 2018). It can also act as pro-oxidants that could suppress division of malignant cells by distracting epidermal growth factor receptor (EGFR) or mitogen activated protein kinase (EGFR/MAPK), phosphatidylinositide 3-kinases (PI3K), protein kinase B (Akt), along with nuclear factor kappa-light-chain-enhancer of stimulated NF-κ B cells (Abotaleb et al., 2018; Neagu et al., 2019). Moreover, several flavonoids were reported for their effectiveness against specific cancer targets or signaling pathways. Flavonoid genistein hinders apoptosis protein 1, NF-κB and survivin by inhibiting Bcl 2, xL, and c-inhibitor complexes in A2780 and also in C200 cell lines (Solomon et al., 2008), triggered intrinsic apoptotic signaling cascade in HCT-116 and LoVo cell lines (Qin et al., 2016), and amplified activity of caspase-3 in HT-29 cancer cell line (Shafiee et al., 2016). Moreover, daidzein, an isoflavonoid also acts as phytoestrogen by promoting delivery of mitochondrial cytochrome c promoting activation of caspase 7 and 9 and also changing the ratio of Bax to Bcl-2 ratio in MCF-7 breast cancer cell line (Liu et al., 2012; Danciu et al., 2018). Along with this, in HCCSK-HEP-1 cell line, daidzein persuaded apoptosis by augmenting Bak and diminishing anti-apoptotic proteins, ensuing in release of cytochrome c from mitochondria and stimulating successive apoptotic cascade involving caspases 9 and 3 (Park et al., 2013).

The release of cytochrome c is also mediated by another flavanone hesperetin, which on other hand, also triggers caspases 3, caspases 9, and diminish Bax/Bcl-2 ratio in cancerous gastric cell lines, along with MCF-7, HT-29, Eca109, MDA-MB-231 (Kopustinskiene et al., 2020) and gastric cancer cells (Zhang et al., 2015). In H522 cell lines, the aforesaid flavanone hesperetin induced extrinsic pathway as a result of caspase-9 activation, elevated TNF-proteins expression, and diminution in p53 level (Elango et al., 2018). Furthermore, in PC-3 cell lines, it suppresses the NF κB signaling cascade and reduce Bcl-2 expression (Sambantham et al., 2013).

Another flavanone naringenin induced apoptosis by elevated p53 expression, Bax and caspase-3 stimulation, along with downregulation of survivin and Bcl-2 proteins, as studied in SGC-7901 cell line (Bao et al., 2016). Moreover, also mediates overexpression of TNF-family proteins which was recently reported to be associated with extrinsic apoptotic pathway (Kopustinskiene et al., 2020). Likewise, epigallocatechin gallate (EGCG), activated apoptosis, arrested cell-cycle, and hindered NF-β, leading to elevated expression of cyclooxygenase-2 (COX) enzyme (Shirakami et al., 2016). It also upsurges Bax/Bcl-2 ratio along with caspases 3, caspases 9, p53, p21, and reduced Bcl-2, PI3K, Akt, in HFF and T47D cells (Moradzadeh et al., 2017). Another phytoestrogen flavonol quercetin, initiated intrinsic apoptotic cascade by promoting regulation of Bax and caspase-3 and on other hand hindered the regulation of Bcl-2 in MCF-7 breast cancer cells (Chou et al., 2010). It also induced apoptosis in LNCaP, DU145 and PC-3 cells by upregulating the p53 signaling cascade (Wang et al., 2021), by initiating the intrinsic pathway by restraining Bad, Bax, Bcl-2 and COX-2 expression, stimulating caspase-3 and including release of mitochondrial cytochrome c, in HL-60 cell line (Niu et al., 2011). It also triggered caspase activation, downregulated Akt/PI-3-kinase and extracellular signal regulated kinase (ERK) cascade to initiate apoptosis in human hepatoma cell lines (Granado-Serrano et al., 2006; Sun et al., 2018). Similar potent alternation in expression of antiapoptotic and proapoptotic genes was also mediated by catechins (Sutherland et al., 2006), while cocoa flavanols was reported to moderately upregulate the apoptotic cascade in HepG2 (Granado-Serrano et al., 2007) and Caco-2 cells (Martin et al., 2016). In a recent study, flavonol kaempferol, promoted the activation of caspases 3 and 7, enhanced regulation of Bax, Bad and p53, along with downregulation of Bcl-xL protein induced intrinsic apoptosis in CP70/A2780, A2780WT, and OVCAR-3 cells (Luo et al., 2011; Imran et al., 2019).

Flavone apigenin also elevated the regulation of p53 gene in ACHN, RCC, and Caki-1 cells (Meng et al., 2017). It also inhibited the PI3K/Akt signaling cascade, triggered the intrinsic and extrinsic pathway and mitochondrial cytochrome c release, along with repressed Bcl-xL in T24 cells (Zhu et al., 2013; Liu et al., 2018). Another flavone chrysin initiated intrinsic pathway triggered by caspases 9 and 3 activation which on other hand mediated by release of mitochondrial cytochrome c in M17 and SP6.5 cells (Xue et al., 2016). It also upregulated caspase 3 in U937 cells (Khoo et al., 2010). In HT-29 colon cells, an estrogen receptor agonist pelargonidin (7-hydroxyflavonoids) initiated the release of mitochondrial cytochrome c, triggered caspases 9,3, Bax and Bid, along with inhibition in Bcl-2 and Bcl-xL gene expression (Lopez de Las Hazas et al., 2017). It also checked the PI3K/Akt signaling cascade and diminished the proliferation of osteosarcoma U2OS cells (Chen et al., 2021). Moreover, cyanidin flavonoid activated Bax and p53 gene expression and also mediated the release of cytochrome c along with reduction in expression of Bcl-2 gene in glioblastoma U87 cell lines (Hosseini et al., 2017; Kopustinskiene et al., 2020). Another anthocyanidin derivative delphinidin stimulated caspases 9 and 3 while diminished the PI3K/Akt and ERK1/2/MAPK signaling cascade in NSCLC cell lines (Pal et al., 2013; Lim & Song, 2017).

9.3.2 EFFECT OF FLAVONOIDS IN CELL CYCLE REGULATION

The cell contains a sequence of events for its division and duplication into daughter cells known as the cell cycle (Dalton, 2015; Tuli et al., 2019). The cell cycle involves four stages as G_1, S, G_2, and M – in which the cell builds its size, copies its hereditary material, make set-up for division, and afterward isolates. Each progression is efficient and customized (Dalton, 2015). A few atoms manage the cell cycle, including the cyclins proteins that advance the initiation and restrict to Cyclin dependable kinases (CDKs). CDKs are another protein buildings that coordinate the headway and upkeep of the cell cycle (Tamura, 2015). Then again, there were some positive focuses in the cell cycle, named as checkpoints, which can cease the cell cycle to explore the integrity of genetic material, targeting the assembly of daughter cells with precise genetic material (Tamura, 2015). In a few sorts of diseases, this fine guideline of these checkpoints is upset which drives the cells to perform excessive expansion and division (De Felice, 2019).

Several flavonoids are reported to suppress malignant cell growth which is mediated by diverse signaling molecules that arrest cell cycle at above-mentioned checkpoints and induce apoptosis (Tuli et al., 2019). Few *ex-vivo* research have revealed that many flavonoids can facilitate cell cycle arrest due to their tendency to hinder CDKs (Casagrande & Darbon, 2001). It can also control the activity of MAPKs as a feasible mechanism for their significant anti-neurodegenerative activity and protection against allergic, autoimmune, and cardiovascular diseases (CVDs) (Schroeter et al., 2002; Ahn et al., 2004). Several flavonoids as apigenin, luteolin, and quercetin also inhibited cell cycle at either the G_1 or G_2 phase as studied *ex vivo* in numerous cell lines (Kobayashi et al., 2002). An isoflavone genistein inhibited the cell cycle at G_2/M phases and triggers apoptosis in breast, gastric, lung, and prostate cancer cells (Kobayashi et al., 2002). It also hindered the G_0/G_1 phase of the cell cycle in fibroblasts and melanoma cells (Lian et al., 1998).

In some studies, cyclins as cdk4 and cdk2 kinases were activated by binding of cyclin D and cyclin E/A, in G_1 and S phase, while in other cases only cdc2 kinase was activated by binding to cyclin B in G_2 and M phase (Dalton, 2015). Out of these, the binding of cdk2 and cdk4 and activities were repressed by the binding of p27 and p21 inhibitors (Sherr, 1994). It has also been revealed that p21 is effectively transactivated by p53 gene product and suppresses tumorigenicity *in vivo* in experimental systems (Kobayashi et al., 2002; Dalton, 2015). Quercetin treated HeLa cells derived from cervical cancer showed lower p21 expression (Ponte et al., 2021). The same was observed in HCT116 human colon cancer cells when treated with genistein (Zhang et al., 2020), MDA-MB-231 breast cancer cells treated with luteolin, and HCT116 colorectal cancer cells treated with taxifolin (Huang et al., 2019; Ponte et al., 2021). Other flavonoids as isorhamnetin and genkwanin, inhibited cell cycle at G_2/M phase and initiated apoptosis in breast cancer cells. Moreover, these flavonoids down-regulated the PI3Kγ-p110, phospho-AKT, phospho-PI3K, phospho-p70S6K, phospho-ULK, and phospho-mTOR, along with their signaling pathway, which also promotes cell cycle arrest (Zhang et al., 2018). In human colon cancer cells (HT-29), kaempferol treatment downregulated the expression of cyclin D, E, and A (Ponte et al., 2021). Another isoflavone genistein arrested LNCaP prostate cancer cells at the G2/M phases, with the suppression of cyclin B expression. Moreover, genistein stimulated p21, a cyclin-dependent kinase inhibitor, which does not rely on p53 initiation. Along with this, other flavonoid, as apigenin and luteolin both elevated p21 levels, but apigenin mediates it through p53-dependent cascade, and luteolin mediates it but in p53-independent cascade (Mahmoud et al., 2013).

In addition, some flavonoids as 3-hydroxyflavone, 6-hydroxyflavone and 5,7-dihydroxyflavone inhibited the G_0/G_1 phase cells in SF295 cell line at growth inhibitory concentrations. Other flavonoids as 3,6-dihydroxyflavone and 5,6,7-trihydroxyflavone increased G_2/M or S phase cells. In other hand, 3-hydroxyflavone and 6-hydroxyflavone reduced G_2/M and G_0/G_1 phase cells, and increased S and G_2/M phase cells as studied in HCT15 cell lines. For the meantime, in HCT15/CL02 cells, 3-hydroxyflavone and 6-hydroxyflavone reduced G_0/G_1 phase cells and elevated the S phase cells (Choi et al., 1999). Some flavonoids have been reported to mediate G_2/M cell cycle arrest by the elevating the expression of p21 which is stimulated by p53 induce inhibition of cyclin B and E (Dalton, 2015). Other flavonoids such as daidzein, cyanidin, luteolin, and taxifolin have also been found to reduce expression of these cyclins (Huang et al., 2019).

9.3.3 EFFECT OF FLAVONOIDS IN ROS-SCAVENGING AND CELL SIGNALING

ROS are mostly produced in the electron transport chain in mitochondria as the results of oxidative aggregate. The measure of ROS produced causes oxidative stress which thusly harm DNA, proteins, and layer lipids (causing irritation) prompting malignant growth along with other degenerative infections (Moskaug et al., 2005). These ROS, containing hydrogen peroxide (H_2O_2) and nitric oxide (NO), are endogenously orchestrated in a few cells and take an interest, as second messengers, in cytokine or potentially development factor signals (Aboet et al., 2021). It can either straightforwardly or in a roundabout way, directs the activity of key flagging chemicals including guanylyl cyclase, phospholipase A2, phospholipase C, phospholipase D, actuating protein-1 (AP-1), atomic element kappa B (NF-κB), insulin receptor, Jun N-terminal kinase, and p38 kinase (MAPK) (Hehner et al., 2000) (*see* Figure 9.2).

These ROS interceded oxidative stress emerging from exogenous and endogenous beginnings put the body under strange physiological conditions (Ozben, 2007). It was accounted for to be firmly connected with a few parts of carcinogenesis, similar to growth bearing state, just as its treatment (Klaunig et al., 2010). The expanded oxidative stress brings about an imbalanced cell oxidation-decrease game plan, which can lead to further alterations such as gene mutations and genetic instability and can affect intracellular signal transduction and transcription factors (Schieber & Chandel, 2014). It has been seen that cancers have strange articulation levels of different cell reinforcement proteins including superoxide dismutase (SOD), catalase (CAT), glutathione peroxidase (GPx), glutathione reductase (GR), glutathione-S-transferase (GST), manganese superoxide dismutase (MnSOD), and copper-zinc superoxide dismutase (CuZnSOD) (Yanishlieva-Maslarova, 2001; Martindale & Holbrook, 2002). Nuclear factor erythroid 2-related element 2 (Nrf 2), a notable transcriptional factor that controls the declaration of previously mentioned cancer prevention agent particles, has been viewed as smothered during the carcinogenesis (Ma, 2013). Moreover, based anti-cancer specialists and radiation treatment utilized for malignant growth therapy were likewise known to apply oxidative pressure, which might result as one reason for drug opposition and disappointment of disease treatment (Rendic & Peter Guengerich, 2012).

Flavonoids functioning as pro-oxidants, have been shown to reduce cancer cell proliferation by inhibiting EGFR/MAPK, PI3K, protein kinase B, and nuclear factor kappa-light-chain-enhancer of activated B cells (NF-κB) (Kopustinskiene, 2020). Moreover, in MCF-7 and 3T3-L1 cells, genistein was shown to influence estrogen receptor activity and affect the Bax/Bcl-2 ratio, therefore suppressing proliferation, differentiation, and triggering of apoptosis (Choi et al., 1999). Another flavonoid baicalin induced apoptosis accompanied with the generation of intracellular ROS and elevation in release of the mitochondrial cytochrome c (Ueda et al., 2002). This release of cytochrome c and mitochondrial transmembrane potential disruption leads the apoptosoma activation in MCF-7 cells treated with 200 μM of the stilbene (Filomeni et al., 2007). Contrasting results have been achieved when cells were treated with quercetin (Chou et al., 2010).

Phosphorylation of proteins at OH groups of serine, threonine, and tyrosine deposits is a significant system of intracellular sign transduction engaged with different cell reactions including the guideline of cell development and multiplication (Birt et al., 2001). The response utilizes ATP as a phosphate contributor and is catalyzed by protein kinases. For example, development factor chemicals tie to extracellular areas of enormous transmembrane receptors show a tyrosine kinase moiety in their intracellular piece (Andersen & Markham, 2005). As an outcome of chemical receptor restricting, the receptor dimerizes and becomes dynamic in the phosphorylation of proteins near the film, along these lines setting off an enormous number of flagging pathways themselves including other PKs, like PKC, a Ser/Thr PK, and mitogen-actuated PKs (MAPKs) (Andersen & Markham, 2005).

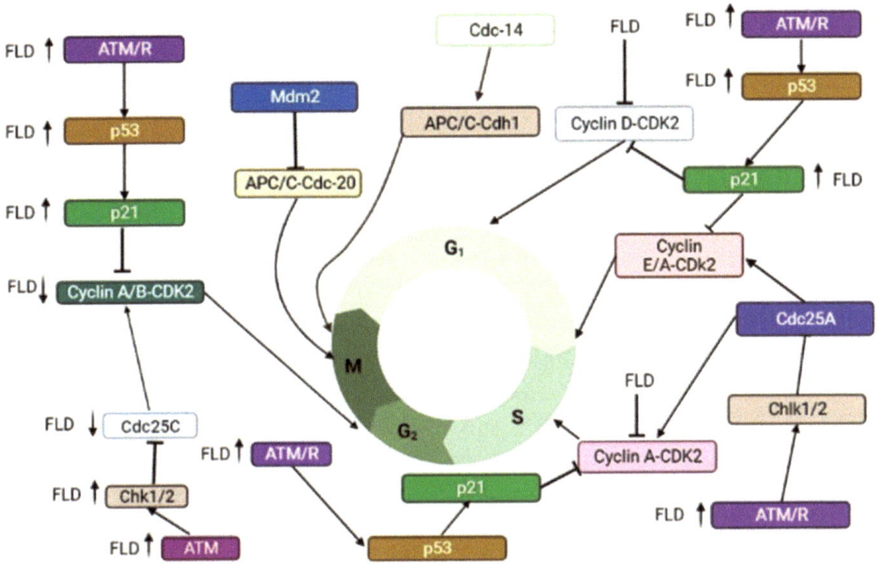

FIGURE 9.2 Cell cycle inhibition by flavonoids, mediated through multiple signaling molecules and cascades.

Previous evidences have revealed that flavonoid-induced ARE-dependent transcriptional gene expression which was reported to be mediated by the initiation of intracellular signaling pathway, including the MAPK. It has been also shown that, among MAPK signaling cascade, translocation of Erk-mediated Quer-derived Nrf-2 and p38 into the nuclei was responsible for consequent initiation of HO-1 expression (Yao et al., 2007; Lee & Kim, 2016). In addition, treatment with naringenin significantly reduced lipid peroxidation, pro-inflammatory cytokines such as IL-6, IL-1β, TNF-α, and triggers the antioxidant enzymes and vitamin C. Furthermore, naringenin effectively repressed CYP1A1, PCNA, and

NF-κB expression, adding for the chemopreventive potential against lung cancer in mice (Bodduluru et al., 2015). Similarly, luteolin also diminished cell viability by increasing the level of GSH and the expression of GSH synthetase in in HT-29 human colon cancer cells (Kang et al., 2014). The declined activities of SOD, CAT, GPx, and GR antioxidant enzymes in mouse colon cancer cells which were also found to be enhanced after luteolin treatment (Balamurugan & Karthikeyan, 2012; Balamurugan et al., 2019). Furthermore, Akt downstream effectors along with Fork-head box O transcription factors, are important regulators of cell growth. The various adverse processes activated upon FOXO suppression include increased generation of ROS. In squamous carcinoma cells, quercetin repressed the cell growth by hindering the EGFR/Akt activation with associated initiation of FOXO1 activation (Huang et al., 2019). Apigenin also revealed an anti-cancer effect on transgenic adenocarcinoma mouse prostate mice by FOXO3a- and FOXO-responsive proteins BIM p27/Kip1 (Shukla et al., 2014; Liu et al., 2018). Additionally, inhibition of FOXO and associated proteins has been studied in pancreatic cancer model *in vivo*, along with Bim and caspase-3 activated (–)-epigallocatechin-3-gallate induced apoptosis. (–)-epigallocatechin-3-gallate also modulated of cell cycle markers as p27/KIP1, angiogenesis as CD31, IL-6, IL-8, HIF1α and VEGF, and metastasis MMP2 and 7 (Shankar et al., 2013).

9.3.4 EFFECT OF FLAVONOIDS IN MITOCHONDRIAL FUNCTIONS

Tumor-cell metabolic activity differs from normal cell due to significantly abnormal mitochondrial functioning (Porporato et al., 2018). As a result, currently there has been utmost interest for naturally derived molecules that might restore normal mitochondrial functioning (Gorlach et al., 2015; Abotaleb et al., 2018). In malignancies, voltage-gated anion channel (VDAC) and hexokinase coupling hinders apoptotic initiation (Romani et al., 1999). In MDA-MB-231 and MCF-7 cells originated from human breast carcinoma, flavone oroxylin A was revealed to promote the release of hexokinase from mitochondria, ensuing in the inhibition of glycolysis (Fattorusso et al., 2002). Overexpression of mitochondrial antiapoptotic proteins of the BCL-2 family results in obstruction in apoptotic pathways and subsequently initiation of cancer (Gorlach et al., 2015).

In MDA-MB-231 and MCF-7 cells started from human bosom carcinoma, flavone oroxylin an uncovered to advance the arrival of hexokinase from mitochondria, following in the restraint of glycolysis (Fattorusso et al., 2002). The raised articulation of mitochondrial antiapoptotic proteins of the BCL-2 family brings about protection from apoptotic pathways (Gorlach et al., 2015). A phytoestrogen isoflavonoid daidzein likewise acted to invigorate mitochondrial cytochrome c delivery, prompting caspase 7 and 9 setting off alongside change in Bax/Bcl-2 proportion as revealed in MCF-7 cells (Jin et al., 2010). Another flavonoid epigallocatechin-3-gallate and naringenin diminished the BCL-2 quality articulation in SGC-7901 gastric disease and in HuCC-T1 cholangiocarcinoma cell lines (Kwak et al., 2017). Mitochondrial adenine nucleotide translocase protein, implanted in the internal mitochondrial layer is answerable for ATP/ADP trade (Klingenberg, 2008). It is a part of MPTP (Mitochondrial Permeability Transition Pore) complex, which is a key component setting off apoptosis (Halestrap, 2004).

In mitochondria extracted from rat renal cortex, quercetin at a dose of 50 M obstructed 46% adenine nucleotide translocase, but apigenin at a dose of 20 M reduced it as studied in human prostate cancer DU145 cells (Ortega & Garcia, 2009; Oishi et al., 2013).

The electron transport chain derived from rat hepatic mitochondria and sub-mitochondrial particles was reported to be hindered from a procyanidin-rich pine (*Pinus pinaster*) bark extract, influencing complexes I, II, and III (Moini et al., 1999). Due to elevated ROS production in complex III of the mitochondrial electron transport chain, the isoflavone genistein triggered permeability transition in the mitochondrial membrane as studied in isolated rat hepatic mitochondria (Salvi et al., 2002). Epigallocatechin-3-gallate repressed the rapid development in malignant pleural mesothelioma cells by controlling electron transport chain complex I and II, along with ATP synthase (Valenti et al., 2013). Furthermore, epigallocatechin-3-gallate controlled the apoptotic signaling pathway and modified mitochondrial bioenergetic activities (Oliveira et al., 2016). Anthocyanins inhibited apoptosis by lowering cytosolic cytochrome c levels and promoting electron transport between NADH dehydrogenase and cytochrome c (Bendokas et al., 2019). Quercetin, naringenin, hesperetin, and kaempferol have been demonstrated to promote the TCA cycle, changing anaerobic glycolysis to phosphorylation, which is generally repressed in cancer cells (Wang et al., 2009; Constantin et al., 2013). Mitochondria in cancer cells are often hyperpolarized, with membrane potentials reaching 220 mV, rendering them resistant to cell death communication (Fantin et al., 2006). Additionally, many flavonoids possess pKa values ranging from 6 to 9, i.e., near to the physiological pH of the cytoplasm and mitochondrial chambers, and hence have the capacity to penetrate the mitochondrial matrix and release a proton in its moderately alkaline environment (pH 7.8) (Van Dijk et al., 2000). This action may be important in cancer chemoprevention since modest mitochondrial uncoupling efficiently protects cells against oxidative damage.

9.4 PROBABLE MOLECULAR MECHANISM FOR ANTI-CANCER ACTIVITY OF FLAVONOIDS

Flavonoids have been frequently marked as *in vitro* enzyme inhibitors which was mediated through several complex signaling cascade. The ligand and receptors involved in these signal transduction can mediate some of their action probably by conjoining radical scavenging and interaction with enzyme functions (Balasuriya & Rupasinghe, 2011; Batra and Sharma, 2013). These non-covalent molecular interactions between flavonoid–protein interaction were mediated by a structural unit known as phenolic nucleus that is associated with proteins (Batra & Sharma, 2013). These interactions can be of electrostatic interactions or Van Der Waals type. In the first type, H-bonding is the key electrostatic interaction, while in the second type, the dispersion interactions with non-polar amino acids develop a non-polar aromatic ring followed by subsequent release of water molecule. This type of oxidative covalent coupling and redox reactions in flavonoid–protein interaction may result from 1–2 electron oxidation carried by oxidation, ROS scavenging and enzymatic oxidation. For some structurally open proteins (e.g., salivary proteins), binding coefficients are slightly low with polyphenols however condensation and polymerization of these molecules produces a constricted affinity (Balasuriya & Rupasinghe, 2011). It can probably be due to nonspecific binding in a stretched protein chain or at the outer surface of proteins.

In terms of cell cycle regulation, the phase of the cell cycle in which the DNA replicated, and the chromosomes separation takes place is coordinated by intense phosphorylation which on other hand controlled by extremely regulated kin cyclin-dependent kinases (CDKs) (Batra &

Sharma, 2013). The most likely mechanism involved with the significant anti-carcinogenic properties of flavonoids could be their capacity to hinder several PKs, thereby preventing signal transduction cascade of cell proliferation (Ren et al., 2003; Kopustinskiene et al., 2020).

The pro-oxidant activity of flavonoids appears to be involved mainly in the inhibition of mitochondrial respiration which is recently correlated with its ability to endure auto-oxidation to produce superoxide anions as studied *in vitro*. Then again, transition metal ions connected to proteins that catalyze auto-oxidation, appears to be suspicious for their huge support in the auto-oxidation of polyphenols. As an elective component, supportive of oxidant movement of flavonoids could be dependent on the action of peroxidases. For example, as studied *in vitro*, peroxyl radical is formed when apigenin acts in presence of myeloperoxidase (Kashyap et al., 2021) which leads to the intracellular ROS production and associated apoptotic pathway initiation leading to necrotic cell death (Wu et al., 2020).

9.5 CONCLUSION

The above-mentioned *in vitro* and *in vivo* studies have revealed that food and beverages rich in flavonoids are associated with a decreased risk of several types of diseases and disorders including cancer. These merits of flavonoids are due to their scavenging ability, apoptosis initiation, mitochondrial activity regulation along with cell cycle and cell signaling regulation.

Through different pathways flavonoids can hinder free extremist intervened cytotoxicity, lipid peroxidation, cancer development, and adjust endogenous chemical action. Accordingly, flavonoids in adequate amount may give security against persistent and degenerative sicknesses. It is a result of every one of these well-being advancing properties flavonoids are alluded to as a semi-fundamental part of the food. Current reports show that flavonoids give preferred antioxidant property over usually utilized normal and engineered cancer prevention agents, for example, α-tocopherol, Trolox, ascorbic corrosive, butylated hydroxyanisole, and hydroxytoluene. Nonetheless, comprehensive data about the above targets and related components of activity can likewise be recovered by using frameworks science, transcriptomics, proteomics, and metabolomics devices. Thusly, future examination ought to distract on the synergistic strategies for flavonoids with other existing anti-cancer medications.

KEYWORDS

- **anti-cancer remedies**
- **curcumin**
- **cyclin dependable kinases**
- **cyclooxygenase-2**
- **extracellular signal-regulated kinase**
- **flavonoids**
- **reactive oxygen species**
- **resveratrol**

REFERENCES

Abot, A., Fried, S., Cani, P. D., & Knauf, C., (2021). Reactive oxygen species/reactive nitrogen species as messengers in the gut: Impact on physiology and metabolic disorders. *Antioxid. Redox Signal., 10*, 1089.

Abotaleb, M., Samuel, S. M., Varghese, E., Varghese, S., Kubatka, P., Liskova, A., & Busselberg, D., (2018). Flavonoids in cancer and apoptosis. *Cancers, 11*, 28.

Ahn, S. C., Kim, G. Y., Kim, J. H., Baik, S. W., Han, M. K., Lee, H. J., Moon, D. O., et al., (2004). Epigallocatechin-3-gallate, constituent of green tea, suppresses the LPS-induced phenotypic and functional maturation of murine dendritic cells through inhibition of mitogen-activated-protein kinases and NF-kB. *Biochem Biophys Res Commun., 313*, 148–155.

Balamurugan, K., & Karthikeyan, J., (2012). Evaluation of luteolin in the prevention of N-nitrosodiethylamine-induced hepatocellular carcinoma using animal model system. *Indian J. Clin. Biochem., 27*(2), 157–163.

Balamurugan, K., Mendoza-Villanueva, D., Sharan, S., Summers, G. H., Dobrolecki, L. E., Lewis, M. T., & Sterneck, E., (2019). C/EBPδ links IL-6 and HIF-1 signaling to promote breast cancer stem cell-associated phenotypes. *Oncogene., 38*, 3765–3780.

Bao, L., Liu, F., Guo, H. B., Li, Y., Tan, B. B., Zhang, W. X., & Peng, Y. H., (2016). Naringenin inhibits proliferation, migration, and invasion as well as induces apoptosis of gastric cancer SGC7901 cell line by downregulation of AKT pathway. *Tumour Biol., 37*, 11365–11374.

Batra, P., & Sharma, A. K., (2013). Anti-cancer potential of flavonoids: Recent trends and future perspectives. *3 Biotech., 3*, 439–459.

Béliveau, R., & Gingras, D., (2007). Role of nutrition in preventing cancer. *Can. Fam. Physician., 53*, 1905–1911.

Benavente-García, O., & Castillo, J., (2008). Update on uses and properties of citrus flavonoids: New findings in anti-cancer, cardiovascular, and anti-inflammatory activity. *J. Agric. Food Chem., 56*, 6185–6205.

Bendokas, V., Skemiene, K., Trumbeckaite, S., Stanys, V., Passamonti, S., Borutaite, V., & Liobikas, J., (2019). Anthocyanins: From plant pigments to health benefits at the mitochondrial level. *Crit. Rev. Food Sci. Nutr.,* 1–14.
Birt, D. F., Hendrich, S., & Wang, W., (2001). Dietary agents in cancer prevention: Flavonoids and isoflavonoids. *Pharmacol. Ther., 90,* 157–177.
Blackadar, C. B., (2016). Historical review of the causes of cancer. *World J. Clin. Oncol., 7,* 54–86.
Bock, F. J., & Tait, S. W. G., (2019). Mitochondria as multifaceted regulators of cell death. *Nat. Rev. Mol. Cell Biol., 21,* 85–100.
Bodduluru, L. N., Kasala, E. R., Madhana, R. M. R., & Sriram, C. S., (2015). Natural killer cells: The journey from puzzles in biology to treatment of cancer. *Cancer Lett., 357,* 454–467.
Casagrande, F., & Darbon, J. M., (2001). Effects of structurally related flavonoids on cell cycle progression of human melanoma cells: Regulation of cyclin-dependent kinases CDK2 and CDK1. *Biochem Pharmacol., 61,* 1205–1215.
Cassidy, A., & Minihane, A. M., (2017). The role of metabolism (and the microbiome) in defining the clinical efficacy of dietary flavonoids. *Am. J. Clin. Nutr., 105,* 10–22.
Chen, W., Wu, S., Huang, Y., Zhang, T., Dong, H., Zheng, X., Chen, T., Gong, X., et al., (2021). A c-Met inhibitor suppresses osteosarcoma progression via the ERK1/2 pathway in human osteosarcoma cells. *Onco. Targets Ther., 14,* 4791.
Chiche, J., Brahimi-Horn, M. C., & Pouyssegur, J., (2010). Tumor hypoxia induces a metabolic shift causing acidosis: A common feature in cancer. *J. Cell. Mol. Med., 14,* 771–794.
Choi, S. U., Ryu, S. Y., Yoon, S. K., Jung, N. P., Park, S. H., Kim, K. H., Choi, E. J., & Lee, C. O., (1999). Effects of flavonoids on the growth and cell cycle of cancer cells. *Anti-cancer Res., 6,* 5229–5233.
Chou, C. C., Yang, J. S., Lu, H. F., Ip, S. W., Lo, C., Wu, C. C., Lin, J. P., Tang, N. Y., Chung, J. G., Chou, M. J., et al., (2010). Quercetin-mediated cell cycle arrest and apoptosis involving activation of a caspase cascade through the mitochondrial pathway in human breast cancer MCF-7 cells. *Arch. Pharmacal Res., 33,* 1181–1191.
Conney, A. H., (2003). Enzyme induction and dietary chemicals as approaches to cancer chemoprevention: The seventh DeWitt S. Goodman lecture. *Cancer Res., 63,* 7005–7031.
Constantin, R. P., Do Nascimento, G. S., Constantin, R. P., Salgueiro, C. L., Bracht, A., Ishii-Iwamoto, E. L., Yamamoto, N. S., & Constantin, J., (2013). Citrus flavanones affect hepatic fatty acid oxidation in rats by acting as prooxidant agents. *Biomed. Res. Int., 2013,* 342973.
Dalton, S., (2015). Linking the cell cycle to cell fate decisions. *Trends Cell Biol., 25,* 592–600.
Danciu, C., Avram, S., Pavel, I. Z., Ghiulai, R., Dehelean, C. A., Ersilia, A., Minda, D., et al., (2018). Main isoflavones found in dietary sources as natural anti-inflammatory agents. *Curr. Drug Targets, 19,* 841–853.
De Felice, F., Marchetti, C., Tombolini, V., & Panici, P. B., (2019). Immune check-point in endometrial cancer. *Int. J. Clin. Oncol., 24,* 910–916.
De Pascual-Teresa, S., Santos-Buelga, C., & Rivas-Gonzalo, J. C., (2000). Quantitative analysis of flavan-3-ols in Spanish foodstuffs and beverages. *J. Agric. Food Chem., 48,* 5331–5337.

Elango, R., Athinarayanan, J., Subbarayan, V. P., Lei, D. K. Y., & Alshatwi, A. A., (2018). Hesperetin induces an apoptosis-triggered extrinsic pathway and a p53-independent pathway in human lung cancer H522 cells. *J. Asian Nat. Prod. Res., 20*, 559–569.

Fantin, V. R., St-Pierre, J., & Leder, P., (2006). Attenuation of LDH-A expression uncovers a link between glycolysis, mitochondrial physiology, and tumor maintenance. *Cancer Cell., 9*, 425–434.

Fattorusso, E., Iorizzi, M., Lanzotti, V., & Taglialatela-Scafati, O., (2002). Chemical composition of shallot (*Allium ascalonicum* Hort.). *J. Agric. Food Chem., 50,* 5686–5690.

Filomeni, G., Graziani, I., Rotilio, G., & Ciriolo, M. R., (2007). Trans-Resveratrol induces apoptosis in human breast cancer cells MCF-7 by the activation of MAP kinases pathways. *Genes Nutr., 2*, 295–305.

Franco, O. H., Bonneux, L., De Laet, C., Peeters, A., Steyerberg, E. W., & Mackenbach, J. P., (2004). The polymeal: A more natural, safer, and probably tastier (than the polypill) strategy to reduce cardiovascular disease by more than 75%. *BMJ., 329*, 1447–1450.

Gorlach, S., Fichna, J., & Lewandowska, U., (2015). Polyphenols as mitochondria-targeted anti-cancer drugs. *Cancer Lett., 366*, 141–149.

Granado-Serrano, A. B., Martin, M. A., Bravo, L., Goya, L., & Ramos, S., (2006). Quercetin induces apoptosis via caspase activation, regulation of Bcl-2, and inhibition of PI-3-kinase/Akt and ERK pathways in a human hepatoma cell line (HepG2). *J. Nutr., 136*, 2715–2721.

Granado-Serrano, A. B., Martin, M. A., Izquierdo-Pulido, M., Goya, L., Bravo, L., & Ramos, S., (2007). Molecular mechanisms of (-)-epicatechin and chlorogenic acid on the regulation of the apoptotic and survival/proliferation pathways in a human hepatoma cell line. *J. Agric. Food Chem., 55*, 2020–2027.

Halestrap, A. P., (2004). Mitochondrial permeability: Dual role for the ADP/ATP translocator? *Nature, 430*, 984.

Hehner, S. P., Hofmann, T. G., Dienz, O., Dröge, W., & Schmitz, M. L., (2000). Tyrosine-phosphorylated Vav1 as a point of integration for T-cell receptor-and CD28-mediated activation of JNK, p38, and interleukin-2 transcription. *Int. J. Biol. Chem., 275*(24), 18160–18171.

Hollman, P. C., & Katan, M. B., (1997). Absorption, metabolism and health effects of dietary flavonoids in man. *Biomed. Pharmacother., 51,* 305–310.

Hosseini, M. M., Karimi, A., Behroozaghdam, M., Javidi, M. A., Ghiasvand, S., Bereimipour, A., Aryan, H., et al., (2017). Cytotoxic and apoptogenic effects of cyanidin-3-glucoside on the glioblastoma cell line. *World Neurosurg., 108*, 94–100.

Huang, L., Jin, K., & Lan, H., (2019). Luteolin inhibits cell cycle progression and induces apoptosis of breast cancer cells through downregulation of human telomerase reverse transcriptase. *Oncol. Lett., 17*, 3842–3850.

Imran, M., Rauf, A., Shah, Z. A., Saeed, F., Imran, A., Arshad, M. U., Ahmad, B., Bawazeer, S., Atif, M., Peters, D. G., et al., (2019). Chemo-preventive and therapeutic effect of the dietary flavonoid kaempferol: A comprehensive review. *Phytother. Res., 33*, 263–275.

Jan, R., & Chaudhry, G. E., (2019). Understanding apoptosis and apoptotic pathways targeted cancer therapeutics. *Adv. Pharm. Bull., 9,* 205–218.

Jaramillo, M. C., & Zhang, D. D., (2013). The emerging role of the Nrf2–Keap1 signaling pathway in cancer. *Genes Dev., 27*, 2179–2191.

Johnson, I. T., (2004). New approaches to the role of diet in the prevention of cancers of the alimentary tract. *Mutat Res., 551*, 9–28.

Kang, M. H., Kim, I. H., & Nam, T. J., (2014). Phloroglucinol induces apoptosis via apoptotic signaling pathways in HT-29 colon cancer cells. *Oncol. Rep., 32*, 1341–1346.

Kashyap, P., Shikha, D., Thakur, M., & Aneja, A., (2021). Functionality of apigenin as a potent antioxidant with emphasis on bioavailability, metabolism, action mechanism and *in vitro* and *in vivo* studies: A review. *J. Food Biochem.*, e13950.

Kasiri, N., Rahmati, M., Ahmadi, L., Eskandari, N., & Motedayyen, H., (2020). Therapeutic potential of quercetin on human breast cancer in different dimensions. *Inflammopharmacology., 28*, 39–62.

Kawabata, K., Yoshioka, Y., & Terao, J., (2019). Role of intestinal microbiota in the bioavailability and physiological functions of dietary polyphenols. *Molecules, 24*, 370.

Keservani, R. K., & Sharma, A. K., (2014). Flavonoids: Emerging trends and potential health benefits. *Journal of Chinese Pharmaceutical Sciences, 23*(12), 815.

Keservani, R. K., Kesharwani, R. K., Sharma, A. K., Vyas, N., & Chadoker, A., (2010b). Nutritional supplements: An overview. *International Journal of Current Pharmaceutical Review and Research, 1* (1), 59–75.

Keservani, R. K., Kesharwani, R. K., Vyas, N., Jain, S., Raghuvanshi, R., & Sharma, A. K., (2010a). Nutraceutical and functional food as future food: A review. *Der Pharmacia Lettre, 2*(1), 106–116.

Keservani, R. K., Sharma, A. K., & Kesharwani, R. K., (2020). "*Nutraceuticals and Dietary Supplements: Applications in Health Improvement and Disease Management*". CRC Press. ISBN: 9781771888738.

Khoo, B. Y., Chua, S. L., & Balaram, P., (2010). Apoptotic effects of chrysin in human cancer cell lines. *Int. J. Mol. Sci., 11*, 2188–2199.

Klaunig, J. E., Kamendulis, L. M., & Hocevar, B. A., (2010). Oxidative stress and oxidative damage in carcinogenesis. *Toxicol. Pathol., 38*, 96–109.

Klingenberg, M., (2008). The ADP and ATP transport in mitochondria and its carrier. *Biochim. Biophys. Acta, 1778,* 1978–2021.

Kobayashi, T., Nakata, T., & Kuzumaki, T., (2002). Effect of flavonoids on cell cycle progression in prostate cancer cells. *Cancer Lett., 176*, 17–23.

Kopustinskiene, D. M., Jakstas, V., Savickas, A., & Bernatoniene, J., (2020). Flavonoids as anti-cancer agents. Nutrients, *122*, 457.

Kroemer, G., & Pouyssegur, J., (2008). Tumor cell metabolism: Cancer's Achilles' heel. *Cancer Cell, 13*, 472–482.

Kwak, T. W., Park, S. B., Kim, H. J., Jeong, Y. I., & Kang, D. H., (2017). Anti-cancer activities of epigallocatechin-3-gallate against cholangiocarcinoma cells. *Oncotargets Ther., 10*, 137–144.

Lebelo, M. T., Joubert, A. M., & Visagie, M. H., (2019). Warburg effect and its role in tumourigenesis. *Arch. Pharmacal Res., 42*, 833–847.

Lee, N., & Kim, D., (2016). Cancer metabolism: Fueling more than just growth. *Mol. Cells, 39*, 847–854.

Li, J., Lai, Y., Cao, Y., Du, T., Zeng, L., Wang, G., Chen, X., et al., (2021). Correction to: SHARPIN overexpression induces tumorigenesis in human prostate cancer LNCaP, DU145 and PC-3 cells via NF-κB/ERK/Akt signaling pathway. *Med. Onc., 9*, 104.

Lian, F., Bhuiyan, M., Li, Y. W., Wall, N., Kraut, M., & Sarkar, F. H., (1998). Genistein-induced G2-M arrest, p21WAF1 upregulation, and apoptosis in a non-small-cell lung cancer cell line. *Nutr. Cancer., 31*, 184–191.

Lim, W., & Song, G., (2017). Inhibitory effects of delphinidin on the proliferation of ovarian cancer cells via PI3K/AKT and ERK 1/2 MAPK signal transduction. *Oncol. Lett., 14*, 810–818.

Liskova, A., Koklesova, L., Samec, M., Smejkal, K., Samuel, S. M., Varghese, E., & Kubatka, P., (2020). Flavonoids in cancer metastasis. *Cancers, 12*, 1498.

Lopez De, L. H. M. C., Mosele, J. I., Macia, A., Ludwig, I. A., & Motilva, M. J., (2017). Exploring the colonic metabolism of grape and strawberry anthocyanins and Their *in vitro* apoptotic effects in HT-29 colon cancer cells. *J. Agric. Food Chem., 65*, 6477–6487.

Lotito, S. B., & Frei, B., (2006). Consumption of flavonoid-rich foods and increased plasma antioxidant capacity in humans: Cause, consequence, or epiphenomenon? *Free Radic. Biol. Med., 41*, 1727–1746.

Lu, S., & Wang, Y., (2018). Nonmetabolic functions of metabolic enzymes in cancer development. *Cancer Commun., 38*, 63.

Luo, H., Rankin, G. O., Li, Z., Depriest, L., & Chen, Y. C., (2011). Kaempferol induces apoptosis in ovarian cancer cells through activating p53 in the intrinsic pathway. *Food Chem., 128*, 513–519.

Martin, M. A., Goya, L., & Ramos, S., (2016). Preventive effects of cocoa and cocoa antioxidants in colon cancer. *Diseases, 4*, 1–6.

Martindale, J. L., & Holbrook, N. J., (2002). Cellular response to oxidative stress: Signaling for suicide and survival. *J. Cell. Physiol., 192*, 1–15.

McArthur, K., & Kile, B. T., (2018). Apoptotic caspases: Multiple or mistaken identities? *Trends Cell Biol., 28*, 475–493.

Mejri, F., Selmi, S., Martins, A., Benkhoud, H., Baati, T., Chaabane, H., Njim, L., et al., (2018). Broad bean (*Vicia faba* L.) pods: A rich source of bioactive ingredients with antimicrobial, antioxidant, enzyme inhibitory, anti-diabetic and health-promoting properties. *Food Funct., 9*, 2051–2059.

Meng, S., Zhu, Y., Li, J. F., Wang, X., Liang, Z., Li, S. Q., Xu, X., Chen, H., Liu, B., Zheng, X. Y., et al., (2017). Apigenin inhibits renal cell carcinoma cell proliferation. *Oncotarget., 8*, 19834–19842.

Moini, H., Arroyo, A., Vaya, J., & Packer, L., (1999). Bioflavonoid effects on the mitochondrial respiratory electron transport chain and cytochrome c redox state. *Redox Rep., 4*, 35–41.

Moradzadeh, M., Hosseini, A., Erfanian, S., & Rezaei, H., (2017). Epigallocatechin-3-gallate promotes apoptosis in human breast cancer T47D cells through down-regulation of PI3K/AKT and telomerase. *Pharmacol. Rep., 69*, 924–928.

Moskaug, J. Ø., Carlsen, H., Myhrstad, M. C., & Blomhoff, R., (2005). Polyphenols and glutathione synthesis regulation. *Am. J. Clin. Nutr., 81*, 277S–283S.

Mullie, P., Clarys, P., Deriemaeker, P., & Hebbelinck, M., (2008). Estimation of daily human intake of food flavonoids. *Int. J. Food Sci. Nutr., 59*, 291–298.

Murota, K., Nakamura, Y., & Uehara, M., (2018). Flavonoid metabolism: The interaction of metabolites and gut microbiota. *Biosci. Biotechnol. Biochem., 82*, 600–610.

Nabavi, S. M., Samec, D., Tomczyk, M., Milella, L., Russo, D., Habtemariam, S., Suntar, I., Rastrelli, L., Daglia, M., Xiao, J., et al., (2018). Flavonoid biosynthetic pathways in plants: Versatile targets for metabolic engineering. *Biotechnol. Adv., 38*, 107316.

Neagu, M., Constantin, C., Popescu, I. D., Zipeto, D., Tzanakakis, G., Nikitovic, D., Fenga, C., et al., (2019). Inflammation and metabolism in cancer cell-mitochondria key player. *Front. Oncol., 9*, 348.

Niu, G., Yin, S., Xie, S., Li, Y., Nie, D., Ma, L., Wang, X., & Wu, Y., (2011). Quercetin induces apoptosis by activating caspase-3 and regulating Bcl-2 and cyclooxygenase-2 pathways in human HL-60 cells. *Acta Biochim. Biophys. Sin., 43*, 30–37.

Oishi, M., Iizumi, Y., Taniguchi, T., Goi, W., Miki, T., & Sakai, T., (2013). Apigenin sensitizes prostate cancer cells to Apo2L/TRAIL by targeting adenine nucleotide translocase-2. *PLoS ONE., 8*, e55922.

Oliveira, M. R., Nabavi, S. F., Daglia, M., Rastrelli, L., & Nabavi, S. M., (2016). Epigallocatechin gallate and mitochondria-A story of life and death. *Pharmacol.Res., 104*, 70–85.

Ortega, R., & Garcia, N., (2009). The flavonoid quercetin induces changes in mitochondrial permeability by inhibiting adenine nucleotide translocase. *J. Bioenerg. Biomembr., 41*, 41–47.

Ozben, T., (2007). Oxidative stress and apoptosis: Impact on cancer therapy. *J. Pharm. Sci., 96*(9), 2181–2196.

Pal, H. C., Sharma, S., Strickland, L. R., Agarwal, J., Athar, M., Elmets, C. A., & Afaq, F., (2013). Delphinidin reduces cell proliferation and induces apoptosis of non-small-cell lung cancer cells by targeting EGFR/VEGFR2 signaling pathways. *PLoS ONE., 8*, e77270.

Pandjaitan, N., Howard, L. R., Morelock, T., & Gil, M. I., (2005). Antioxidant capacity and phenolic content of spinach as affected by genetics and maturation. *J. Agric. Food Chem., 53*, 8618–8623.

Park, H. J., Jeon, Y. K., You, D. H., & Nam, M. J., (2013). Daidzein causes cytochrome c-mediated apoptosis via the Bcl-2 family in human hepatic cancer cells. *Food Chem. Toxicol., 60*, 542–549.

Passamonti, S., Terdoslavich, M., Franca, R., Vanzo, A., Tramer, F., Braidot, E., Petrussa, E., & Vianello, A., (2009). Bioavailability of flavonoids: A review of their membrane transport and the function of bilitranslocase in animal and plant organisms. *Curr. Drug Metab., 10*, 369–394.

Patel, M. J., (2008). A review of potential health benefits of flavonoids. *Lethbridge Undergrad Res J., 3*.

Ponte, L. G. S., Pavan, I. C. B., Mancini, M. C. S., Da Silva, L. G. S., Morelli, A. P., Severino, M. B., Bezerra, R. M. N., & Simabuco, F. M., (2021). The hallmarks of flavonoids in cancer. *Molecules, 26*, 2029.

Porporato, P. E., Filigheddu, N., Pedro, J. M. B., Kroemer, G., & Galluzzi, L., (2018). Mitochondrial metabolism and cancer. *Cell Res., 28*, 265–280.

Qin, J., Teng, J., Zhu, Z., Chen, J., & Huang, W. J., (2016). Genistein induces activation of the mitochondrial apoptosis pathway by inhibiting phosphorylation of Akt in colorectal cancer cells. *Pharm. Biol., 54*, 74–79.

Rendic, S., & Guengerich, F. P., (2012). Contributions of human enzymes in carcinogen metabolism. *Chem. Res. Toxicol., 25*, 1316–1383.

Rodriguez-Garcia, C., & Sanchez-Quesada, C., (2019). Dietary flavonoids as cancer chemopreventive agents: An updated review of human studies. *Antioxidants, 8*, 137.

Romani, A., Mulinacci, N., Pinelli, P., Vincieri, F. F., & Cimato, A., (1999). Polyphenolic content in five Tuscany cultivars of *Olea europaea* L. *J. Agric. Food Chem., 47*, 964–967.

Salvi, M., Brunati, A. M., Clari, G., & Toninello, A., (2002). Interaction of genistein with the mitochondrial electron transport chain results in the opening of the membrane transition pore. *Biochim. Biophys. Acta, 1556*, 187–196.

Sambantham, S., Radha, M., Paramasivam, A., Anandan, B., Malathi, R., Chandra, S. R., & Jayaraman, G., (2013). Molecular mechanism underlying hesperetin-induced apoptosis by in silico analysis and in prostate cancer PC-3 cells. *Asian Pac. J. Cancer Prev., 14*, 4347–4352.

Scheline, R. R., (1991). *Handbook of Mammalian Metabolism of Plant Compounds.* CRC Press.

Schieber, M., & Chandel, N. S., (2014). ROS function in redox signaling and oxidative stress. *Curr. Biol., 24*, R453-R462.

Schroeter, H., Boyd, C., Spencer, J. P., Williams, R. J., Cadenas, E., & Rice-Evans, C., (2002). MAPK signaling in neurodegeneration: Influences of flavonoids and of nitric oxide. *Neurobiol Aging, 23*, 861–880.

Shafiee, G., Saidijam, M., Tavilani, H., Ghasemkhani, N., & Khodadadi, I., (2016). Genistein induces apoptosis and inhibits proliferation of HT29 colon cancer cells. *Int. J. Mol. Cell. Med., 5*, 178–191.

Sherr, C. J., (1993). Mammalian G1 cyclins. *Cell, 73*, 1059–1065.

Shirakami, Y., Sakai, H., Kochi, T., Seishima, M., & Shimizu, M., (2016). Catechins and its role in chronic diseases. *Adv. Exp. Med. Biol., 929*, 67–90.

Solomon, L. A., Ali, S., Banerjee, S., Munkarah, A. R., Morris, R. T., & Sarkar, F. H., (2008). Sensitization of ovarian cancer cells to cisplatin by genistein: The role of NF-kappa *B. J. Ovarian Res., 1*, 9.

Sutherland, B. A., Rahman, R. M., & Appleton, I., (2006). Mechanisms of action of green tea catechins, with a focus on ischemia-induced neurodegeneration. *J. Nutr. Biochem., 17*, 291–306.

Tamura, K., (2015). Development of cell-cycle checkpoint therapy for solid tumors. *Jpn. J. Clin. Oncol., 45*, 1097–1102.

Thilakarathna, S. H., & Rupasinghe, H. P., (2013). Flavonoid bioavailability and attempts for bioavailability enhancement. *Nutrients., 5*, 3367–3387.

Tuli, H. S., Tuorkey, M. J., Thakral, F., Sak, K., Kumar, M., Sharma, A. K., Sharma, U., Jain, A., Aggarwal, V., & Bishayee, A., (2019). Molecular mechanisms of action of genistein in cancer: Recent advances. *Front. Pharmacol., 10*, 1336.

Ueda, S., Nakamura, H., Masutani, H., Sasada, T., Takabayashi, A., Yamaoka, Y., & Yodoi, J., (2002). Baicalin induces apoptosis via mitochondrial pathway as prooxidant. *Mol. Immunol., 38*, 781–791.

Valenti, D., De Bari, L., Manente, G. A., Rossi, L., Mutti, L., Moro, L., & Vacca, R. A., (2013). Negative modulation of mitochondrial oxidative phosphorylation by epigallocatechin-3 gallate leads to growth arrest and apoptosis in human malignant pleural mesothelioma cells. *Biochim. Biophys. Acta, 1832*, 2085–2096.

Van, D. C., Driessen, A. J., & Recourt, K., (2000). The uncoupling efficiency and affinity of flavonoids for vesicles. *Biochem. Pharmacol., 60*, 1593–1600.

Vegliante, R., Di Leo, L., Ciccarone, F., & Ciriolo, M. R., (2018). Hints on ATGL implications in cancer: Beyond bioenergetic clues. *Cell Death Dis., 9*, 316.

Vrhovsek, U., Rigo, A., Tonon, D., & Mattivi, F., (2004). Quantitation of polyphenols in different apple varieties. *J. Agric. Food Chem., 52*, 6532–6538.

Wang, J., Fang, F., Huang, Z., Wang, Y., & Wong, C., (2009). Kaempferol is an estrogen-related receptor alpha and gamma inverse agonist. *FEBS Lett., 583*, 643–647.

Weinberg, F., Ramnath, N., & Nagrath, D., (2019). Reactive oxygen species in the tumor microenvironment: An overview. *Cancers, 11*, 1191.

Wu, J., Ye, J., Kong, W., Zhang, S., & Zheng, Y., (2020). Programmed cell death pathways in hearing loss: A review of apoptosis, autophagy and programmed necrosis. *Cell Prolif., 53*, e12915.

Xiao, Z. P., Peng, Z. Y., Peng, M. J., Yan, W. B., Ouyang, Y. Z., & Zhu, H. L., (2011). Flavonoids health benefits and their molecular mechanism. *Med Chem., 11*, 169–177.

Xue, C., Chen, Y., Hu, D. N., Iacob, C., Lu, C., & Huang, Z., (2016). Chrysin induces cell apoptosis in human uveal melanoma cells via intrinsic apoptosis. *Oncol. Lett., 12*, 4813–4820.

Yanishlieva-Maslarova, N. V., (2001). Inhibiting oxidation. Antioxidants in food: Practical applications, 22–70.

Yao, P., Nussler, A., Liu, L., Hao, L., Song, F., Schirmeier, A., & Nussler, N., (2007). Quercetin protects human hepatocytes from ethanol-derived oxidative stress by inducing heme oxygenase-1 via the MAPK/Nrf2 pathways. *J. Hepatol., 47*, 253–261.

Zaidi, N., Lupien, L., Kuemmerle, N. B., Kinlaw, W. B., Swinnen, J. V., & Smans, K., (2013). Lipogenesis and lipolysis: The pathways exploited by the cancer cells to acquire fatty acids. *Prog. Lipid Res., 52*, 585–589.

Zhang, J., Wu, D., Vikash, Song, J., Wang, J., Yi, J., & Dong, W., (2015). Hesperetin induces the apoptosis of gastric cancer cells via activating the mitochondrial pathway by increasing reactive oxygen species. *Dig. Dis. Sci., 60*, 2985–2995.

Zhong, H., Xiao, M., Zarkovic, K., Zhu, M., Sa, R., Lu, J., Tao, Y., Chen, Q., Xia, L., Cheng, S., et al., (2017). Mitochondrial control of apoptosis through modulation of cardiolipin oxidation in hepatocellular carcinoma: A novel link between oxidative stress and cancer. *Free Radic. Biol. Med., 102*, 67–76.

Zhu, Y., Mao, Y., Chen, H., Lin, Y., Hu, Z., Wu, J., Xu, X., et al., (2013). Apigenin promotes apoptosis, inhibits invasion and induces cell cycle arrest of T24 human bladder cancer cells. *Cancer Cell Int., 13*, 54.

CHAPTER 10

PLANT METABOLITES AS IMMUNOMODULATORS

SAUMYA DAS[1], MANAS KUMAR DAS[2], URMILA JAROULIYA[3], and EKNATH D. AHIRE[4]

[1]Associate Professor, Pharmacy Institute, Noida Institute of Engineering and Technology, Knowledge Park-2, Greater Noida, Uttar Pradesh, India

[2]Director, Orlean College of Pharmacy, Knowledge Park-3, Greater Noida, Uttar Pradesh, India

[3]SOS in Biochemistry, Jiwaji University, Gwalior, Madhya Pradesh, India

[4]Department of Pharmaceutics, METs, Institute of Pharmacy, Affiliated Savitribai Phule Pune University, BKC, Adgoan, Nashik, Maharashtra, India

ABSTRACT

Numerous synthetically prepared drugs come up with various adverse effects and contraindications. The patients' immune system started worsening as a consequence of the adverse effects caused due to synthetic drugs. They started encountering one over the other pathological conditions, be it autoimmune disorders, hypersensitivity reactions, cancers, other metabolic disorder, etc. These worsening conditions led to the awareness among the people about the need to modulate the immune system steadily to achieve

Advances in Flavonoids for Human Health and Prevention of Diseases. Nisha Sharma, Deepika Saini, Rajesh K. Kesharwani, Prakash Chandra Gupta, & Raj K. Keservani (Eds.)

the desired effects, rather than treating the disorders in an advanced manner. This review chapter tries to focus on the plant metabolites that have potential immunotherapeutic effects. Many plant metabolites and some phytoconstituents accountable for immunomodulation have been explained. These plant metabolites can either suppress or stimulate the immune response as desired. Moreover, these plant metabolites are also known to possess other activities too, such as antioxidant, antidiabetic, and anti-inflammatory activity. Some of the plant metabolites discussed are – phenols, phytophenols, tannins, alkaloids, sterols, etc. This chapter emphasizes the use of several natural products/drugs in the augmentation and suppression of impaired immune system under unfavorable conditions. The principle accent of this chapter is to emphasize the capability of Indian Ayurvedic medication as adjuvants to work against the side consequences of contemporary therapies.

10.1 INTRODUCTION

Ethnobotany is the study of traditional plants for their medicinal properties and is an effective method to discover future medicines. The discovery of drugs in plants involves a variety of methods including botanical, ethnobotanical, phytochemical, and biological methods. They carry on offering us with new chemicals (lead molecules) to develop drugs against various drug targets including cancer, HIV/AIDS, malaria, Alzheimer's disease (AD) and pain (Archana et al., 2011). Immunity is the natural protection mechanism of the body against the pathogens/foreign particles (Nader, 2017). There are various internal stimuli (such as immunization, previous infection) and external stimuli whose kick start the immune response of the body. An immunomodulator is a chemical constituent which either enhances or suppresses the immune response of the body. The intrinsic nature of immunomodulation is that a pharmacological substance (here plant metabolite), which acts under various dose-time regimens displays an immunomodulatory effect (Dham, 1995). The substance which enhances the immune reaction is called immunostimulant while the one which suppresses/retards the immune reaction is called immunosuppressant. Immunomodulation by using herbal/medicinal plants is an optional traditional therapy for certain diseases where the host immune response is to be repaired.

1. **Active Immunity:** It is apparent that vaccination is the only ultimate strategy to provide good immunity to the patients (Trujillo et al., 2005).
2. **Passive Immunity:** The principle is to use the neutralizing antibodies (Nab) against the virus. Nab can either be engineered in labs or obtained from the sera of recovered patients. This is known as plasma therapy.

10.1.1 IMMUNOMODULATION

It is a technique that may modify the immune system of an organism by interfering with its functions; if an agent consequence in an augmentation of immune reactions, its miles termed an immune-stimulative drug, which typically implies stimulation of nonspecific device (i.e., granulocytes, macrophages, positive T-lymphocytes, and distinctive effector substances). Immunosuppression implies especially to lessen resistance in opposition to infections or stress and can be because of environmental or chemotherapeutic factors (Labro, 2012). Immunostimulation and immunosuppression each need to be reconnoitered so that you can regulate normal immunological functioning. Hence, both immunostimulant agents and immunosuppressing agents have their personal significance, and is looking for better retailers exerting these sports is turning into a discipline of essential hobby all around the world (Patwardhan et al., 1990). Natural adjuvants, artificial mediator, and antibody reagents are applied as immunosuppressive and immunostimulant agents (Doherty, 1981). But there are essential obstacles to the overall use of those agents, which include the improved threat of contamination and comprehensive impact for the duration of the immune system (Diasio & LoBuglio, 1996).

10.1.2 IMMUNOMODULATORS

The application of plant crop as immunomodulators remains withinside the development platform. There are numerous treatments utilized in conventional medicinal drug programs that may modify the immune system (Sultana & Chandy, 2018). A kind of plant extract, including polysaccharides, lectins, peptides, flavonoids, and tannins, was mentioned to regulate the immune system (Lutsiak, 2006). Traditional systems of Indian medicine, which include Siddha and Ayurveda, have recommended a way to boom the body's natural resistance to disease. Many Indian medicinal vegetation and numerous Rasayanas are stated to have a defensive effect (Wen et al., 2012). Rasayana organization is assumed to promote health, infection, and longevity. Permitting to Ayurveda, they strengthen all body tissues, prevent aging, promote immunity, and protect from various diseases (Shivprasad et al., 2006).

10.1.3 DIFFERENT ASPECTS OF IMMUNE BOOSTER

The immune system is the power of our body to offer resistance or safety in opposition to infections and sicknesses because of viruses, microorganisms, and different dangerous microorganisms. So obviously, antibodies are meals, fruits, veggies, or artificial dietary supplements (Keservani et al., 2010a, b) that assist us to expand this capacity to combat disease. All the components of our body that assist us to combat in opposition to contagions and assist in growing our immune system collectively are referred to as immune system. Our immune system has the capacity to discover dangerous viruses and viruses that attack our bodies–these cells, additionally referred to as antigens, assault them through producing 'antibodies.' Immune boosters are an excellent benefit to adults due to the fact they assist in reinforcing the body internally (Sharma et al., 2017). Immune marketers can assist defend the body in opposition to numerous diseases and ailments. They are made, maintaining in thought the needs of your body. Using the immune system can top off many essential nutrients, which includes vitamins, minerals, etc., withinside the body. They assist to speed up the restoration process of your body. It keeps you prompted and will increase your productivity. They additionally assist in preserving weight, blood pressure (BP), digestion, growth, healthy cholesterol, and lots of different elements of your health. There are many meals, fruits, and vegetables that could naturally improve your immune system. One of the perfect approaches to enhance your immune machine is to have a healthful weight loss plan. Fruits and vegetables rich in beta carotene, Vitamin C, Vitamin E, and zinc are suitable for reinforcement the immune system. These encompass broccoli, kiwi, kale, red, green, or yellow peppers, orange juice, papaya, cauliflower, sweet potatoes, tomatoes, strawberries, avocados, almonds, peanuts, spinach, eggs, and milk, etc. All types of berries, in addition to foods rich in omega-3 fatty acids including beans, flax seeds, or even nuts may be eaten to strengthen the immune system. Other herbs (Keservani et al., 2017, 2020; Sharma et al., 2020; Rane et al., 2020a–c) that improve the immune system are garlic, black cumin, and licorice add these to your weight loss plan to have a healthy and satisfying immune system (https://www.seniority.in/list-of-immunity-boosters-to-help-fighting-and germs).

10.2 PRIMARY AND SECONDARY METABOLITES

Metabolism can be summarized as the sum of all the biochemical reactions that obtain position in the body. Medium metabolites and metabolism products, and are usually restricted to small molecules (Thirumurugan et al., 2018).

10.2.1 PRIMARY METABOLITES

Primary metabolites are the molecules made throughout the growth phase. The procedure is known as primary metabolism or trophophase. They are important in the cell growth, reproduction, and development. For example, the nucleosides, vitamins, amino acids, etc. They are of two types (Table 10.1):

1. **Primary Essential Metabolites:** They are required to sustain the process of cell growth. Example: vitamins, nucleosides, amino acids, etc.
2. **Primary Metabolic End Products:** They are the normal end products of the fermentation procedure of primary metabolism. Example: ethanol, acetone, lactic acid, butanol, etc.

10.2.2 SECONDARY METABOLITES

Secondary metabolites (idiolites) are the products formed at the conclusion of metabolism. Soon after the tropophase, the process enters into another phase known as idiophase or secondary metabolism. This phase occurs when there is an accumulation of waste products. The secondary metabolites just support the cell survival and do not benefit the cell growth directly. Example-steroids, essential oils, alkaloids, phenols, etc.

TABLE 10.1 The Comparisons between Primary and Secondary Plant Metabolites

Basis for Comparison	Primary Metabolite	Secondary Metabolite
Also known as	Tropophase	Idiophase
It occurs at	Growth phase	Stationary phase
Production	They are formed in enormous amounts and their extraction is easy.	They are formed in small amounts and their extraction is difficult.
Occurrence	Same in each species, which means the yield the same products.	Differs in dissimilar species.
Importance	1. These products are used in industries. 2. They play important character in cell growth, reproduction, and development.	1. Secondary metabolites such as antibiotics, gibberellins are also important. 2. They helps in the cell survival and do not directly benefit the cell growth.
Examples	Carbohydrates, Vitamins, lipids, and proteins.	Steroids, Phenolics, essential oils, alkaloids, etc.

Source: https://biodifferences.com/difference-between-primary-metabolites-and-secondary-metabolites.html.

Second metabolites also vary from basic metabolites in that they have a limited allocation in the plant kingdom. That is, definite secondary metabolites are typically found in only one plant type or in an associated group of species, while basic metabolites are originate through the plant kingdom. Likewise, they are usually only produced during certain dates of plant growth (Hashemi & Davoodi, 2012).

10.2.3 TYPES OF SECONDARY PLANT METABOLITES (KIJJOA, 2002)

1. **Phenolics:** Simple phenolics (gallic acid, salicylaldehyde), tannins (geranin, tellimagrandin), coumarins (umbelliferone, scopoletin), flavonoids (Keservani & Sharma, 2014; Keservani et al., 2016a, b) (quercetin, kaempferol), chromones and xanthones (eugenin), stilbenes (resveratrol), lignans.
2. **Alkaloids:** Nicotine, caffeine, vinblastine.

3. **Saponins:** Triterpenoidal saponin (soyasaponin I), steroidal saponin (dioscin).
4. **Terpenes:** Monoterpenes (linalool, limonene, citronellal), sesquiterpenes (farnesol, caryophyllene), diterpenes (gibberellane, labdane), triterpene (quassin).
5. **Lipids:** Fixed oils, waxes, essential oils.
6. **Carbohydrates:** Mucilages.

10.2.4 CHEMICAL STRUCTURE OF SOME PLANT SECONDARY METABOLITES (MOHIUDDIN, 2019) (Figures 10.1 and 10.2)

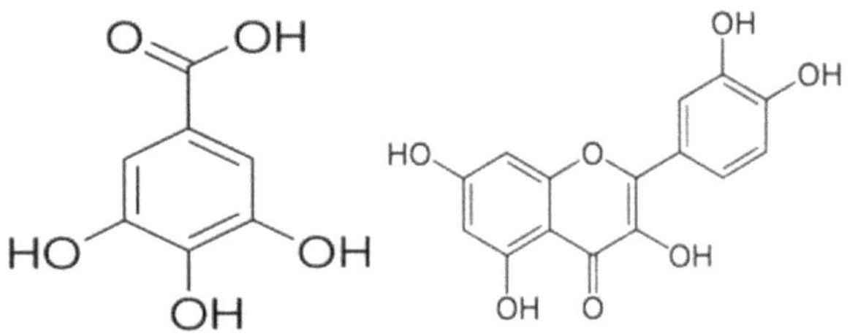

FIGURE 10.1 Gallic acid quercetin.

Source: Structure drawn by using ChemBioDraw Ultra 14.0.

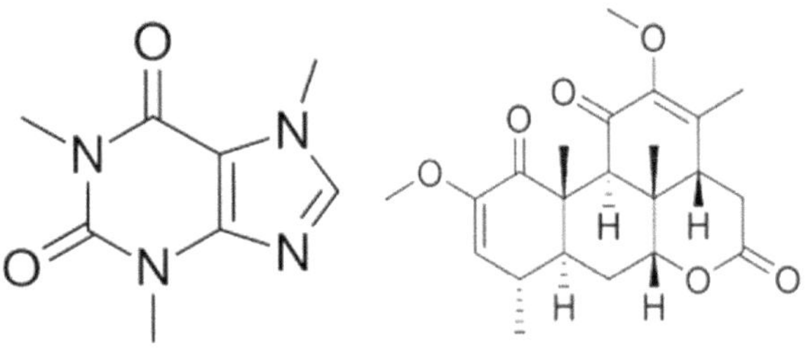

FIGURE 10.2 Caffeine quassin.

Source: Structure drawn by using ChemBioDraw Ultra 14.0.

10.2.5 PRODUCTION/EXTRACTION OF SECONDARY METABOLITES FROM PLANTS

Victorious extraction commences with cautious assortment and preparation of plant samples. During plant extraction, it is important to minimize disturbances from compounds that may be released in conjunction with targeted metabolites, and to avoid extract contamination, as well as to prevent decay of essential metabolites or artifact formation due to extraction conditions or soluble impurities (Pagare et al., 2015).

1. **Conventional:** This depends on the release of metabolites from plant tissues by various means such as solvent, steam, and extraction.
2. **Immobility:** Cells or biocatalyst are trapped within the matrix through insertion, ad or joint communication. In addition to the suitable substrate, the required secondary metabolites are synthesized.
3. ***In-Vitro* Tissue, Organ, and Cell Culture:** Plant culture and tissue culture can be established below germ-free environment from plants, such as plant roots, leaves, stems, meristems, etc., both by replication and secretion of secondary metabolites. Shooting, callus root, cell suspension and hairy root culture are used to bind the metabolite of interest.

10.2.6 MAJOR SECONDARY METABOLITE PATHWAYS

Mainly, there are three main pathways: the shikimate pathway, the polyketide pathway and the isoprenoid pathway. The major source of aromatic compounds is the shikimate pathway. The phenylpropanoid pathway is one of the most important pathways in terms of carbon flux in plants, since more than 20% of the metabolism can go through this pathway. This pathway is important because it leads to other metabolites such as lignins, flavonoids, lignans, and anthocyanins (Hussein & Anssary, 2018).

For terpenoids, there is isoprenoid pathway. Terpenoids constitute almost one-third of the secondary metabolites. Moreover, the C_5-building block is also incorporated in many other skeletons such as anthraquinones, naphthoquinones, cannabinoids, furanocoumarins, and terpenoidindole alkaloids (Isha & Stress, 2019).

10.2.7 BIOACTIVITY OF SECONDARY METABOLITES

1. **Antioxidant Activity:**
   i. *Limonia acidissima* (flavonoids).
   ii. *Murraya koenigii* (kaempferol, quercetin, luteolin).

2. **Antimicrobial Activity:**
   i. *Sisymbrium irio* (against *A. niger, F. oxysporum, E. coli, P. aeruginosa*).
   ii. *Duranta erecta* (against *A. flavus, P. chrysogenum, Rhizopus* spp.).
3. **Antidiabetic Activity:**
   i. *T. cucumerina* (methanol, flavonoids, steroids, alkaloids).
   ii. *Azadirachta indica* (flavonoids, alkaloids).

10.3 IMMUNOMODULATORY EFFECTS OF SECONDARY METABOLITES ON PERSISTENT NEUROINFLAMMATION AND COGNITIVE AGING

The cognitive impairment is a complex process involving multiple mechanisms-overexpression of neuro-inflammatory processes, BBB leakage and overactivation of microglia and astrocytes. Reactive oxygen species (ROS) is a by-product of energy production.

The microglial cells of the brain, along with the intracellular mitochondria, produce ROS in order to keep homeostasis; and are also involved in antioxidant defense mechanisms. Both underproduction and overproduction of ROS results in dysfunctional cells.

10.3.1 RESVERATROL

It is a polyphenol—plant secondary metabolite, which has the potential to defend and annoyed the BBB. It also inhibits interleukin beta (IL-1β) and BV-2 microglial cells in order to play a role in neuroprotection.

10.3.2 CURCUMIN

It is the yellow pigment of *Curcuma longa* and is a powerful antioxidant. It acts by the protection of BBB permeability/integrity, decreasing the microglial cells activation in the brain. It also inhibits free radicals (nitric oxide (NO)) and inflammatory markers (TNF-α, IL-1α and IL-6) produced by microglial cells.

10.3.3 BERBERINE

They are the alkaloids. Berberine is obtained from *Tinospora cordifolia, Berberis aristata, Berberis lyceum*, etc. Berberine has BBB-protective effects.

10.3.4 CAFFEINE

Caffeine leads to the leakage of inflammatory marker – IgG and Evans blue, to the brain, thus protecting the BBB integrity. Caffeine suppresses or inhibits the production of TNF-α and IL-10. Caffeine has BBB-protective effects.

10.4 SECONDARY METABOLITES AS ANTI-INFLAMMATORY AGENTS

Inflammation is the basic response to any injury. Inflammation is a process mediated by the signaling molecules produced by mast cells, leukocytes, macrophages as well as by the activation of factors which cause edema due to the accumulation of leukocytes at the site of inflammation. Anti-inflammatory drugs are separated into two categories–steroidal and non-steroidal anti-inflammatory medicine (NSAIDs). Steroidal drugs act on the cells and mediators responsible for inflammation, while NSAIDs act on cyclooxygenases-1 and 2 (COX-1 and COX-2). Plant secondary metabolites are a great source of drugs and are used widely in dispensing as anti-inflammatory drugs.

10.4.1 PHENOL AND PHYTOPHENOL COMPOUNDS

Phytophenols are secondary metabolites based on a common carbon skeleton structure-the C_6-C_3 phenylpropanoid unit (Grigore, 2017). Phenolic compounds include – flavonoids, condensed tannins, gallotannins. These mentioned phenolic compounds inhibit the molecular targets of pro-inflammatory mediators in inflammatory responses.

10.4.2 GALLOTANNINS

They have various effects—such as anticancer, antiviral, and anti-inflammatory properties. The mechanism of anti-inflammatory activity is to scavenge the free radicals and to inhibit some inflammatory mediators such as cytokines, COX-2, etc.

10.4.3 ALKALOIDS

Isoquinoline, indole, and diterpene alkaloids possess good anti-inflammatory activity. Bisbenzylisoquinoline alkaloids possess antimalarial, immunomodulatory activities along with anti-inflammatory properties (Njoroge & Bussmann, 2006).

10.4.4 STEROLS

They are helpful in reducing plasma total and LDL cholesterol levels. Sterols also possess antiatherosclerotic activities since they show effects on antioxidant, coagulation, and hepatic systems too. *Lepidium sativum* contained the highest number of sterols. Strong anti-inflammatory activity was detected in *Picris hieracioides, Foenicum vulgare, Cichorium intybus* leaves and *Cynara cardunculus, P. hieracioides* contained only γ-sitosterol. -Sitosterol and stigmasterol were the predominant sterols since they were widely distributed (Mona, 2014).

10.5 CHEMISTRY OF PLANT-DERIVED IMMUNOMODULATORS (VENKATALAKSHMI, 2016)

10.5.1 GLYCOSIDES

These organic compounds of plant and animal origin produce one or more sugar molecules on enzymatic or acid hydrolysis. Chemically, these are acetals or ethers of sugars produced by the dealings of sugar hydroxyls with non-sugar fragments with loss of water molecules. Many glycosides have been revealed to have desired immunomodulatory effects.

10.5.2 FLAVONOIDS

Chemically, flavonoids have a 15-carbon skeleton (C_6–C_3–C_6) which consists of two phenyl rings connected by a three-carbon bridge. Several types of flavonoids exert immunomodulatory activities, including apigenin (Kure et al., 2017), oligomeric proanthocyanidins, isoflavonoids, flavones, and anthocyanidins (Jain et al., 2019).

10.5.3 VOLATILE OILS AND TERPENOIDS

The term terpene represents the hydrocarbons (C_5H_8), while terpenoids include the hydrocarbon as well as their oxygenated derivatives.

10.5.4 VITAMIN D

Vitamin D has been known for over 20 years to have significant effects on immune cells. Vitamin D is a broken-ring steroid, so it is called a seco-steroid. Vitamin D2 (ergocalciferol) and vitamin D3 (cholecalciferol) are the two main forms of vitamin D. Vitamin D2 is obtained from plants, whereas vitamin D3 is produced photochemically in the epidermis of animals (Lemire & Archer, 1991). When 7-dihydrocholesterol is exposed to UVB radiation (295–310 nM), previtamin D is formed, which undergoes thermal conversion and two separate hydroxylations (the P450 enzymes 25-hydroxylase and 1α-hydroxylase), respectively, and then provide activation. 1,25(OH) 2D (Feronet et al., 2009). Vitamin D synthesis peaks at wavelengths between 295 and 297 nm (UV index 3 or higher), and exposure to the sun for 15 minutes at least twice a week produces enough vitamin D. When exposed to UV-B rays for longer periods of time, the body breaks down previtamin D as soon as it produces it and reaches equilibrium (Kumar et al., 2012).

10.5.5 SAPONIN

Saponin's auxiliary activity was found 50 years ago, so after the veterinary vaccine, especially the disease, he discovered a wide polyp. This is a powerful auxiliary for the antibody response to the strong antigen of the mouse due to IgGL and IgG_2 antibodies. In addition to immune competition, which can explain the success of the experimental vaccine for parasitic protozoal diseases (Bukhari et al., 2015), it is very widely spread in the plant kingdom. They are glycosides which include a sterol or triterpene ring with connected sugars (Price, 1987). Numerous show strong fungicidal activity, and it's been projected that this will gain the plant as a protection mechanism in opposition to fungal attacks.

10.5.6 IMMUNOSTIMULATORY POLYSACCHARIDES

A great numerous glucan from fungi, lichens, and algae have been originating to hold immunostimulatory possessions (Wagner, 1995; Bomford, 1988). The glucans have been extracted from mushrooms which feature in traditional folk remedies, or are edible, e.g., *Lentinus edodes* (Petrovsky, 2006; Chihara et al., 1969).

10.5.7 SIGNALING MECHANISM

Immunomodulator plants activate macrophages to inhibit the growth of tumor cells *in vitro* and also induces them to release interleukin (IL)-1. Thus, by acting at the level of the macrophages, it is likely that function as a specific immunostimulant through IL-1 secretion, which may help to trigger T lymphocytes (Agarwal, 1999).

10.5.8 COW URINE AS IMMUNOMODULATOR

Nevertheless of the use of Ayurveda or Siddhi or using cow urine distillate Immunomodulator are based on classical texts and organisms, the conventional machine of medicine is based on Charak, Sushrute, Vagabhatta, Ashtangardia, and Samhitas, based on classical textual content and systems, doctors, and products. The cow urine becomes defined because the simplest material/secretion of an animal begins with the handiest treatment price Srushti Harita Samhita and Ashtagandha. In Ayurveda, cows became projected to enhance overall health. Panchavalee (Cowpathy) has been an old age-associated system with a completely high region in Ayurveda for 5,000 years, and this kind of treatment method did now no longer acquire the value of the past period. Ayurvedic Animal Medicines is particularly prepared as a small leather-based product and means 5 elements resulting from cows with milk, cottage cheese, foam, urine, and manure. Augmented Camden Track B, Blastogenesis B, and T lymphocyte, which turned into cooked withinside the urine of the bottle was found. IgG, IGA, and IgM antibody mouse. Cow urine has been found to increase the phagocytic interest of macrophages, which facilitates withinside the anticipation and manipulation of bacterial contaminations. It additionally will increase the secretion of IL-1 and IL-2. Accordingly, cow urine has been proven to have an effective impact on immunity, which may be used as an immunomodulatory candidate in animals (Ganguly & Prasad, 2010).

10.6 ALKALOIDS AS IMMUNOMODULATORS

Alkaloid, a category of certainly taking place natural nitrogen-containing bases. Alkaloids have numerous and crucial physiological outcomes on human beings and different animals. Well-recognized alkaloids consist of strychnine, morphine, ephedrine, quinine, and nicotine. Alkaloids are discovered typically in plants and are particularly ordinary in definite families of flowering plants. In fact, as numerous as one-quarter of higher plants are anticipated to comprise alkaloids, of which numerous 1,000 differing types were recognized. The chemical structures of alkaloids are extraordinarily variable. Usually, an alkaloid consists of as a minimum one nitrogen atom in an amine-type structure, i.e., one derived from ammonia by changing hydrogen atoms with hydrogen-carbon groups referred to as hydrocarbons. This or some other nitrogen atom may be active as a base in acid-base reactions. The name alkaloid (alkali-like) was initially carried out to the materials due to the fact, just like the inorganic alkalis, they react with acids to shape salts. Most alkaloids have one or more in their nitrogen atoms as a part of a ring of atoms, often referred to as a cyclic system. Alkaloid names usually end withinside the suffix -ine, a reference to their chemical type as amines. In their natural form most alkaloids are colorless, nonvolatile, crystalline solids. They additionally have a tendency to have a bitter taste. Alkaloids are frequently categorized on the basis of their chemical structure. For example, the ones alkaloids that comprise a ring system known as indole are called indole alkaloids. On this foundation, the essential types of alkaloids are the pyrrolidines, tropanes, pyridines, isoquinolines, pyrrolizidines, quinolines, indoles, and the terpenoids and steroids. Alternatively, alkaloids may be categorized in step with the organic system wherein they arise. For example, the opium alkaloids arise withinside the opium poppy (*Papaver somniferum*). This dual type system absolutely produces little confusion because there's a hard association between the chemical kinds of alkaloids and their organic distribution (https://www.britannica.com/science/alkaloid). The impact of 33 ergot alkaloids (natural and semisynthetic) at the introduction of antibody development in quarantined spleen cells by antigen in tissue cultures. The effects are quantitatively articulated as the variety of antibody-forming cells. The group of herbal ergot alkaloids (ergopeptines) employed a powerful inhibitory action at the antibody response. Compounds resulting from lysergic acid are absolutely inactive in this esteem. A new, very vigorous semisynthetic drug (Table 10.1, compounds 2) resulting from inactive lysergol changed into designated. The compounds interrelating with alpha-adrenergic-receptors on lymphocytes (methoxamine, prazosin, phentolamine, act synergistically with the inhibitory impact of ergopeptine alkaloids (Sterzl et al., 1987).

10.6.1 CLASSIFICATION OF ALKALOIDS

Alkaloids are massive class of nitrogen comprising compounds that vary in terms of structure, synthesis, and biological activity. Consequently, distinctive classifications are available in alkaloids based on a few criteria. Usually, alkaloids are categorized based on:

- Biosynthetic pathway based at the precursor applied for synthesis;
- Chemical type-based on chemical entity;
- Pharmacological type-based on particular pharmacological properties; and
- Taxonomic type based on allocation of alkaloids in distinct plant groups.

Alkaloids are assembly into six based on the character of amino acid from which it begins its biosynthesis (particularly ornithine, anthranilic acid, histidine, lysine, phenylalanine, and tryptophan). Alternative system of class includes 9 classes based on chemical structure: acridines, amines, amides, canthinones, benzylisoquinolines, imidazoles, furoquinolines, indolquinazolines, and quinazolines. Supplementary, a chemotaxonomic method in alkaloid type was additionally to be had. Another class of alkaloids was true, pseudo, and proto alkaloids characterized with heterocyclic nitrogenic bases resulting from amino acids or heterocyclic nitrogen base isn't resulting from amino acids or they're primary amines resulting from amino acids, however nitrogen isn't part of fragrant system (e.g., vanillylamide). An extra realistic system of class consists of a heterocyclic and non-heterocyclic organization similar to whether or not nitrogen is part of the cyclic ring system.

10.6.2 DIFFERENT TYPES OF ALKALOIDS

10.6.2.1 GINSENOSIDES

Ginsenosides are the essential ingredients of ginseng. Ginsenosides have the particular biological interest and medicinal value, including anti-inflammatory, antioxidation, antitumor, and inhibition of cell apoptosis. With the augment of stress in life, the occurrence of anxious system illnesses is likewise increasing ginsenosides, because the immunomodulatory drugs. Ginseng has been nicely referred to as an immune modulator. Roots (mostly), leaves of ginseng, stems, and their extracts had been used for preserving immune homeostasis and improving resistance to contamination or microbial assaults through results on the immune system. The immune system consists of numerous kinds of cells with their personal specialized features, and every kind of the immune cells differentially responds to ginseng treatment. This assessment abridges the current understanding of the results of ginseng on the immune system. Amongst the loads of literatures on immune responses, the most effective comprised the reviews which used mice, rats, or people as *in vivo* model organisms, murine or human-derived cell traces as *in vitro* experimental models, and *P. ginseng* as a supply of ginseng (Kang & Min, 2012). NK cells are a kind of cytotoxic lymphocytes of the innate immune system which are worried withinside the first line of protection in opposition to freshly springing up malignant cells and inflamed cells. Distinguished from cytotoxic T lymphocytes (Amirghofran et al., 2009), they are able to understand target cells withinside the absence of antibodies and MHC and do now no longer involve activation to kill goal cells thus allowing a quick immune response. In addition to killing goal cells non-specifically, NK cells launch pro-inflammatory cytokines including IFN-γ which turns on macrophages to damage phagocytized microbes. Many literatures have mentioned that overall ginseng extract complements NK cell features. Oral administration of *P. ginseng* extract for prolonged natural killing activity in mice, and ginsenoside Rg1-treated mice additionally confirmed an improved natural killing activity of the splenocytes. Excitingly, the stimulatory outcomes of ginseng on NK cells were determined irrespective of disorder situation of the host. Consequently, the features of NK cells inspired through ginseng were similar among immunosuppressed mice and immunocompetent mice. The experimental use of blood sample of human patients who had persistent fatigue syndrome and bought immunodeficiency syndrome (AIDS) confirmed purposeful augmentation of NK cells remoted from peripheral blood mononuclear cells. *In vivo* test carried out via way of means of (Scaglione et al., 1990) verified that NK cell feature in people captivating ginseng turned into encouraged in comparison to the placebo group. Co-operatively, those effects suggest that ginseng can also additionally enhance NK cell activity regardless of host immune conditions (Kang & Min, 2012). Structure of National Center for Biotechnology Information (https://pubchem.ncbi.nlm.nih.gov/compound/Ginsenoside-RG1) (Figure 10.3).

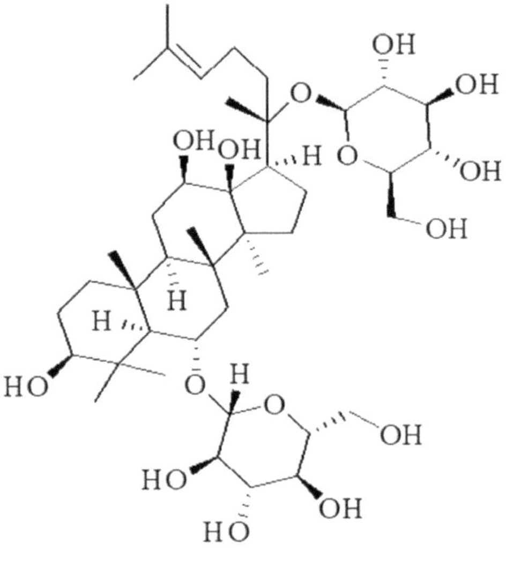

Ginsenosides

FIGURE 10.3 Structure of ginsenoside.

Source: Structure drawn by using ChemBioDraw Ultra 14.0.

10.6.2.2 BERBERINE

Berberine is a plant alkaloid with an extended records of medicinal use in each Ayurvedic and Chinese medicine. It is found in *Hydrastis canadensis* (goldenseal), *Coptis chinensis* (Coptis or goldenthread), *Berberis aquifolium* (Oregon grape), *Berberis vulgaris* (barberry), and *Berberis aristata* (tree turmeric). The berberine alkaloid may be discovered withinside the roots, rhizomes, and stem bark of the plants. Berberine extracts and decoctions have shown tremendous antimicrobial interest in opposition to plenty of organisms which incorporates bacteria, viruses, fungi, protozoans, helminths, and chlamydia. In China, berberine is an over-the-counter drug for the treatment of bacterial diarrhea. In 1988, the hypoglycemic impact of berberine was first of all mentioned while berberine turned into prescribed to deal with diarrhea in diabetic patients (Sharma et al., 2017). Moreover, numerous clinical and preclinical research demonstrate the ameliorative impact of berberine in opposition to numerous disorders such as metabolic, neurological, and cardiological problems. This evaluation presents a precis concerning the pharmacokinetic and pharmacodynamic capabilities of berberine, with a focal point at the special mechanisms underlying its multispectrum pastime. Nevertheless, several literatures have been posted via way of means of diverse authors determining the phytochemical and pharmaceutical factors in conjunction with conventional makes use of but there's no a lot more literature concerning thus far the significance of Berberine, which is an essential constituent of this species (Chander et al., 2017). As an immunity supplement, berberine is normally used to relieve allergies in addition to an extensive variety of issues because of autoimmunity, which includes irritable bowel disease (IBD), type-1 diabetes, rheumatoid arthritis, and multiple sclerosis (MS). While human examine evidence for these makes use of is presently lacking, animal studies are incredibly supportive of berberine's capacity to lessen inflammation and autoimmune responses. The study was conducted to examine the usage of berberine in mice with tubule interstitial nephritis (TIN), a kind of human autoimmune kidney conditions. The mice had been given berberine at doses of 10 mg/kg body weight for three days and had been examined 2 months later. The treatment shows lessen damage and enhance condition of the kidneys (https://supplementsinreview.com/immune/berberine-immune/) (Figure 10.4).

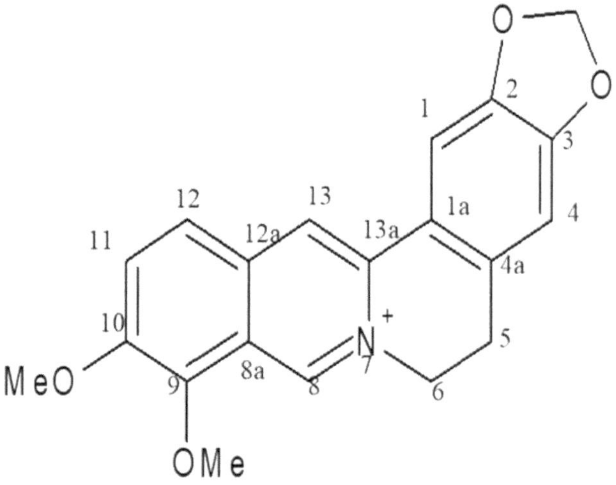

FIGURE 10.4 Structure of berberine.

Source: Structure drawn by using ChemBioDraw Ultra 14.0.

Berberine is proposed to have the ability as a therapeutic agent for lipid lowering. In this pilot study, berberine decreased serum cholesterol, triglycerides, and LDL-C. This activity is just like that said somewhere else *in vivo* (https://pubchem.ncbi.nlm.nih.gov/compound/Berberine). However, in addition, studies consisting of final results in humans are needed to verify this activity and its benefit. The method of berberine adaptable lipid metabolism has been examined through many groups. In hamsters with hyperlipidemia, berberine decreased serum cholesterol and LDL-C, and improved LDL receptor mRNA in addition to protein withinside the liver. These results have been partially because of stabilization of LDL receptor mRNA mediated through the ERK signaling pathway. In accumulation to up-regulation of the LDL receptor, berberine inhibit lipid synthesis in human hepatocytes through activation of AMPK pathway (Yin, 2008). Scientists have raised several hypotheses wherein berberine may also assist with diabetes on a cellular level, which includes all of the following:

- Directly growing AMPK pathway, which stimulates glucose uptake in muscle cells and facilitates balance in high blood sugar level (rat and cell studies).
- Delay the breakdown of carbohydrates into simple sugars (rat study).

- Enhancing glucagon secretion (rat study).
- Mimics insulin action through growing the capacity of the body to soak up glucose (cell study) (Singh & Chaudhuri, 2017).
- Decreasing glucose absorption through the intestinal lining (cell study).
- Increasing adiponectin, a protein which helps regulate blood sugar levels (through AMPK) (cell study) Activating the blood sugar transport activity of glucose transporter 1 (GLUT1) (cell study).

Some proof indicates that berberine can also additionally enhance metabolic health in ladies with PCOS and insulin resistance (Wu, 2016).

10.6.2.3 CORDIFOLIOSIDE A

Cordifolioside A is remoted from *Tinospora cordifolia. T. cordifolia* is a shrub this is local to India. Its root stems and leaves are utilized in Ayurvedic medicine. It is applied for diabetes, allergic rhinitis, excessive cholesterol, (hay fever), disappointed stomach, gout, lymphoma, rheumatoid arthritis (RA), and different cancers, hepatitis, peptic ulcer disease (PUD), fever, gonorrhea, syphilis, and enhance the immune system (Sonavale, 2019). Clerodanefuranoditerpenoid glycosides like cordioside, cordiofolioside A and cordiol had been stated, and non-specific immune responses. Cordifolioside A is an active immunostimulant used withinside the excellent control and standardization of the plant's preparations. A cadinene sesquiterpenoid glycoside, tinocordiside, enjoys an immunomodulatory activity (https://pubchem.ncbi.nlm.nih.gov/compound/45359937). The polysaccharide fraction from *T. cordifolia* effectively decreased the metastatic ability of B16F-10 cancer cells because of its antioxidant activity towards DPPH and superoxide radicals. G1-4A, an arabinogalactan polysaccharide remoted from *T. cordifolia*, establishes galactose for their macrophage activation, which performs a vital function in particular (Figure 10.5) (Singh & Chaudhuri, 2017).

10.7 CONCLUSION

Plant based secondary metabolites in the type of alkaloids, terpenoids, flavonoids, and polysaccharides, glycoside products cause variation in the immunomodulatory properties. Presently flavonoids, alkaloids, and steroids from different plants were found to possess significant biological activity (Jayakumar & Murugan, 2016). Various plant-based extracts protect the BBB integrity and function. These extracts are originated from different plants and food supply (e.g., tea polyphenols, resveratrol, antioxidants, and plant alkaloids). Subsequently microglial cells donate to neurodegenerative ailments by triggering neuro-inflammatory progressions and oxidative stress, plant secondary metabolites are helpful in preventing cognitive aging in older individuals. Research shows that plant based secondary metabolites exert beneficial effects on cardiovascular protection, anti-cancer activity as well as anti-inflammatory effects. A lot of synthetically prepared drugs come up with a variability of adverse effects and contraindications. The immune system of the patients started worsening as an end result of these adverse effects. They started encountering one over the other pathological conditions. As a result of these deteriorating conditions, more people are now aware of the importance of gradually altering the immune system rather than aggressively treating disorders to produce the desired results. As a result, there is much greater demand for plant-derived products/drugs and daily remedies.

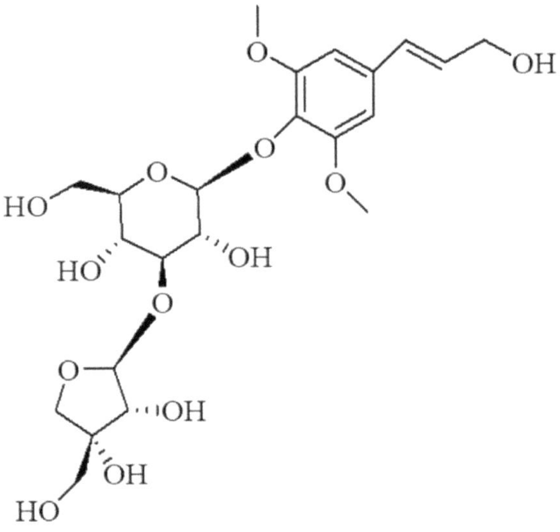

Cordifolioside A

FIGURE 10.5 Structure of cordifolioside A.

Source: Structure drawn by using ChemBioDraw Ultra 14.0.

COMPETING INTEREST

The authors declare that there are no competing interests.

KEYWORDS

- **alkaloids**
- **glucose transporter 1**
- **immune system**
- **inhibits interleukin beta**
- **irritable bowel disease**
- **non-steroidal anti-inflammatory medicine**
- **phytoconstituents**
- **phytophenols**
- **plant metabolites**
- **reactive oxygen species**

REFERENCES

Agarwal, S. S., (1999). Immunomodulation: A review of studies on Indian medicinal plants and synthetic peptides, part 1: Medicinal plants. *Proc. Ind. Natl. Sci. Acad., 65*(3), 79–204.

Amirghofran, Z., Bahmani, M., Azadmehr, A., & Javidnia, K., (2009). Immunomodulatory activities of various medicinal plant extracts: Effects on human lymphocytes apoptosis. *Immunological Investigations, 38*, 181–192.

Bomford, R., (1988). Immunomodulators from plants and fungi. *Phytotherapy Research, 2*(4), 159–164.

Bukhari, S., Abbas, N., Jantan, I., & Waqas, A., (2015). Plant-derived immunomodulators: An insight on their preclinical evaluation and clinical trials. *Frontiers in Plant Science, 4*(3), 23–34.

Chander, V., Aswal, J. S., Dobhal, R., & Uniyal, D. P., (2017). A review on pharmacological potential of berberine; an active component of Himalayan *Berberis aristata. J. Phytopharmacol., 6*(1), 53–55.

Chih, C. W., Hui, M. C., & Ning, S. Y., (2012). In: Lie-Fen, S., & Allan, S. Y. L., (eds.), *Chapter 6 – Developing Phytocompounds from Medicinal Plants as Immunomodulators* (Vol. 62, pp. 197–272). Advances in Botanical Research, Academic Press ISSN 0065-2296, ISBN 9780123945914.

Chihara, G., Maeda, Y., Hamuro, J., Sasaki, T., & Fukuoka, F., (1969). Inhibition of mouse sarcoma 180 by polysaccharides from *Lentinus edodes* (Berk.) sing. *Nature, 222*(5194), 687–688.

Christybapita, D., Divyagnaneswari, M., & Dinakaran, M. R., (2007). Oral administration of *Ecliptaalba* leaf aqueous extract enhances the non-specific immune responses and disease resistance of *Oreochromis mossambicus. Fish Shellfish Immun., 23*, 840–852.

Dham, S. K., (1995). Immunomodulation–clinical applications. *Medical Journal, Armed Forces India, 51*(3), 149, 150.

Diasio, R. B., & LoBuglio, A. F., (1996). Immunomodulators: Immunosuppresive agents and immunostimulants. In: *Goodman and Gilman's the Pharmacological Basis of Therapeutics* (9th edn., pp. 1291–1307). New York, McGraw-Hill.

Doherty, N. S., (1981). Selective effects of immunosuppressive agents against the delayed hypersensitivity responses and humoral response to sheep red blood cell in mice. *Agent Action, 11*, 237–242.

EBSCOhost | 48587936 | *Immunomodulatory Activity of Methanolic Extract of Morusalba Linn. (Mulberry) Leaves.* https://web.a.ebscohost.com/abstract? (accessed on 13 February 2023).

Feron, F., Fernandes, D. A., Abreu, D., Eyles, D., & Féron, F., (2009). Vitamin, D, a neuro-immunomodulator: *Implications for Neurodegenerative and Autoimmune Diseases, 5*, 56.

Ganguly, S., & Prasad, A., (2010). Role of plant extracts and cow urine distillate as immunomodulator in comparison to levamisole-a review. *Journal of Immunology and Immunopathology, 12*(2), 91–94.

Garcia, D., Leiro, J., Delgado, R., Sanmartin, M. L., & Ubeira, F. M., (2003). *Mangifera indica* L. extract (Vimang) and mangiferin modulate mouse humoral immune responses. *Phytother Res., 17*, 1182–1187.

Grigore, A., (2017). In: Soto-Hernandez, M., Palma-Tenango, M., & Garcia-Mateos, M. D. R., (eds.), *Plant Phenolic Compounds as Immunomodulatory Agents, Phenolic Compounds–Biological Activity*. Intech Open.

Hashemi, S. R., & Davoodi, H., (2012). *Herbal plants as new immuno-stimulator in poultry industry: A review. Asian Journal of Animal and Veterinary Advances, 7*(2), 105–116.

Hodge, G., Sandy, H., & Han, P., (2002). *Allium sativum* (Garlic) suppresses leukocyte inflammatory cytokine production *in vitro*: Potential therapeutic use in the treatment of inflammatory bowel disease. *Cytometry, 48*, 209–215.

Hong, Y. K., Wu, H. T., Ma, T., Liu, W. J., & He, X. J., (2009). Effects of *Glycyrrhiza glabra* polysachharides on immune and antioxidant activities in high-fat mice. *Int. J. BioMacro, 45*, 61–64.

https://biodifferences.com/difference-between-primary-metabolites-and-secondary-metabolites.html (accessed on 13 February 2023).

https://supplementsinreview.com/immune/berberine-immune/ (accessed on 13 February 2023).

https://www.britannica.com/science/alkaloid (accessed on 13 February 2023).

https://www.seniority.in/list-of-immunity-boosters-to-help-you-fight-infections/ (accessed on 13 February 2023).

Hussein, R. A., & Anssary, A. A. E., (2018). In: Philip, F. B., (ed.), *Plants Secondary Metabolites: The Key Drivers of the Pharmacological Actions of Medicinal Plants, Herbal Medicine.* Intech Open, https://www.intechopen.com/books/herbal-medicine/

plants-secondary-metabolites-the-key-drivers-of-the-pharmacological-actions-of-medicinal-plants (accessed on 13 February 2023).

Isah, T., (2019). Stress and defense responses in plant secondary metabolites production. *Biol. Res., 52*, 39.

Jain, C., Khatana, S., & Vijayvergia, R., (2019). Bioactivity of secondary metabolites of various plants: A review. *Int. J. Pharm. Sci. & Res., 10*(2), 494–504.

Jatawa, A. S., Paul, R., & Tiwari, A., (2011). Indian medicinal plants: A rich source of natural immuno-modulator. *International Journal of Pharmacology, 7*, 198–205.

Jayakumar, K., & Murugan, K., (2016). Solanum alkaloids and their pharmaceutical roles: A review. *J. Anal. Pharm. Res., 3*(6), 00075.

Kang, S., & Min, H., (2012). Ginseng, the 'immunity boost': The effects of panax ginseng on immune system. *J. Ginseng Res., 36*(4), 354–368.

Keservani, R. K., & Sharma, A. K., (2014). Flavonoids: Emerging trends and potential health benefits. *Journal of Chinese Pharmaceutical Sciences, 23*(12), 815.

Keservani, R. K., Kesharwani, R. K., Sharma, A. K., Vyas, N., & Chadoker, A., (2010b). Nutritional supplements: An overview. *International Journal of Current Pharmaceutical Review and Research, 1*(1), 59–75.

Keservani, R. K., Kesharwani, R. K., Vyas, N., Jain, S., Raghuvanshi, R., & Sharma, A. K., (2010a). Nutraceutical and functional food as future food: A review. *Der Pharmacia Lettre, 2*(1), 106–116.

Keservani, R. K., Sharma, A. K., & Kesharwani, R. K., (2016a). Nutraceutical and functional foods for cardiovascular health. In: *Food Process Engineering* (pp. 257–278, 291–312). Apple Academic Press, CRC Press. ISBN: 9781771884020.

Keservani, R. K., Sharma, A. K., & Kesharwani, R. K., (2016b). Medicinal effect of nutraceutical fruits for the cognition and brain health. *Scientifica,* 10. Article ID 3109254. doi: 10.1155/2016/3109254.

Keservani, R. K., Sharma, A. K., & Kesharwani, R. K., (2017). An overview and therapeutic applications of nutraceutical and functional foods. *Recent Advances in Drug Delivery Technology*, 160–201. ISBN:9781522507543.

Keservani, R. K., Sharma, A. K., & Kesharwani, R. K., (2020). *Nutraceuticals and Dietary Supplements: Applications in Health Improvement and Disease Management.* CRC Press. ISBN: 9781771888738.

Kijjoa, A., (2002). Plant secondary metabolites with immunomodulatory activity. In: Rauter, A. P., Palma, F. B., Justino, J., Araújo, M. E., & Dos, S. S. P., (eds.), *Natural Products in the New Millennium: Prospects and Industrial Application* (p. 47). Proceedings of the Phytochemical Society of Europe, Springer, Dordrecht.

Kumar, D., Arya, V., Kaur, R., Bhat, Z. A., Gupta, V. K., & Kumar, V., (2012). A review of immunomodulators in the Indian traditional health care system. *Journal of Microbiology, Immunology and Infection, 45*(3), 165–184.

Kure, C., Timmer, J., & Stough, C., (2017). The immunomodulatory effects of plant extracts and plant secondary metabolites on chronic neuroinflammation and cognitive aging: A mechanistic and empirical review. *Front Pharmacol., 8*, 117.

Kyo, E., Uda, N., Kasuga, S., & Itakura, Y., (2001). Immunomodulatory effects of aged garlic extract. *JNUTR, 131*, 1075S–1079S.

Lemire, J. M., & Archer, D. C., (1991). 1,25-dihydroxyvitamin D3 prevents the *in vivo* induction of murine experimental autoimmune encephalomyelitis. *J. Clin. Invest., 87*(3), 1103–1107.

Lutsiak, M. E., Kwon, G. S., & Samuel, J., (2006). Biodegradable nanoparticle delivery of a Th2-biased peptide for induction of Th1 immune responses. *J. Pharm. Pharmacol., 58*, 739–747.

Marie, T. L., (2012). Immunomodulation and infection: Back to the future. *Expert Review of Anti-Infective Therapy, 10*(3), 245–247.

Mohammed, M. S., Osman, W. J. A., Garelnabi, E. A. E., Osman, Z., Osman, B., Khalid, H. S., & Mohamed, M. A., (2014). Secondary metabolites as anti-inflammatory agents. *The Journal of Phytopharmacology, 3*(4), 275–285.

Mohiuddin, A. K., (2019). Chemistry of secondary metabolites. *Ann. Clin. Toxicol., 2*(1), 10–14.

Mukherjee, P. K., Nema, N. K., Bhadra, S., Mukherjee, D., Braga, F. C., & Matsabisa, M. G., (2014). Immunomodulatory leads from medicinal plants. *Indian Journal of Traditional Knowledge, 13*(2), 235–256.

Nader, D., (2017). Immunological mechanisms in disease and in the surgical patient. *Immunological Investigations, 46*(8), 765–768.

National Center for Biotechnology Information. PubChem. Database. *Ginsenoside RG1*. CID=441923, https://pubchem.ncbi.nlm.nih.gov/compound/Ginsenoside-RG1 (accessed on 13 February 2023).

National Center for Biotechnology Information. PubChem. Database. *Berberine*. CID=2353, https://pubchem.ncbi.nlm.nih.gov/compound/Berberine (accessed on 13 February 2023).

National Center for Biotechnology Information. PubChem. Database. CID=45359937. https://pubchem.ncbi.nlm.nih.gov/compound/45359937 (accessed on 13 February 2023).

Njoroge, G. N., & Bussmann, R. W., (2006). Diversity and utilization of antimalarial ethnophytotherapeutic remedies among the Kikuyus (Central Kenya). *J. Ethnobiol. Ethnomed., 2*, 8.

Pagare, S., Bhatia, M., Tripathi, N., & Bansal, Y. K., (2015). Secondary metabolites of plants and their role: Overview. *Current Trends in Biotechnology and Pharmacy, 9*, 293–304.

Pandit, S., Bandyopadhyay, A., Ponnusankar, S., Ota, S., & Mukherjee, P. K., (2011). Exploring the possible metabolism mediated interaction of *Glycyrrhiza glabra* extract with CYP3A4 AND CYP2D6. *Phytotherapy Res., 25*, 1429–1434.

Patwardhan, B., Kalbag, D., Patki, P. S., & Nagsampagi, B. A., (1990). Search of immunomodulatory agents: A review. *Indian Drugs, 28*(2), 348–358.

Petrovsky, N., (2006). Novel human polysaccharide adjuvants with dual Th1 and Th2 potentiating activity. *Vaccine, 24*(2), S2-26-9.

Price, K. R., Johnson, I. T., & Fenwick, G. R., (1987). The chemistry and biological significance of saponins in foods and feeding stuffs. *CRC Crit. Rev. Food Sci. Nutr., 26*(1), 27–135.

Rana, A. C., & Gupta, G. L., (2007). *Withania somnifera* (Ashwagandha): A review. *Pharmacognosy Reviews, 1*(1), 129–136.

Rane, B. R., Bharath, M. S., Patil, R. R., Keservani, R. K., & Jain, A. S., (2020b). Novel approaches in nutraceuticals. In: Rajesh, K. K., Raj, K. K., & Anil, K. S., (eds.), *Enhancing the Therapeutic Efficacy of Herbal Formulations Through Novel Drug Delivery Systems* (pp. 241–266). IGI Global International Publisher, Pennsylvania, USA, Chapter 11. ISBN13: 9781799844532.

Rane, B. R., Patil, A. S., Keservani, R. K., & Jain, A. S., (2020a). Novel approaches in herbal formulation. In: Rajesh, K. K., Raj, K. K., & Anil, K. S., (eds.), *Enhancing the Therapeutic Efficacy of Herbal Formulations Through Novel Drug Delivery Systems* (pp. 43–68). IGI Global International Publisher, Pennsylvania, USA, Chapter 2. ISBN13: 9781799844532.

Rane, B. R., Tadavi, S. A., & Keservani, R. K., (2020c). Naturopathy. In: Anil, K. S., Raj, K. K., & Surya, P. G., (eds.), *Herbal Product Development* (pp. 321–347). Apple Academic Press, CRC Press, Taylor & Francis Group, Chapter 12. ISBN: 9781771888776.

Scaglione, F., Ferrara, F., Dugnani, S., Falchi, M., Santoro, G., & Fraschini, F., (1990). Immunomodulatory effects of two extracts of *Panax ginseng*, Meyer CA. *Drugs Exp. Clin. Res., 16*, 537–542.

Sharma, P., Kumar, P., Sharma, R., Gupta, G., & Chaudhary, A., (2017). Immunomodulators: Role of medicinal plants in immune system. *Natl. J. Physiol. Pharm. Pharmacol., 7*(6), 552–556.

Shivaprasad, H. N., Kharya, M. D., Rana, A. C., & Mohan, S., (2006). Preliminary immunomodulatory activities of the aqueous extract of *Terminalia chebula. Pharmaceutical Biology, 44*(1), 32–34.

Singh, D., & Chaudhuri, P. K., (2017). Chemistry and pharmacology of *Tinospora cordifolia. Natural Product Communications, 12*(2), 299–308.

Sonavale, R., Narkhede, A., Mulik, A., Kunchiraman, B., & Jagtap, S. D., (2019). Effect of edaphic factors on major secondary metabolites of *Tinospora cordifolia* and neem guduchi with respect to their immunomodulatory effect. *Int. J. Pharm. Sci. & Res., 10*(7), 3386–3394.

Sterzl, J., Rehácek, Z., & Cudlín, J., (1987). Regulation of the immune response by ergot alkaloids. *Czech Med., 10*(2), 90–98.

Sultana, R., & Chandy, V., (2018). Immunomodulators from plant origin: A review. *International Journal of Pharmaceutical Erudition., 8*(1), 31–44.

Sunila, E. S., & Kuttan, G., (2004). Immunomodulatory and antitumor activity of *Piper longum* Linn. and piperine, *J. Ethnopharmacol., 90*, 339–346.

Thirumurugan, D., Cholarajan, A., Raja, S. S. S., & Vijayakumar, R., (2018). In: Vijayakumar, R., & Raja, S. S. S., (eds.) *An Introductory Chapter: Secondary Metabolites, Secondary Metabolites – Sources and Applications*. Intech Open, doi: 10.5772/intechopen.79766.

Trujillo-Vargas, C. M., Mayer, K. D., Bickert, T., Palmetshofer, A., Grunewald, S., & Ramirez-Pineda, J. R., (2005). Vaccinations with T-helper type 1 directing adjuvants have different suppressive effects on the development of allergen-induced T-helper type 2 responses. *Clin. Exp. Allergy, 35*, 1003–1013.

Vaghasiya, J., Vaghasiya, J., Datani, M., Nandkumar, K., Malaviya, S., & Jivani, N., (2010). *Comparative Evaluation of Alcoholic and Aqueous Extracts of Ocimum Sanctum for Immunomodulatory Activity, 43*, 51–57.

Venkatalakshmi, P., Vadivel, V., & Brindha, P., (2016). Role of phytochemicals as immunomodulatory agents: A review. *Int. J. Green Pharm., 10*, 1–18.

Wagner, H. K. M., (1995). Immunostimulants and adaptogens from plants. In: *Phytochemistry of Medicinal Plants* (pp. 1–18). Springer US.

Wu, X. K., Wang, Y. Y., & Liu, J. P., (2016). Randomized controlled trial of letrozole, berberine, or a combination for infertility in the polycystic ovary syndrome. *Fertil. Steril., 106*(3), 757–765.

Yin, J., Xing, H., & Ye, J., (2008). Efficacy of berberine in patients with type 2 diabetes mellitus. *Metabolism, 57*(5), 712–717.

INDEX

I

J

K

L

N

O

P